Finite Element Methods and Their Applications

Edited by Mahboub Baccouch

Published in London, United Kingdom

IntechOpen

Supporting open minds since 2005

Finite Element Methods and Their Applications
http://dx.doi.org/10.5772/intechopen.83274
Edited by Mahboub Baccouch

Contributors
Mahboub Baccouch, Sutisna Ali Nanang Ali, Tu Anh Do, Dang Quoc Quoc Vuong, Bui Minh Dinh, Suresh Babu Baluguri, G. Srinivas, Jieling Zhao, RameshBabu Chandran, Chad Pope, Edward Lum, Younis Abid Abid Sabawi, Ubaidillah, Bhre Wangsa Wangsa Lenggana, Vinod Bandela, Saraswathi Kanaparthi, Prantasi Harmi Harmi Tjahjanti, Septia Hardy Sujiatanti, Antonio Bilotta, Raghuvir Pai, S.M. Abdul Khader, Nitesh Kumar

First published in London, United Kingdom, 2021 by IntechOpen
IntechOpen is the global imprint of INTECHOPEN LIMITED, registered in England and Wales, registration number: 11086078, 5 Princes Gate Court, London, SW7 2QJ, United Kingdom
Printed in Croatia

British Library Cataloguing-in-Publication Data
A catalogue record for this book is available from the British Library

Additional hard and PDF copies can be obtained from orders@intechopen.com

Finite Element Methods and Their Applications
Edited by Mahboub Baccouch
p. cm.
Print ISBN 978-1-83962-341-7
Online ISBN 978-1-83962-342-4
eBook (PDF) ISBN 978-1-83962-356-1

For EU product safety concerns:
IN TECH d.o.o., Prolaz Marije Krucifikse Kozulić 3, 51000 Rijeka, Croatia, info@intechopen.com or visit our website at intechopen.com.

Meet the editor

Mahboub Baccouch is a professor in the Mathematics Department, University of Nebraska at Omaha (UNO). He received his Ph.D. in Mathematics from Virginia Tech in 2008. He joined UNO in 2008 as a tenure-track assistant professor. He was promoted to associate professor with tenure in 2014. In 2019, he was promoted to full professor. In 2017, he received the College of Arts and Sciences Excellence in Research Award. His research interest is computational mathematics and he actively investigates topics related to numerical methods for deterministic and stochastic differential equations using high-order methods including finite element and discontinuous Galerkin methods. He is also studying numerical solutions to many mathematical problems with broader applications.

Contents

Preface **XI**

Chapter 1 **1**
A Brief Summary of the Finite Element Method for Differential Equations
by Mahboub Baccouch

Chapter 2 **43**
Fluid Structure Interaction Study of Stenosed Carotid Artery
Considering the Effects of Blood Pressure and Altered Gravity
by S.M. Abdul Khader, Nitesh Kumar and Raghuvir Pai

Chapter 3 **63**
The Finite Element Method Applied to the Magnetostatic
and Magnetodynamic Problems
by Dang Quoc Vuong and Bui Minh Dinh

Chapter 4 **93**
A Combination of Finite Difference and Finite Element Methods
for Temperature and Stress Predictions of Early-Age Concrete Members
by Tu Anh Do

Chapter 5 **109**
Convective Heat and Mass Transfer of Two Fluids in a Vertical Channel
by Suresh Babu Baluguri and G. Srinivas

Chapter 6 **131**
A Dynamic Finite Element Cellular Model and Its Application
on Cell Migration
by Jieling Zhao

Chapter 7 **149**
Nuclear Reactor Thermal Expansion Reactivity Effect Determination
Using Finite Element Analysis Coupled with Monte Carlo Neutron
Transport Analysis
by Chad Pope and Edward Lum

Chapter 8 **169**
Finite Element Analysis and Its Applications in Dentistry
by Vinod Bandela and Saraswathi Kanaparthi

Chapter 9 193
Rolling Resistance Estimation for PCR Tyre Design Using
the Finite Element Method
by Sutisna Nanang Ali

Chapter 10 211
Finite Element Analysis in Nanotechnology Research
by RameshBabu Chandran

Chapter 11 221
Finite Element Method for Ship Composite-Based on Aluminum
by Prantasi Harmi Tjahjanti and Septia Hardy Sujiatanti

Chapter 12 245
A MATLAB-Based Symbolic Approach for the Quick Developing
of Nonlinear Solid Mechanics Finite Elements
by Antonio Bilotta

Chapter 13 265
A Posteriori Error Analysis in Finite Element Approximation
for Fully Discrete Semilinear Parabolic Problems
by Younis Abid Sabawi

Chapter 14 285
Finite Element Magnetic Method for Magnetorheological Based Actuators
by Ubaidillah and Bhre Wangsa Lenggana

Preface

The finite element method (FEM) is a widely used technique for numerical simulations in many areas of science and engineering. The method has gained increased popularity over the years for the solution of complex mathematical problems. It is now a powerful and popular numerical method for solving partial differential equations, with flexibility in dealing with complex geometric domains and various boundary conditions. Although the method has been extensively used in the field of structural mechanics, it has also been successfully applied to solve several other types of engineering problems, such as heat conduction, fluid dynamics, seepage flow, and electric and magnetic fields. In particular, FEM has been successfully applied to fluid-structure interaction, thermomechanical, thermochemical, and thermo-chemo-mechanical problems, biomechanics, biomedical engineering, piezoelectricity, ferroelectricity, electromagnetics, and more.

An important advantage of FEM, and the main reason for its popularity among academics and industrial developers, is the ability to handle mathematical problems on domains with arbitrary geometry. An attractive feature is the ability to generate solutions to problems governed by linear and nonlinear differential equations. Moreover, FEM enjoys a firm theoretical foundation that is mostly free of ad hoc schemes and heuristic numerical approximations, thereby inspiring confidence in the physical relevance of the solution.

This book provides several applications of FEM for solving real-world problems. It is a useful resource for students in science and engineering, researchers with diverse educational background, practicing scientists and engineers, computational scientists, and applied mathematicians.

Chapter 1 introduces the method for several one-dimensional and two-dimensional model problems. The remaining chapters consider applications of FEM to several problems. These applications include fluid problems, magnetostatic and magneto-dynamic problems, stress predictions of early-age concrete members, application on cell migration, dentistry, nanotechnology research, ship composite-based on aluminum, and nonlinear solid mechanics. The emphasis of the text is on the simulation of several physical phenomena of FEM, but many mathematical and numerical aspects to important problems are also given.

Chapter 1 provides a summary of FEM. Since the remaining chapters of this textbook are based on FEM, we present it in the first chapter as a general method for approximating solutions of ordinary differential equations (ODEs) and partial differential equations (PDEs). To be more specific, we use simple one-dimensional and two-dimensional model problems to introduce FEM.

Chapter 2 studies the pulsatile flow of blood with different physiological pressure conditions and altered gravity. It summarizes the investigation on the effects of hypertension in comparison with normal blood pressure on normal and stenosed carotid artery bifurcation. In addition, it discusses the effects of altered gravity during the change of posture from sleeping to standing under normal blood pressure conditions.

Chapter 3 applies FEM to solve magnetostatic and magnetodynamic problems.

Chapter 4 presents a two-dimensional finite difference scheme for thermal analysis of a concrete element. FEM is then used to calculate the thermal stresses in the concrete. The analysis results are compared with measurements of actual concrete elements. The combined approach can be a simple and useful tool for analyzing temperatures and thermal stresses in early-age concrete elements.

Chapter 5 presents a mathematical model for convective heat and mass transfer of two immiscible fluids in a vertical channel of variable width with thermo-diffusion, diffusion-thermal effects. The governing boundary layer equations generated for momentum, angular momentum, energy, and species concentration are solved with appropriate boundary conditions using Galerkin FEM. The effects of the pertinent parameters are studied in detail. Furthermore, the chapter analyzes the rate of heat transfer, mass transfer, and shear stress near both walls.

Chapter 6 introduces a newly developed finite element cellular model to simulate collective cell migration and explore the effects of mechanical feedback between cells and between cells and substrate. The viscoelastic model represents one cell with many triangular elements. Intercellular adhesions between cells are represented as linear springs. Furthermore, the chapter includes a mechano-chemical feedback loop between cell-substrate mechanics and cell migration. The results reproduce a set of experimental observations of patterns of collective cell migration during epithelial wound healing. In addition, the chapter demonstrates that cell-substrate-determined mechanics play an important role in regulating persistent and oriented collective cell migration. It also illustrates that our finite element cellular model can be applied to study a number of tissue-related problems regarding cellular dynamic changes at the subcellular level.

Chapter 7 describes FEM coupled with Monte Carlo analysis as a methodology for quantification of a particularly important nuclear parameter that is primarily influenced by thermal and mechanical phenomena present in nuclear reactors. FEM described in this chapter is used to evaluate the reactivity coefficient associated with the thermal expansion-driven spacing of the assemblies along with the much more complicated reactivity coefficient associated with the thermal expansion and mechanical interaction of the fuel assembly hexagonal ducts.

Chapter 8 presents a brief application of FEM in dentistry. It provides an overview of several methods.

Chapter 9 presents rolling resistance estimation in the design process of passenger car radial tire by using FEM to digitally simulate the tire. The simulation firstly computes the deformation of several alternative designs of tires under certain loading and then calculates the value of deformation force in each tire component during deformation. The total force of deformation is considered as energy loss or hysteresis loss resulting in tire rolling resistance. The experiment was carried out on three different tire designs: two grooves, three grooves, and four grooves. The four-groove tire design gave the smallest rolling resistance coefficient. Finally, the simulation was continued to compare different crown radii of the tires and the results show that the largest crown radius generates the lowest rolling resistance.

Chapter 10 elaborates the applications of FEM in varied applications of nanotechnology including carbon nanotubes (CNTs), nanobeams, nanorods, nanobiomaterials, graphene-coated materials, nanosensors, nanotips, and curved nanobeams.

Chapter 11 explains the use of alternative materials for ship building, namely aluminum-based composite material, which is an aluminum alloy AlSi10Mg (b) ship-building material based on the European Nation (EN) Aluminum Casting (AC)-43,100, with silicon carbide reinforcement. Composite ship models use ANSYS software to determine the distribution of stress.

Chapter 12 proposes a symbolic mathematical approach for the rapid early-phase development of finite elements. The algebraic manipulator adopted is MATLAB and the applicative context is the analysis of hyperelastic solids or structures under the hypothesis of finite deformation kinematics. The work has been finalized through the production, in an object-oriented programming style, of three MATLAB classes implementing a truss element, tetrahedral element, and plane element. The approach proposed, starting with the mathematical formulation and finishing with the code implementation, is described and its effectiveness, in terms of minimization of the gap between the theoretical formulation and its actual implementation, is highlighted.

Chapter 13 investigates the error estimation of numerical approximation to a class of semi-linear parabolic problems. More specifically, the time discretization uses the backward Euler Galerkin method, and the space discretization uses FEM for which the meshes are allowed to change in time. The key idea in the analysis is to adapt the elliptic reconstruction technique enabling us to use the a posteriori error estimators derived for elliptic models and to obtain optimal order of convergence for Lipschitz and non-Lipschitz nonlinearities. This chapter also addresses some challenges dealing with the nonlinear term by employing a continuation argument.

Chapter 14 presents a finite element magnetic method for magnetorheological-based actuators. We consider several discussions such as necessary magnetostatic using free software finite element magnetic method; design consideration for the magnetic circuit of the device and case studies of several types of simulation in magnetorheological material-based devices. During the design process, magnetostatic simulation using the finite element magnetic method is carried out to make a better magnetic circuit.

We thank all the authors who contributed to this book with their studies that provide accessible and excellent explanations of the applications of FEM. This work would not have been possible without our excellent contributors. Finally, we express our thanks to Author Service Manager Mr. Josip Knapić and the staff at IntechOpen for their invaluable support and editorial assistance.

Mahboub Baccouch
Department of Mathematics,
University of Nebraska at Omaha,
Omaha, Nebraska, USA

Chapter 1

A Brief Summary of the Finite Element Method for Differential Equations

Mahboub Baccouch

Abstract

The finite element (FE) method is a numerical technique for computing approximate solutions to complex mathematical problems described by differential equations. The method was developed in the 1950s to solve complicated problems in engineering, notably in elasticity and structural mechanics modeling involving elliptic partial differential equations and complicated geometries. But nowadays the range of applications is quite extensive. In particular, the FE method has been successfully applied to many problems such as fluid–structure interaction, thermomechanical, thermochemical, thermo-chemo-mechanical problems, biomechanics, biomedical engineering, piezoelectric, ferroelectric, electromagnetics, and many others. This chapter contains a summary of the FE method. Since the remaining chapters of this textbook are based on the FE method, we present it in this chapter as a method for approximating solutions of ordinary differential equations (ODEs) and partial differential equations (PDEs).

Keywords: the finite element method, initial-value problems, boundary-value problems, Laplace equation, heat equation, wave equation

1. Introduction

1.1 An overview of the finite element method

Differential equations arise in many disciplines such as engineering, mathematics, sciences, economics, and many other fields. Unfortunately solutions to differential equations can rarely be expressed by closed formulas and numerical methods are needed to approximate their solutions. There are many numerical methods for approximating the solution to differential equations including the finite difference (FD), finite element (FE), finite volume (FV), spectral, and discontinuous Galerkin (DG) methods. These methods are used when the mathematical equations are too complicated to be solved analytically.

The FE method has become the standard numerical scheme for approximating the solution to many mathematical problems; see [1–9] and the references therein just to mention a few. In simple words, the FE method is a numerical method to solve differential equations by discretizing the domain into a finite mesh. Numerically speaking, a set of differential equations are converted into a set of algebraic equations to be solved for unknown at the nodes of the mesh. The FE method originated from the need to solve complex elasticity and structural analysis problems in civil and aeronautical engineering. The first development can be traced back to the work by Hrennikoff in 1941 [10] and Courant in 1943 [11]. Although these pioneers used different perspectives in their FE approaches, they each identified the one common and essential characteristic: mesh discretization of a continuous domain into a set of discrete sub-domains, usually called elements. Another fundamental mathematical contribution to the FE method is represented by Gilbert Strang and George Fix [12]. Since then, the FE method has been generalized for the numerical modeling of physical systems in many engineering disciplines including electromagnetism, heat transfer, and fluid dynamics.

The advantages of this method can be summarized as follows:

1. Numerical efficiency: The discretization of the calculation domain with finite elements yields matrices that are in most cases sparse and symmetric. Therefore, the system matrix, which is obtained after spatial and time discretization, is sparse and symmetric too. Both the storage of the system matrix and the solution of the algebraic system of equations can be performed in a very efficient way.

2. Treatment of nonlinearities: The modeling of nonlinear material behavior is well established for the FE method (e.g., nonlinear curves, hysteresis).

3. Complex geometry: By the use of the FE method, any complex domain can be discretized by triangular elements in 2D and by tetrahedra elements in 3D.

4. Applicable to many field problems: The FE method is suited for structural analysis, heat transfer, electrical/magnetical analysis, fluid and acoustic analysis, multi-physics, etc.

COMSOL Multiphysics (known as FEMLAB before 2005) is a commercial FE software package designed to address a wide range of physical phenomena. It is widely used in science and industry for research and development. It excels at modeling almost any multi-physics problem by solving the governing set of PDEs via the FE method. This software package is able to solve one, two and three-dimensional problems. It comes with a modern graphical user interface to set up simulation models and can be scripted from Matlab or via its native Java API.

In this chapter, we introduce the FE method for several one-dimensional and two-dimensional model problems. Although the FE method has been extensively used in the field of structural mechanics, it has been successfully applied to solve several other types of engineering problems, such as heat conduction, fluid dynamics, seepage flow, and electric and magnetic fields. These applications prompted mathematicians to use this technique for the solution of complicated problems. For illustration, we will use simple one-dimensional and two-dimensional model problems to introduce the FE method.

2. The FE method for ODEs

2.1 The FE method for first-order linear IVPs

We first present the FE method as an approximation technique for solving the following first-order initial-value problem (IVP) using piecewise linear polynomials

$$u' = f(x), \quad x \in [a, b], \quad u(a) = u_0. \tag{1}$$

In order to apply the FE method to solve this problem, we carry out the following process.

1. Derive a weak form (variational formulation). This can be done by multiplying the ODE in (1) by a test function $v(x) \in V_0 = \left\{ v \in L^2[a,b] : \|v\|^2 + \|v'\|^2 < \infty,\ v(a) = 0 \right\}$, where $\|v\|^2 = \int_a^b v^2(x)dx$, integrating from a to b, using integration by parts, and applying $v(a) = 0$, to get
$$\int_a^b fvdx = \int_a^b u'vdx = -\int_a^b uv'dx + u(b)v(b) - u(a)v(a) = -\int_a^b uv'dx + u(b)v(b).$$

2. Generate a triangulation (also called a mesh) of the computational domain $[a,b]$. For a one-dimensional problem, a mesh is a set of points in the interval $[a,b]$, say, $a = x_0 \leq x_1 \leq \cdots \leq x_N = b$. The point x_i is called a node or nodal point. The length of the interval (called an element) $I_i = [x_{i-1}, x_i]$ is $h_i = x_i - x_{i-1}$. Let $h = \max_{1 \leq i \leq N} h_i$ (called a mesh size that measures how fine the partition is). If the mesh is uniformly distributed, then $x_i = a + ih, i = 0, 1, \ldots, N$, where $h = \frac{b-a}{N}$.

3. Define a finite dimensional space over the triangulation: Let the solution u be in the space V. For the model problem (1), the solution space is $V = C^1[a,b]$. We wish to construct a finite dimensional space (subspace) $V_h \subset V$ based on the mesh. When the FE space is a subspace of the solution space, the method is called conforming. It is known that in this case, the FE solution converges to the true solution provided the FE space approximates the given space in some sense [3]. Different finite dimensional spaces will generate different FE solutions.

   Define the FE space as the set of all continuous piecewise linear polynomials $V_h = \{v : v|_{I_i} \in P^1(I_i),\ i = 1, 2, \ldots, N,\ v(a) = 0\}$, where $P^1(I_i)$ is the space of polynomials of degree ≤ 1 on I_i. Functions in V_h are linear on each I_i, and continuous on the whole interval $[a,b]$. An example of such a function is shown in **Figure 1**.

   We remark that any function $v \in V_h$ is uniquely determined by its nodal values $v(x_i)$.

4. Construct a set of basis functions based on the triangulation. Since V_h has finite dimension, we can find one set of basis functions. A basis for V_h is $\left\{\phi_j\right\}_{j=0}^{N}$, where $\phi_j \in V_h$ are linearly independent. Then

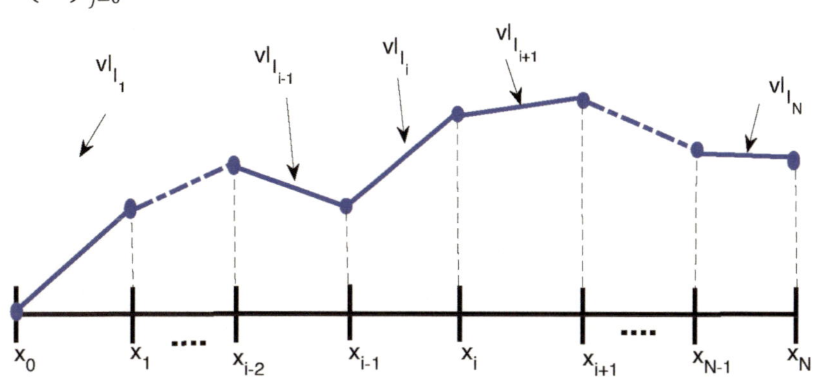

Figure 1.
A continuous piecewise linear function v.

$V_h = \left\{ v_h(x) \in V,\ v_h(x) = \sum_{j=0}^{N} c_j \phi_j(x) \right\}$ is the space spanned by the basis functions $\{\phi_i\}_{i=0}^{N}$. The simplest finite dimensional space is the piecewise continuous linear function space defined over the triangulation.

$$V_h = \{v_h(x) \in V,\ v_h(x) \text{ is piecewise continuous linear over } [\mathrm{a}, \mathrm{b}] \text{ with } v_h(a) = 0\}.$$

There are infinite number of sets of basis functions. We should choose a set of basis functions that are simple, have compact (minimum) support (that is, zero almost everywhere except for a small region), and meet the regularity requirement, that is, they have to be continuous, and differentiable except at nodal points. The simplest ones are the so-called hat functions satisfying $\phi_i(x_i) = 1$ and $\phi_i(x_j) = 0$ for $i \neq j$. The analytic form is (see **Figure 2**)

$$\phi_0(x) = \begin{cases} \dfrac{x_1 - x}{h}, & x \in I_1, \\ 0, & \text{else}, \end{cases}, \quad \phi_N(x) = \begin{cases} \dfrac{x - x_{N-1}}{h}, & x \in I_N, \\ 0, & \text{else}, \end{cases},$$

$$\phi_i(x) = \begin{cases} \dfrac{x - x_{i-1}}{h}, & x \in I_i, \\ \dfrac{x_{i+1} - x}{h}, & x \in I_{i+1}, \\ 0, & \text{else}. \end{cases}$$

5. Approximate the exact solution u by a continuous piecewise linear function $u_h(x)$. The FE method consists of finding $u_h \in V_h$ such that

$$-\int_a^b u_h v' dx + u_h(b) v(b) = \int_a^b f v dx, \quad \forall\ v \in V_h.$$

This type of FE method (with similar trial and test space) is sometimes called a Galerkin method, named after the famous Russian mathematician and engineer Galerkin.

Implementation: The FE solution is a linear combination of the basis functions. Writing $u_h(x) = \sum_{j=0}^{N} c_j \phi_j(x)$, where $c_0, c_1, \ldots, c_N$ are unknowns, and choosing $v = \phi_i,\ i = 1, 2, \ldots, N$ to get

$$-\sum_{j=0}^{N} c_j \int_a^b \phi_j \phi_i' dx + c_N \phi_i(b) = \int_a^b f \phi_i dx, \quad i = 1, 2, \ldots, N,$$

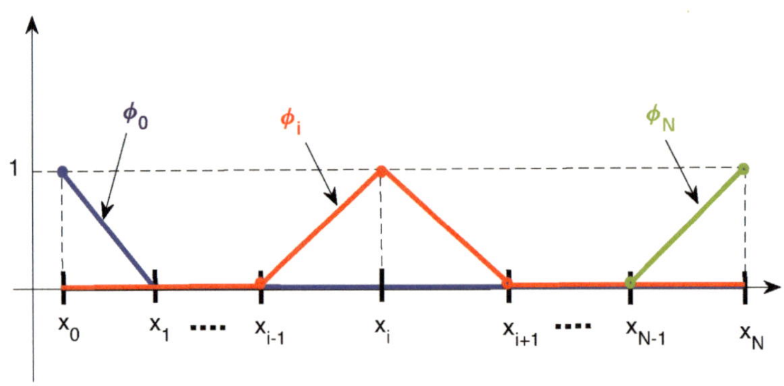

Figure 2.
A typical hat function ϕ_i on a mesh. Also shown is the half hat functions ϕ_o and ϕ_N.

since $u_h(b) = c_N$. Note that using the hat functions, we have $u_h(x_0) = 0$ and $u_h(x_i) = \sum_{j=0}^{N} c_j\phi_j(x_i) = c_i\phi_i(x_i) = c_i$ for $i = 1, 2, \dots, N$. Thus, we get the following linear system

$$-\sum_{j=1}^{N} c_j \int_a^b \phi_j \phi_i' dx + c_N \phi_i(b) = \int_a^b \phi_0 \phi_i' dx, \quad i = 1, 2, \dots, N.$$

Finally, we solve the linear system for $c_1, \dots, c_N$. We note that for $i = 1, 2, \dots, N-1$, we have

$$\int_a^b \phi_i \phi_i' dx = \int_{x_{i-1}}^{x_{i+1}} \phi_i \phi_i' dx = \frac{1}{h_i} \int_{x_{i-1}}^{x_i} \left(\frac{x - x_{i-1}}{h_i}\right) dx - \frac{1}{h_i} \int_{x_i}^{x_{i+1}} \left(\frac{x_{i+1} - x}{h_i}\right) dx = 0.$$

However, for $i = N$, we have

$$\int_a^b \phi_N {\phi_N}' dx = \int_{x_{N-1}}^{x_N} \phi_N {\phi_N}' dx = \int_{x_{N-1}}^{x_N} \left(\frac{x - x_{N-1}}{h_N}\right)\left(\frac{x - x_{N-1}}{h_N}\right) dx$$
$$= \frac{1}{h_N} \int_{x_{N-1}}^{x_N} \frac{x - x_{N-1}}{h_N} dx = \frac{1}{2}.$$

Similarly, for $i = 1, 2, \dots, N$, we have

$$\int_a^b \phi_{i-1} \phi_i' dx = \int_{x_{i-1}}^{x_i} \phi_{i-1} \phi_i' dx = \int_{x_{i-1}}^{x_i} \left(\frac{x_i - x}{h_i}\right)\left(\frac{x - x_{i-1}}{h_i}\right) dx = \frac{1}{h_i} \int_{x_{i-1}}^{x_i} \frac{x_i - x}{h_i} dx = \frac{1}{2},$$
$$\int_a^b \phi_{i+1} \phi_i' dx = \int_{x_i}^{x_{i+1}} \phi_{i+1} \phi_i' dx = \int_{x_i}^{x_{i+1}} \left(\frac{x - x_i}{h_{i+1}}\right)\left(\frac{x_{i+1} - x}{h_{i+1}}\right) dx = -\frac{1}{h_{i+1}} \int_{x_i}^{x_{i+1}} \frac{x - x_i}{h_{i+1}} dx$$
$$= -\frac{1}{2}.$$

We next calculate $\int_a^b f\phi_i dx$. Since it depends on f, we cannot generally expect to calculate it exactly. However, we can approximate it using a quadrature rule. Using the Trapezoidal rule $\int_a^b f(x)dx \approx \frac{b-a}{2}(f(a) + f(b))$ and using $\phi_i(x_{i-1}) = \phi_i(x_{i+1}) = 0$ and $\phi_i(x_i) = 1$, we get

$$\int_a^b f\phi_i dx = \int_{x_{i-1}}^{x_i} f\phi_i dx + \int_{x_i}^{x_{i+1}} f\phi_i dx \approx \frac{h_i + h_{i+1}}{2} f(x_i), \quad i = 1, 2, \dots, N-1,$$

$$\int_a^b f(x)\phi_N dx = \int_{x_{N-1}}^{x_N} f(x)\phi_N dx \approx \frac{h_N}{2}(f(x_{N-1})\phi_N(x_{N-1}) + f(x_N)\phi_N(x_N)) = \frac{h_N}{2} f(x_N).$$

Thus, we obtain the following linear system of equations

$$\begin{bmatrix} 0 & \frac{1}{2} & 0 & \cdots & 0 \\ -\frac{1}{2} & 0 & \frac{1}{2} & \ddots & 0 \\ \vdots & \ddots & \ddots & \ddots & \vdots \\ 0 & \cdots & -\frac{1}{2} & 0 & \frac{1}{2} \\ 0 & \cdots & 0 & -\frac{1}{2} & \frac{1}{2} \end{bmatrix} \begin{bmatrix} c_1 \\ c_2 \\ \vdots \\ c_{N-1} \\ c_N \end{bmatrix} = \begin{bmatrix} \frac{h_1 + h_2}{2} f(x_1) \\ \frac{h_2 + h_3}{2} f(x_2) \\ \vdots \\ \frac{h_{N-1} + h_N}{2} f(x_{N-1}) \\ \frac{h_N}{2} f(x_N) \end{bmatrix}.$$

The determinant of the above matrix is $\frac{1}{2^N}$. Thus, the system has a unique solution $c_1, c_2, \ldots, c_N$.

Remark 2.1 *Suppose that* $u(a) = u_0$, *then we let* $u_h(x) = \sum_{j=0}^{N} c_j \phi_j(x)$. *Since* $u_0 = u_h(x_0) = \sum_{j=1}^{N} c_j \phi_j(x_0) = c_0 \phi_0(x_0) = c_0$, *we only need to find* $c_1, c_2, \ldots, c_N$. *Choosing* $v = \phi_i,\ i = 1, 2, \ldots, N$, *we get the following linear system*

$$-\sum_{j=1}^{N} c_j \int_a^b \phi_j \phi_i' dx + c_N \phi_i(b) = \int_a^b f \phi_i dx + u_0 \int_a^b \phi_0 \phi_i' dx, \quad i = 1, 2, \ldots, N.$$

Finally, we solve the linear system for $c_1, \ldots, c_N$. We note that $\int_a^b \phi_0 \phi_i' dx = 0$ for $i = 2, \ldots, N$ and

$$\int_a^b \phi_0 {\phi_1}' dx = \int_{x_0}^{x_1} \left(\frac{x_1 - x}{h_1}\right)\left(\frac{x - x_0}{h_1}\right)' dx = \frac{1}{h_1} \int_{x_0}^{x_1} \frac{x_1 - x}{h_1} dx = \frac{1}{2}.$$

Following the same steps used for the case $u(a) = 0$, we obtain the following linear system of equations

$$\begin{bmatrix} 0 & \frac{1}{2} & 0 & \cdots & 0 \\ -\frac{1}{2} & 0 & \frac{1}{2} & \ddots & 0 \\ \vdots & \ddots & \ddots & \ddots & \vdots \\ 0 & \cdots & -\frac{1}{2} & 0 & \frac{1}{2} \\ 0 & \cdots & 0 & -\frac{1}{2} & \frac{1}{2} \end{bmatrix} \begin{bmatrix} c_1 \\ c_2 \\ \vdots \\ c_{N-1} \\ c_N \end{bmatrix} = \begin{bmatrix} \frac{h_1 + h_2}{2} f(x_1) + \frac{u_0}{2} \\ \frac{h_2 + h_3}{2} f(x_2) \\ \vdots \\ \frac{h_{N-1} + h_N}{2} f(x_{N-1}) \\ \frac{h_N}{2} f(x_N) \end{bmatrix}.$$

2.2 The FE method for first-order nonlinear IVPs

Here, we extend the FE method for the nonlinear IVP using piecewise linear polynomials

$$u' = f(x, u), \quad x \in [a, b], \quad u(a) = u_0. \tag{2}$$

The FE method consists of finding $u_h \in V_h = \{v : v|_{I_i} \in P^1(I_i),$ $i = 1, 2, \ldots, N,\ v(a) = 0\}$, such that

$$u_h(b) v(b) - \int_a^b u_h v' dx = \int_a^b f(x, u_h) v dx, \quad \forall\ v \in V_h.$$

Writing $u_h(x) = \sum_{j=0}^{N} c_j \phi_j(x)$ and choosing $v = \phi_i,\ i = 1, 2, \ldots, N$, we get

$$c_N \left(\phi_i - \int_a^b \phi_N \phi_i' dx\right) - \sum_{j=0}^{N-1} c_j \int_a^b \phi_j \phi_i' dx - \int_a^b f\left(x, \sum_{j=0}^{N} c_j \phi_j\right) \phi_i dx = 0,$$
$$i = 1, 2, \ldots, N,$$

where $u_h(x_0) = c_0 = u_0$. Finally, we solve the nonlinear system for $c_1, c_2, \ldots, c_N$ using *e.g.*, Newton's method for systems of nonlinear equations. The system can be written as $F_i(c_1, c_2, \ldots, c_N) = 0,\ i = 1, 2, \ldots, N$, where

$$F_i = c_N\left(\phi_i - \int_a^b \phi_N \phi_i' dx\right) - \sum_{j=0}^{N-1} c_j \int_a^b \phi_j \phi_i' dx - \int_a^b f\left(x, \sum_{j=0}^{N} c_j \phi_j\right)\phi_i dx,$$
$$i = 1, 2, \ldots, N.$$

Let $\alpha_i = \sum_{j=0}^{N} c_j \int_a^b \phi_j \phi_i' dx$ and $\beta_i = \int_a^b f\left(x, \sum_{j=0}^{N} c_j \phi_j\right)\phi_i dx$. Then, for $i = 1, 2, \ldots, N-1$,

$$\begin{aligned}
\alpha_i &= c_{i-1}\int_{x_{i-1}}^{x_i} \phi_{i-1}\phi_i' dx + c_i\left(\int_{x_{i-1}}^{x_i} \phi_i\phi_i' dx + \int_{x_i}^{x_{i+1}} \phi_i\phi_i' dx\right) + c_{i+1}\int_{x_i}^{x_{i+1}} \phi_{i+1}\phi_i' dx \\
&= c_{i-1}\int_{x_{i-1}}^{x_i} \frac{x_i - x}{h_i^2} dx + c_i\left(\int_{x_{i-1}}^{x_i} \frac{x - x_{i-1}}{h_i^2} dx - \int_{x_i}^{x_{i+1}} \frac{x_{i+1} - x}{h_{i+1}^2} dx\right) - c_{i+1}\int_{x_i}^{x_{i+1}} \frac{x - x_i}{h_{i+1}^2} dx \\
&= \frac{1}{2}c_{i-1} + c_i\left(\frac{1}{2} - \frac{1}{2}\right) - \frac{1}{2}c_{i+1} = \frac{1}{2}c_{i-1} - \frac{1}{2}c_{i+1},
\end{aligned}$$

$$\begin{aligned}
\alpha_N &= c_{N-1}\int_{x_{N-1}}^{x_N} \phi_{N-1}{\phi_N}' dx + c_N\int_{x_{N-1}}^{x_N} \phi_N {\phi_N}' dx \\
&= c_{N-1}\int_{x_{N-1}}^{x_N} \frac{x_N - x}{h_N^2} dx + c_N\int_{x_{N-1}}^{x_N} \frac{x - x_{N-1}}{h_N^2} dx = \frac{1}{2}c_{N-1} + \frac{1}{2}c_N.
\end{aligned}$$

Similarly,

$$\beta_i = \int_{x_{i-1}}^{x_{i+1}} f\left(x, \sum_{j=0}^{N} c_j\phi_j\right)\phi_i dx = \int_{x_{i-1}}^{x_i} f\left(x, \sum_{j=0}^{N} c_j\phi_j\right)\phi_i dx + \int_{x_i}^{x_{i+1}} f\left(x, \sum_{j=0}^{N} c_j\phi_j\right)\phi_i dx.$$

Using Simpson's Rule $\int_a^b f(x)dx \approx \frac{b-a}{6}\left(f(a) + 4f\left(\frac{a+b}{2}\right) + f(b)\right)$, and using $\phi_i(x_{i-1}) = \phi_i(x_{i+1}) = 0$, $\phi_i(x_i) = 1$, $\sum_{j=0}^{N} c_j\phi_j\left(x_{i-1} + \frac{h_i}{2}\right) = \frac{c_{i-1}+c_i}{2}$, $\phi_i\left(x_{i-1} + \frac{h_i}{2}\right) = \frac{1}{2}$, $\sum_{j=0}^{N} c_j\phi_j(x_i) = c_i$, we have, for $i = 1, 2, \ldots, N-1$,

$$\beta_i \approx \frac{h_i}{3} f\left(x_{i-1} + \frac{h_i}{2}, \frac{c_{i-1} + c_i}{2}\right) + \frac{h_i + h_{i+1}}{6} f(x_i, c_i) + \frac{h_{i+1}}{3} f\left(x_i + \frac{h_{i+1}}{2}, \frac{c_i + c_{i+1}}{2}\right).$$

However, for $i = N$, we have

$$\beta_N \approx \frac{h_N}{6}\left(2f\left(x_{N-1} + \frac{h}{2}, \frac{c_{N-1} + c_N}{2}\right) + f(x_N, c_N)\right).$$

Next, we compute the Jacobian matrix with entries

$$J_{i,j} = \frac{\partial F_i}{\partial c_j} = \int_a^b \phi_j\phi_i' dx - \int_a^b f_u\left(x, \sum_{j=0}^{N} c_j\phi_j\right)\phi_j\phi_i\, dx = a_{i,j} - b_{i,j}, \quad i = 1, 2, \ldots, N.$$

We already computed the entries $a_{i,j}$ as

$$a_{i,i-1} = \int_a^b \phi_{i-1}\phi_i' dx = \frac{1}{2}, \quad a_{i,i} = \int_a^b \phi_i\phi_i' dx = 0, \quad i = 1, 2, \ldots, N-1,$$

$$a_{N,N} = \int_a^b \phi_N {\phi_N}' dx = \frac{1}{2}, \quad a_{i,i+1} = \int_a^b \phi_{i+1}\phi_i' dx = -\frac{1}{2}.$$

Using Simpson's Rule, we get

$$b_{i,i-1}=\int_{x_{i-1}}^{x_i}\phi_{i-1}\phi_i f_u\left(x,\sum_{j=0}^{N}c_j\phi_j\right)dx\approx\frac{h_i}{6}f_u\left(x_{i-1}+\frac{h_i}{2},\frac{c_{i-1}+c_i}{2}\right),$$

$$b_{i,i+1}=\int_{x_i}^{x_{i+1}}\phi_{i+1}\phi_i f_u\left(x,\sum_{j=0}^{N}c_j\phi_j\right)dx\approx\frac{h_{i+1}}{6}f_u\left(x_i+\frac{h_{i+1}}{2},\frac{c_i+c_{i+1}}{2}\right),$$

$$b_{i,i}=\int_{x_{i-1}}^{x_i}\phi_i^2 f_u\left(x,\sum_{j=0}^{N}c_j\phi_j\right)dx+\int_{x_i}^{x_{i+1}}\phi_i^2 f_u\left(x,\sum_{j=0}^{N}c_j\phi_j\right)dx$$

$$\approx\frac{h_i}{6}f_u\left(x_{i-1}+\frac{h_i}{2},\frac{c_{i-1}+c_i}{2}\right)+\frac{h_i+h_{i+1}}{6}f_u(x_i,c_i)+\frac{h_{i+1}}{6}f_u\left(x_i+\frac{h_{i+1}}{2},\frac{c_i+c_{i+1}}{2}\right),$$

$$b_{N,N}=\int_{x_{N-1}}^{x_N}\phi_N^2 f_u\left(x,\sum_{j=0}^{N}c_j\phi_j\right)dx\approx\frac{h_N}{6}\left(f_u\left(x_{N-1}+\frac{h}{2},\frac{c_{N-1}+c_N}{2}\right)+f_u(x_N,c_N)\right).$$

2.3 The FE method for two-point BVPs

Here, we shall study the derivation and implementation of the FE method for two-point boundary-value problems (BVPs). For easy presentation, we consider the following model problem: Find $u\in C^2[a,b]$ such that

$$-u''+q(x)u=f(x),\quad x\in\Omega=(a,b),\quad u(a)=u(b)=0,\qquad(3)$$

where $u:\overline{\Omega}=[a,b]\to\mathbb{R}$ is the sought solution, $q(x)\geq 0$ is a continuous function on $[a,b]$, and $f\in L^2[a,b]$. Under these assumptions, (3) has a unique solution $u\in C^2[a,b]$. For general $q(x)$, it is impossible to find an explicit form of the solution. Therefore, our goal is to obtain a numerical solution via the FE method.

2.3.1 Different mathematical formulations for the 1D model

The model problem (3) can be reformulated into three different forms:

(D)-form: the original differential equation (3).

(V)-form: the variational form or weak form: $\int_a^b u'v'dx+\int_a^b quvdx=\int_a^b fvdx$, for any test function v in the Sobolev space $H_0^1[a,b]=\{v\in L^2[a,b]:\|v\|^2+\|v'\|^2<\infty,\ v(a)=v(b)=0\}$, where $\|v\|^2=\int_a^b v^2(x)dx$. The corresponding FE method is often called the *Galerkin method*. In other words, a Galerkin FE method is a FE method obtained from the variational form.

(M)-form: the minimization form: $\min_{v(x)\in H_0^1[a,b]}\int_a^b\left(\frac{1}{2}(v')^2+\frac{1}{2}qv^2-fv\right)dx$. The corresponding FE method is often called the *Ritz method*.

Under some assumptions, the three different forms are equivalent, that is, they have the same solution as will be explained in the following theorem.

Theorem 2.1 (Mathematical equivalences) *Suppose that u″ exists and continuous on $[a,b]$. Then we have the following mathematical equivalences.*

(D) is equivalent to (V), (V) is equivalent to (M), and (M) is equivalent to (D).

2.3.2 Galerkin method of the problem

To solve (3) using the FE method, we carry out the process described below. Usually, a FE method is always derived from the weak or variational formulation of the problem at hand.

Weak formulation of the problem: The Galerkin FE method starts by rewriting (3) in an equivalent variational formulation. To this end, let us define the vector space $H_0^1 = \left\{ v \in L^2(a,b) : \|v\|^2 + \|v'\|^2 < \infty,\ v(a) = v(b) = 0 \right\}$. Multiplying (3) by a test function $v \in H_0^1$, integrating from a to b, and using integration by parts, we get

$$\int_a^b fv dx = \int_a^b -u''v dx + \int_a^b quv dx = \int_a^b u'v' dx + \int_a^b quv dx,$$

since $v(a) = v(b) = 0$. Hence, the weak (or variational) form of (3) reads: Find $u \in H_0^1$, such that

$$\int_a^b u'v' dx + \int_a^b quv dx = \int_a^b fv dx, \quad \forall\ v \in H_0^1. \tag{4}$$

We want to find $u \in H_0^1$ that satisfies (4). We note that a solution u to (4) is less regular than the solution u (3). Indeed, (4) has only u' whereas (3) contains u''. Furthermore, we can easily verify the following:

1. If u is strong solution (*i.e.*, solves (3)) then u is also weak solution (*i.e.*, solves (4)).

2. Conversely, if u is a weak solution with $u \in C^2[a,b]$, it is also strong solution.

3. Existence and uniqueness of weak solutions is obtained by the Lax-Milgram Theorem.

4. We can consider solutions with lower regularity using the weak formulation.

5. FE method gives an approximation of the weak solution.

From now on, we use the notation $\|v\| = \|v\|_\Omega$, where $\Omega = [a,b]$.

The FE formulation: The FE method is based on the variational form (4). We note that the space H_0^1 contains many functions and it is therefore just as hard to find a function $u \in H_0^1$ which satisfies the variational Eq. (4) as it is to solve the original problem (3). Next, we study in details a special Galerkin method called the FE method. Let $a = x_0 < x_1 < \cdots < x_N = b$ be a regular partition of $[a,b]$. Suppose that the length of $I_i = [x_{i-1}, x_i]$ is $h_i = x_i - x_{i-1}$. We define $h = \max_{i=1,2,\ldots,N} h_i$ to be the mesh size. We wish to construct a subspace $V_h \subset V = H_0^1$. Since V_h has finite dimension, we can find one set of basis functions $\left\{\phi_j\right\}_{j=1}^{N-1}$ for V_h, where $\phi_j \in V_h,\ j = 1, 2, \ldots, N-1$ are linearly independent. We remark that V_h is the space spanned by the basis functions *i.e.*, $V_h = \left\{ v_h(x),\ v_h(x) = \sum_{j=1}^{N-1} c_j \phi_j(x) \right\}$. The FE method consists of choosing a basis for the subspace V_h that satisfies the following properties

1. The matrix A must be sparse (*e.g.* traditional or banded matrix). In this case, iterative methods for solving linear systems can be adapted to obtain an efficient solution.

2. u_h must converge to the solution u of the original problem as $h \to 0$.

It is natural to obtain an approximation u_h to u as follows: Find $u_h \in V_h$ such that

$$\int_a^b u_h' v' dx + \int_a^b q u_h v dx = \int_a^b f v dx, \quad \forall\ v \in V_h. \tag{5}$$

We call u_h the FE approximation of u. We say that (5) is the Galerkin approximation of (4) and the method used to find $u_h \in V_h$ is called Galerkin method.

FE approximation using Lagrange $\mathbb{P}_1$ elements: The simplest finite dimensional space is the piecewise continuous linear function space defined over the triangulation

$$V_{h,0}^1 = \{v_h \in V,\ v_h \text{ is piecewise continuous linear over [a, b] with } v_h(a) = v_h(b) = 0\}.$$

It is easy to show that $V_{h,0}^1$ has a finite dimension even although there are infinite number of elements in $V_{h,0}^1$. The approximation of the FE method is therefore to look for an approximation u_h within a small (finite dimensional) subspace $V_{h,0}^1 = \{v \in V_h^1 |\ v(a) = v(b) = 0\}$ of H_0^1, consisting of piecewise linear polynomials, where $V_h^1 = \left\{v \in C^0[a,b] |\ v|_{I_i} \in P^1(I_i)\right\}$.

Let $V_{h,0}^1$ be the space of all continuous piecewise linear functions, which vanish at the end points a and b. There are many types of basis functions $\{\phi_i\}_{i=1}^{N-1}$. The simplest ones are the so-called hat functions satisfying $\phi_i(x_j) = \delta_{ij}$, where δ_{ij} is the Kronecker symbol. Note especially that there is no need to construct hat functions ϕ_0 and ϕ_N since any function of $V_{h,0}^1$ must vanish at the end points $x_0 = a$ and $x_N = b$.

The explicit expressions for the hat function $\phi_i(x)$ and its derivative $\phi_i'(x)$ are given by

$$\phi_i(x) = \begin{cases} 0, & a \le x \le x_{i-1}, \\ \dfrac{x - x_{i-1}}{h_i}, & x_{i-1} \le x \le x_i, \\ \dfrac{x_{i+1} - x}{h_{i+1}}, & x_i \le x \le x_{i+1}, \\ 0, & x_{i+1} \le x \le b, \end{cases} \quad \phi_i'(x) = \begin{cases} 0, & a < x < x_{i-1}, \\ \dfrac{1}{h_i}, & x_{i-1} < x < x_i, \\ -\dfrac{}{h_{i+1}}, & x_i < x < x_{i+1}, \\ 0, & x_{i+1} < x < b, \end{cases}$$

for $i = 1, 2, \ldots, N-1$. The FE approximation of (4) thus reads: Find $u \in V_{h,0}^1$, such that

$$\int_a^b u_h' v' dx + \int_a^b q u_h v dx = \int_a^b f v dx, \quad \forall\ v \in V_{h,0}^1. \tag{6}$$

We call u_h the FE approximation of u. We say that (6) is the Galerkin approximation of (4) and the method used to find $u_h \in V_{h,0}^1$ is called Galerkin method.

It can be shown that (6) is equivalent to the $N-1$ equations

$$\int_a^b u_h' \phi_i' dx + \int_a^b q u_h \phi_i dx = \int_a^b f \phi_i dx, \quad i = 1, 2, \dots, N-1. \tag{7}$$

Derivation of the discrete system: Since $u_h \in V_{h,0}^1$, we can express it as a linear combination of hat functions *i.e.*,

$$u_h = \sum_{j=1}^{N-1} c_j \phi_j(x), \tag{8}$$

where c_j are real numbers to be determined. We note that the coefficients c_j, $j = 1, 2, \dots, N-1$ are the $N-1$ nodal values of u_h to be determined. Note that the index is only from 1 to $N-1$, because of the zero boundary conditions. We remark that $u_h(a) = u_h(b) = 0$ and $u_h(x_i) = c_i$. So c_i is an approximate solution to the exact solution at $x = x_i$.

We can use either the weak/variational form (V), or the minimization form (M), to derive a linear system of equations for the coefficients c_j.

Substituting (8) into (7) yields

$$\sum_{j=1}^{N-1} c_j \left(\int_a^b \phi_i' \phi_j' dx + \int_a^b q \phi_i \phi_j dx \right) = \int_a^b f \phi_i dx, \quad i = 1, 2, \dots, N-1. \tag{9}$$

The problem (7) is now equivalent to the following: Find the real numbers $c_1, c_2, \dots, c_{N-1}$ that satisfy the linear system (9).

We note that the linear system (9) is equivalent to the system in matrix–vector form

$$A\mathbf{c} = \mathbf{b}, \tag{10}$$

where $\mathbf{c} = [c_1, c_2, \dots, c_{N-1}]^t \in \mathbb{R}^{N-1}$ is the unknown vector, A is an $(N-1) \times (N-1)$ matrix, the so-called stiffness matrix when $q = 0$, with entries

$$a_{ij} = \int_a^b \left(\phi_i' \phi_j' + q \phi_i \phi_j \right) dx, \quad i,j = 1, 2, \dots, N-1, \tag{11}$$

and $\mathbf{b} \in \mathbb{R}^{N-1}$, the so-called load vector, has entries

$$b_i = \int_a^b f \phi_i dx; \quad i = 1, 2, \dots, N-1. \tag{12}$$

To obtain the approximate solution we need to solve the linear system for the unknown vector $\mathbf{c}$. We note that $a_{ij} = a\left(\phi_i, \phi_j\right)$ and $b_i = (f, \phi_i)$, where $a(u,v) = \int_a^b (u'v' + quv) dx$ is a bi-linear and $(f, v) = \int_a^b f v dx$ is a linear form.

2.3.3 Ritz method of the problem

The Ritz method is one of the earliest FE methods. However, not every problem has a minimization form. The minimization form for the model problem (3) is

$$\min_{v(x) \in H_0^1[a,b]} F(v), \quad \text{where } F(v) = \int_a^b \left(\frac{1}{2} (v')^2 + \frac{1}{2} q v^2 - f v \right) dx.$$

As before, we look for an approximate solution of the form (8). If we plug this into the functional form above, we get

$$F(u_h) = \int_a^b \left(\frac{1}{2} \left(\sum_{j=1}^{N-1} c_j \phi_j'(x) \right)^2 + \frac{1}{2} q \left(\sum_{j=1}^{N-1} c_j \phi_j(x) \right)^2 - f \left(\sum_{j=1}^{N-1} c_j \phi_j(x) \right) \right) dx,$$

which is a multi-variable function of $c_1, c_2, \dots, c_{N-1}$ and can be written as $F(u_h) = F(c_1, c_2, \dots, c_{N-1})$. The necessary conditions for a global minimum are $\frac{\partial F}{\partial c_i} = 0$, $i = 1, 2, \dots, N-1$. Taking the partial derivatives directly with respect to c_i, we get

$$\int_a^b \left(\phi_i'(x) \sum_{j=1}^{N-1} c_j \phi_j'(x) + q\phi_i(x) \sum_{j=1}^{N-1} c_j \phi_j(x) - f\phi_i(x) \right) dx = 0, \quad i = 1, 2, \dots, N-1.$$

Exchange the order of integration and the summation, we get

$$\sum_{j=1}^{N-1} c_j \int_a^b \left(\phi_i'(x)\phi_j'(x) + q\phi_i(x)\phi_j(x) \right) dx = \int_a^b f\phi_i(x) dx = 0, \quad i = 1, 2, \dots, N-1,$$

which is exactly the same linear system (9) obtained using the Galerkin method.

2.3.4 Computer implementation

It is straightforward to calculate the entries $\hat{a}_{i,j} = \int_a^b \phi_i' \phi_j' dx$. For $|i-j| > 1$, we have $\hat{a}_{i,j} = 0$, since ϕ_i and ϕ_j lack overlapping support. However, if $i = j$, then

$$\hat{a}_{i,i} = \int_a^b (\phi_i')^2 dx = \int_{x_{i-1}}^{x_i} \left(\frac{1}{h_i} \right)^2 dx + \int_{x_i}^{x_{i+1}} \left(-\frac{1}{h_{i+1}} \right)^2 dx = \frac{1}{h_i} + \frac{1}{h_{i+1}}, \quad i,j = 1, 2, \dots, N-1.$$

Furthermore, if $j = i+1$, then

$$\hat{a}_{i,i+1} = \int_a^b \phi_i' \phi_{i+1}' dx = \int_{x_i}^{x_{i+1}} \left(-\frac{1}{h_{i+1}} \right) \left(\frac{1}{h_{i+1}} \right) dx = -\frac{1}{h_{i+1}}, \quad i,j = 1, 2, \dots, N-2. \tag{13}$$

By symmetry, we also have

$$\hat{a}_{i+1,i} = \int_a^b \phi_{i+1}' \phi_i' dx = -\frac{1}{h_{i+1}}, \quad i,j = 1, 2, \dots, N-2.$$

To obtain $\tilde{a}_{i,j} = \int_a^b q\phi_i\phi_j dx$ and $b_i = \int_a^b f\phi_i dx$, we use the composite trapezoidal rule

$$\int_a^b f(x) dx = \sum_{i=1}^{N} \int_{x_{i-1}}^{x_i} f(x) dx \approx \frac{1}{2} \left[h_1 f(x_0) + \sum_{i=1}^{N-1} (h_i + h_{i+1}) f(x_i) + h_N f(x_N) \right].$$

So, we can easily verify that

$$\tilde{a}_{i,j} = \int_a^b q\phi_i\phi_j dx \approx \begin{cases} \frac{q_i}{2}(h_i + h_{i+1}), & i = j \\ 0, & i \neq j \end{cases}, \quad b_i = \int_a^b f\phi_i dx \approx \frac{1}{2}(h_i + h_{i+1}) f_i,$$

where $q_i = q(x_i)$ and $f_i = f(x_i)$. Thus, the matrix $A = \left(\hat{a}_{i,j} + \tilde{a}_{i,j}\right)$ is tridiagonal and has the form

$$A = \begin{bmatrix} \frac{1}{h_1} + \frac{1}{h_2} + \frac{q_1}{2}(h_1 + h_2) & -\frac{1}{h_2} & 0 & \cdots & 0 \\ -\frac{1}{h_2} & \frac{1}{h_2} + \frac{1}{h_3} + \frac{q_2}{2}(h_2 + h_3) & -\frac{1}{h_3} & \ddots & 0 \\ 0 & -\frac{1}{h_3} & \ddots & \cdots & 0 \\ \vdots & \ddots & \ddots & \ddots & -\frac{1}{h_{N-1}} \\ 0 & \cdots & 0 & -\frac{1}{h_{N-1}} & \frac{1}{h_{N-1}} + \frac{1}{h_N} + \frac{q_{N-1}}{2}(h_{N-1} + h_N) \end{bmatrix}.$$

Finally, we obtain the following system: $c_0 = c_N = 0$ and

$$-\frac{1}{h_i} c_{i-1} + \frac{1}{h_i + h_{i+1}} c_i - \frac{1}{h_{i+1}} c_{i+1} + \frac{q_i(h_i + h_{i+1})}{2} c_i = \frac{1}{2}(h_i + h_{i+1}) f_i, \quad i = 1, 2, \dots, N-1,$$

Remark 2.2 *Suppose that the partition is uniform i.e.,* $h_i = h = \frac{b-a}{N}$ *for all* $i = 1, 2, \dots, N$*. Then the stiffness matrix A and the load vector* $\mathbf{b}$ *have the form:*

$$A = \begin{bmatrix} \frac{2}{h} + hq_1 & -\frac{1}{h} & 0 & \cdots & 0 \\ -\frac{1}{h} & \frac{2}{h} + hq_2 & -\frac{1}{h} & \ddots & 0 \\ 0 & -\frac{1}{h} & \ddots & \cdots & 0 \\ \vdots & \ddots & \ddots & \ddots & -\frac{1}{h_{N-1}} \\ 0 & \cdots & 0 & -\frac{1}{h} & \frac{2}{h} + hq_{N-1} \end{bmatrix}, \quad \mathbf{b} = h \begin{bmatrix} f_1 \\ f_2 \\ f_3 \\ \vdots \\ f_{N-1} \end{bmatrix}.$$

Finally, we obtain the following system: $c_0 = c_N = 0$ and

$$\frac{-c_{j-1} + 2c_j - c_{j+1}}{h} + hq_i c_j = hf_i \Rightarrow -\frac{c_{j-1} - 2c_j + c_{j+1}}{h^2} + q_i c_j = f_i, \quad i = 1, 2, \dots, N-1,$$

which is the same system obtained using the finite difference method, where u'' is approximated using the second-order midpoint formula $u''(x_j) \approx \frac{u(x_{j-1}) - 2u(x_j) + u(x_{j+1})}{h^2}$. We conclude that the above FE method using the composite trapezoidal rule is equivalent to the finite difference method of order 2.

2.3.5 *Existence, uniqueness, and basic* a priori *error estimate*

Lemma 2.1 *The matrix A with entries* $a_{i,j} = \int_a^b \phi_i' \phi_j' dx$ *is symmetric positive definite i.e.,* $a_{i,j} = a_{j,i}$ *and*

$$\mathbf{x}^t A \mathbf{x} = \sum_{i,j=1}^{N-1} x_i a_{i,j} x_j > 0, \quad \text{for all nonzero } \mathbf{x} = [x_1, \dots, x_{N-1}]^t \in \mathbb{R}^{N-1}.$$

Theorem 2.2 *The linear system (10) obtained using the FE method has a unique solution. Consequently, the FE method solution* u_h *is unique.*

Next, we state a general convergence result for the Galerkin method. We first define the following norm and semi-norm: For $v \in H_0^1$, we define

$$\|v\| = \left(\int_a^b v^2(x)dx \right)^{1/2}, \quad |v|_1 = \|v'\| = \left(\int_a^b (v'(x))^2 dx \right)^{1/2}.$$

Theorem 2.3 *Suppose that $q(x) \geq 0, \forall \ x \in [a, b]$. Let u be the solution to (4) and u_h be the solution to (6). Then there exists a constant C such that*

$$\|(u - u_h)'\| \leq C\|(u - v_h)'\|, \quad \forall \ v_h \in V_{h,0}^1, \tag{14}$$

where C is given by $C = 1 + \max_{x \in [a,b]} |q(x)|$, which is independent of the choice of $V_{h,0}^1$.

Remark 2.3 *From (14), taking the minimum over all $v_h \in V_{h,0}^1$, we get $\|(u - u_h)'\| \leq C \min\limits_{v_h \in V_{h,0}} \|(u - v_h)'\|$. Thus, $|u - u_h|_1 \leq C \min_{v_h \in V_{h,0}} |u - v_h|_1$, where* $C = 1 + \max_{x \in [a,b]} |q(x)|$.

Next, we study the convergence of u_h to u. Let $u \in H_0^1$. Define the piecewise linear interpolant by

$$\pi u = \sum_{j=1}^{N} u(x_j)\phi_j(x) \in V_{h,0}^1, \quad x \in [a, b].$$

Since $\pi u \in V_{h,0}^1$, the estimate (14) gives

$$\|(u - u_h)'\| \leq C\|(u - \pi u)'\|.$$

This inequality suggest that the error between u and u_h is controlled by the interpolation error $u - \pi u$ in the $|\cdot|_1$-norm.

Theorem 2.4 (A priori error estimate) *Suppose that $q(x) \geq 0 \ \forall \ x \in [a, b]$. Let u be the solution to (4) and u_h be the solution to (6). Then there exists a constant C such that*

$$\|(u - u_h)'\|^2 \leq C \sum_{i=1}^{N} h_i^2 \|u''\|_{I_i}^2,$$

where C is a constant independent of h. Consequently, if $h = \max_i h_i$, then

$$\|(u - u_h)'\|^2 \leq Ch^2 \|u''\|^2.$$

Remark 2.4

1. If the partition is not uniform then we obtain the same error estimate with $h = \max_{i=1,2,\ldots,N}(x_i - x_{i-1})$.

2. The error is expressed in terms of the exact solution u. If it is expressed in terms of the computed solution u_h it is an *a posteriori* error estimate (this yields a computable error bound).

3. $u_h \to u$ in the $\|v'\|$-norm as $h = max_i(h_i) \to 0$. If $\|(u - u_h)'\| = 0$ then $u - u_h$ is constant, but since $u(0) = u_h(0)$ we also have $u - u_h = 0$ and therefore $u_h = u$.

4. u_h is the best approximation within the space $V^1_{h,0}$ with respect to the $\|v'\|$-norm.

5. The norm $\|v'\|$ is referred to as the energy norm and has often a physical meaning.

2.3.6 Boundary conditions

In problem (3) we considered a homogeneous Dirichlet boundary conditions. Here, we extend the FE method to boundary conditions of different types. There are three important types of boundary conditions (BCs):

1. Dirichlet BCs: $u(a) = \alpha$ and $u(b) = \beta$ for two real numbers α and β. This BC is also known as strong BC or essential BC.

2. Neumann BCs: $u'(a) = \alpha$ and $u'(b) = \beta$ for two real numbers α and β. This BC is also known as natural BCs.

3. Robin BCs: $u'(a) = \alpha u(a)$ and $u'(b) = \beta u(b)$ for two real numbers α and β.

Note that any combination is possible at the two boundary points.

Nonhomogeneous Dirichlet boundary conditions: Let us consider the following two-point BVP: find $u \in C^2(a,b)$ such that

$$-u'' = f(x), \quad x \in (a,b), \quad u(a) = \alpha, \quad u(b) = \beta, \tag{15}$$

where α and β are given constants and $f \in C(a,b)$ is a given function. In this case, the admissible function space $H^1_0 = \left\{v : \|v\|^2 + \|v'\|^2 < \infty, \ v(a) = v(b) = 0\right\}$ and the FE space $V^1_{h,0}$ defined earlier remain the same. Multiplying (15) by a test function $v \in H^1_0$ and integrating by parts gives

$$\int_a^b fv dx = \int_a^b -u''v dx = -u'(b)v(b) + u'(a)v(a) + \int_a^b u'v' dx = \int_a^b u'v' dx,$$

since $v(a) = v(b) = 0$. Hence, the weak or variational form of (15) reads: Given $u(a) = \alpha$, $u(b) = \beta$, find $u \in H^1 = \left\{v : \|v\|^2 + \|v'\|^2 < \infty\right\}$, such that

$$\int_a^b u'v' dx = \int_a^b fv dx, \quad \forall \ v \in H^1_0. \tag{16}$$

Let V^1_h and $V^1_{h,0}$, respectively, be the space of all continuous piecewise linear functions and the space of all continuous piecewise linear functions which vanish at the endpoints a and b. We also let $a = x_0 < x_1 < \cdots < x_N = b$ be a uniform partition of the interval $[a,b]$. Moreover let $\{\phi_i\}$ be the set of hat basis functions of V_h associated with the $N+1$ nodes $x_j, \ j = 0, 1, \ldots, N$, such that $\phi_i(x_j) = \delta_{ij}$. The FE approximation of (16) thus reads: Find $u_h \in V^1_h$ such that $u_h(a) = \alpha$, $u_h(b) = \beta$, and

$$\int_a^b u_h'v' dx = \int_a^b fv dx, \quad \forall \ v \in V^1_{h,0}. \tag{17}$$

It can be shown that (17) is equivalent to the $N-1$ equations

$$\int_a^b u_h' \phi_i' dx = \int_a^b f \phi_i dx, \quad i = 1, 2, \ldots, N-1. \tag{18}$$

Expanding u_h as a linear combination of hat functions

$$u_h = \sum_{j=0}^{N} c_j \phi_j(x) = \alpha \phi_0(x) + \sum_{j=1}^{N-1} c_j \phi_j(x) + \beta \phi_N(x), \tag{19}$$

where the coefficients $c_j, \; j = 1, 2, \ldots, N-1$ are the $N-1$ nodal values of u_h to be determined.

Substituting (19) into (18) yields

$$\sum_{j=1}^{N-1} c_j \left(\int_a^b \phi_i' \phi_j' dx \right) = \int_a^b (f \phi_i - \alpha \phi_0' \phi_i' - \beta \phi_N' \phi_i') dx, \quad i = 1, 2, \ldots, N-1,$$

which is a $(N-1) \times (N-1)$ system of equations for c_j. In matrix form we write

$$A\mathbf{c} = \mathbf{b}, \tag{20}$$

where A is a $(N-1) \times (N-1)$ matrix, the so-called stiffness matrix, with entries

$$a_{i,j} = \int_a^b \phi_i' \phi_j' dx, \quad i,j = 1, 2, \ldots, N-1, \tag{21}$$

$\mathbf{c} = [c_1, c_2, \ldots, c_{N-1}]^t$ is a $(N-1)$ vector containing the unknown coefficients $c_j, \; j = 1, 2, \ldots, N-1$, and $\mathbf{b}$ is a $(N-1)$ vector, the so-called load vector, with entries

$$b_i = \int_a^b (f \phi_i - \alpha \phi_0' \phi_i' - \beta \phi_N' \phi_i') dx, \quad i = 1, 2, \ldots, N-1. \tag{22}$$

Computer Implementation: The explicit expression for a hat function $\phi_i(x)$ is given by

$$\phi_i(x) = \begin{cases} 0, & a \le x \le x_{i-1}, \\ \dfrac{x - x_{i-1}}{h_i}, & x_{i-1} < x \le x_i, \\ \dfrac{x_{i+1} - x}{h_{i+1}}, & x_i < x \le x_{i+1}, \\ 0, & x_{i+1} < x \le b, \end{cases} \quad i = 1, 2, \ldots, N-1,$$

$$\phi_0(x) = \begin{cases} \dfrac{x_1 - x}{h_1}, & x_0 < x \le x_1, \\ 0, & x_1 < x \le b, \end{cases} \quad \phi_N(x) = \begin{cases} 0, & x_0 < x \le x_{N-1}, \\ \dfrac{x - x_{N-1}}{h_N}, & x_{N-1} < x \le b. \end{cases}$$

For simplicity we assume the partition is uniform so that $h_i = h$ for $i = 1, 2, \ldots, N$. Hence the derivative $\phi_i'(x)$ is either $-\frac{1}{h}, \frac{1}{h}$, or 0 depending on the interval.

It is straightforward to calculate the entries of the stiffness matrix. For $|i - j| > 1$, we have $a_{i,j} = 0$, since ϕ_i and ϕ_j lack overlapping support. However, if $i = j$, then

$$a_{i,j} = \int_a^b (\phi_i')^2 dx = \int_{x_{i-1}}^{x_i} \left(\frac{1}{h}\right)^2 dx + \int_{x_i}^{x_{i+1}} \left(-\frac{1}{h}\right)^2 dx = \frac{2}{h}, \quad i,j = 1, 2, \ldots, N-1,$$

where we have used that $x_i - x_{i-1} = x_{i+1} - x_i = h$. Furthermore, if $j = i + 1$, then

$$a_{i,i+1} = \int_a^b \phi_i' \phi_{i+1}' dx = \int_{x_i}^{x_{i+1}} \left(-\frac{1}{h}\right)\left(\frac{1}{h}\right) dx = -\frac{1}{h}, \quad i,j = 1, 2, \ldots, N-2.$$

Changing i to $i-1$ we also have

$$a_{i-1,i} = \int_a^b \phi_{i-1}' \phi_i dx = \int_{x_{i-1}}^{x_i} \left(\frac{1}{h}\right)\left(-\frac{1}{h}\right) dx = -\frac{1}{h}, \quad i,j = 2, 3, \ldots, N-1.$$

Thus the stiffness matrix is

$$A = \frac{1}{h}\begin{bmatrix} 2 & -1 & 0 & \cdots & 0 \\ -1 & 2 & -1 & \ddots & 0 \\ 0 & -1 & 2 & \cdots & 0 \\ \vdots & \ddots & \ddots & \ddots & -1 \\ 0 & \cdots & 0 & -1 & 2 \end{bmatrix}.$$

The entries b_i of the load vector must often be evaluated using quadrature, since they involve the function f which can be hard to integrate analytically. For example, using the trapezoidal rule one obtains the approximate load vector entries

$$b_1 = \int_a^b (f\phi_1 - \alpha\phi_0'\phi_1' - \beta\phi_N'\phi_1')dx = \int_{x_0}^{x_1} \left(f\phi_1 - \alpha\left(-\frac{1}{h}\right)\left(\frac{1}{h}\right)\right)dx + \int_{x_1}^{x_2} f\phi_1$$

$$= \frac{\alpha}{h} + \int_{x_0}^{x_2} f\phi_1 \approx \frac{\alpha}{h} + hf(x_1),$$

$$b_i = \int_a^b (f\phi_i - \alpha\phi_0'\phi_i' - \beta\phi_N'\phi_i')dx = \int_{x_{i-1}}^{x_{i+1}} f\phi_i dx \approx hf(x_i), \quad i = 2, \ldots, N-2,$$

$$b_{N-1} = \int_a^b (f\phi_{N-1}' - \alpha\phi_0'\phi_{N-1} - \beta\phi_N'\phi_{N-1}')dx$$

$$= \int_{x_{N-2}}^{x_{N-1}} f\phi_{N-1} dx + \int_{x_{N-1}}^{x_N} \left(f\phi_{N-1} - \beta\left(\frac{1}{h}\right)\left(-\frac{1}{h}\right)\right)dx = \int_{x_{N-2}}^{x_N} f\phi_{N-1} dx + \frac{\beta}{h} \approx hf(x_{N-1}) + \frac{\beta}{h}.$$

Assembly: We rewrite (20), (21), (22) as

$$\frac{1}{h}\begin{bmatrix} 2 & -1 & 0 & \cdots & 0 \\ -1 & 2 & -1 & \ddots & 0 \\ 0 & -1 & 2 & \cdots & 0 \\ \vdots & \ddots & \ddots & \ddots & -1 \\ 0 & \cdots & 0 & -1 & 2 \end{bmatrix}\begin{bmatrix} c_1 \\ c_2 \\ c_3 \\ \vdots \\ c_{N-1} \end{bmatrix} = \begin{bmatrix} hf(x_1) + \frac{\alpha}{h} \\ hf(x_2) \\ hf(x_3) \\ \vdots \\ hf(x_{N-1}) + \frac{\beta}{h} \end{bmatrix}$$

We note that $u_h(a) = \alpha = u(a)$ and $u_h(b) = \beta = u(b)$. Therefore, we see that the system matrix A remains the same, and only the first and last entries of the load vector $\mathbf{b}$ need to be modified because of the definition of the basis functions $\{\phi_0, \dots, \phi_N\}$. An alternative approach is to use all the basis functions $\{\phi_0, \dots, \phi_N\}$ to form a larger system of equation, *i.e.*, and $(N+1) \times (N+1)$ system. The procedure for inserting the boundary conditions into the system equation is: enter zeros in the first and $(N+1)$-th rows of the system matrix A except for unity in the main diagonal positions of these two rows, and enter α and β in the first and $(N+1)$-th rows of the vector $\mathbf{b}$, respectively.

General boundary conditions: Let us consider the following two-point BVP: find $u \in C^2(a,b)$ such that

$$-u'' = f(x), \quad x \in [a,b], \quad u(a) = \alpha, \quad \gamma u(b) + u'(b) = \beta, \tag{23}$$

where α, β and γ are given numbers and $f \in C(a,b)$ is a given function. The boundary condition at $x = b$ is called a Robin boundary condition (combination and u and u' is prescribed at $x = b$). In this case, the admissible function space is modified to

$$H_0^1 = \left\{v : \|v\|^2 + \|v'\|^2 < \infty, \ v(a) = 0\right\}.$$

Multiplying (23) by a function $v \in H_0^1$ and integrating by parts gives

$$\int_a^b f v dx = \int_a^b -u'' v dx = -u'(b)v(b) + u'(a)v(a) + \int_a^b u'v'dx$$
$$= -(\beta - \gamma u(b))v(b) + u'(a)v(a) + \int_a^b u'v'dx.$$

Since $v(a) = 0$, we are left with

$$\int_a^b u'v'dx + \gamma u(b)v(b) = \int_a^b f v dx + \beta v(b).$$

Hence, the weak or variational form of (23) reads: Given $u(a) = \alpha$, find the approximate solution $u \in H_0^1$, such that

$$\int_a^b u'v'dx + \gamma u(b)v(b) = \int_a^b f v dx + \beta v(b), \quad \forall \ v \in H_0^1. \tag{24}$$

The FE space V_h^1 is now the set of all continuous piecewise linear functions which vanish at the end point a. The FE approximation of (24) thus reads: Find the piecewise linear approximation u_h to the solution u satisfies

$$\int_a^b u_h'v'dx + \gamma u_h(b)v(b) = \int_a^b f v dx + \beta v(b), \quad \forall \ v \in V_h^1, \tag{25}$$

with $u_h(a) = \alpha$. As before, (25) can be formulated in matrix form.

2.4 Model problem with coefficient and general Robin BCs

Let us consider the following two-point BVP: find $u \in C^2(a,b)$ such that

$$-(p(x)u')' = f(x), \quad x \in I = [a,b], \quad p(a)u'(a) = \kappa_0(u(a) - \alpha), \\ p(b)u'(b) = \kappa_1(u(b) - \beta), \tag{26}$$

where $p = p(x)$ with $p(x) \geq p_0 > 0, f \in L^2(I)$, $\kappa_0, \ \kappa_1 \geq 0$, and α, β are given numbers. Let

$$V = \left\{ v \in C^0(I) : \|v\|^2 + \|v'\|^2 < \infty \right\}.$$

Multiplying (26) by a function $v \in V$ and integrating by parts gives

$$\int_a^b fvdx = \int_a^b -(pu')'vdx = \int_a^b pu'v'dx - p(b)u'(b)v(b) + p(a)u'(a)v(a)$$
$$= \int_a^b pu'v'dx - \kappa_1(u(b) - \beta)v(b) + \kappa_0(u(a) - \alpha)v(a).$$

We gather all u-independent terms on the left and obtain

$$\int_a^b pu'v'dx - \kappa_1 u(b)v(b) + \kappa_0 u(a)v(a) = \int_a^b fvdx - \kappa_1 \beta v(b) + \kappa_0 \alpha v(a), \quad \forall \ v \in V.$$

The FE method consists of finding $u_h \in V_h = \left\{ v \in C^0(a,b) | \ v|_{I_i} \in P^1(I_i) \right\}$ such that

$$\int_a^b pu_h'v'dx - \kappa_1 u_h(b)v(b) + \kappa_0 u_h(a)v(a) = \int_a^b fvdx - \kappa_1 \beta v(b) + \kappa_0 \alpha v(a), \quad \forall v \in V_h. \tag{27}$$

Implementation: We need to assemble a stiffness matrix A and a load vector b. Substituting $u_h = \sum_{i=0}^N c_i \phi_i$ into (27) and taking $v = \phi_j$ for $j = 0, 1, \dots, N$ yields

$$\sum_{i=0}^N \int_a^b p\phi_i'\phi_j'dx - \kappa_1 \phi_i(b)\phi_j(b) + \kappa_0 \phi_i(a)\phi_j(a) = \int_a^b f\phi_j dx - \kappa_1 \beta \phi_j(b) + \kappa_0 \alpha \phi_j(a),$$
$$\forall j = 0, 1, \dots, N.$$

which is a $(N+1) \times (N+1)$ system of equations for c_i. In matrix form we write $A\mathbf{c} = \mathbf{b}$, where $\mathbf{c} = [c_0, \dots, c_N]^t$ is a $(N+1)$ vector containing the unknown coefficients $c_i, \ i = 0, 1, \dots, N$, A is a $(N+1) \times (N+1)$ matrix with entries

$$a_{i,j} = \int_a^b p\phi_i'\phi_j'dx - \kappa_1 \phi_i(b)\phi_j(b) + \kappa_0 \phi_i(a)\phi_j(a), \quad i,j = 0, 1, \dots, N,$$

and $\mathbf{b}$ is a $(N+1)$ vector with entries

$$b_j = \int_a^b f\phi_j dx - \kappa_1 \beta \phi_j(b) + \kappa_0 \alpha \phi_j(a), \quad j = 0, 1, \dots, N.$$

Let for simplification $p = 1$. Then the matrix A and the vector $\mathbf{b}$ (when using the trapezoidal rule) are given by

$$A = \begin{bmatrix} \kappa_0 + \frac{1}{h_1} & -\frac{1}{h_1} & 0 & \cdots & 0 \\ -\frac{1}{h_1} & \frac{1}{h_1} + \frac{1}{h_2} & -\frac{1}{h_2} & \ddots & 0 \\ 0 & -\frac{1}{h_2} & \ddots & \cdots & 0 \\ \vdots & \ddots & \ddots & \ddots & -\frac{1}{h_N} \\ 0 & \cdots & 0 & -\frac{1}{h_N} & \frac{1}{h_N} - \kappa_1 \end{bmatrix}, \quad \mathbf{b} = \begin{bmatrix} \frac{h_1}{2} f_0 + \kappa_0 \alpha \\ \frac{h_1 + h_2}{2} f_1 \\ \vdots \\ \frac{h_{N-1} + h_N}{2} f_{N-1} \\ \frac{h_N}{2} f_N - \kappa_1 \beta \end{bmatrix}.$$

2.5 The FE method using Lagrange $\mathbb{P}_2$ elements

Let $a = x_0 < x_1 < \cdots < x_N = b$ be a regular partition of the interval $[a, b]$. Suppose that the length of $I_i = [x_{i-1}, x_i]$ is $h_i = x_i - x_{i-1}$. Let $P^k = \left\{ p(x) = \sum_{j=0}^{k} c_j x^j, c_j \in \mathbb{R} \right\}$ denotes the vector space of polynomials in one variable and of degree less than or equal to k. The FE method for Lagrange P^2 elements involves the discrete space:

$$V_h^2 = \{v(x) \in C^0[a, b], \quad v|_{I_i} \in P^2(I_i), \quad i = 1, \ldots, N\},$$

and its subspace $V_{0,h}^2 = \left\{ v \in V_h^2 | \; v(a) = v(b) = 0 \right\}$. These spaces are composed of continuous, piecewise parabolic functions (polynomials of degree less than or equal to 2). The P^2 FE method consists in applying the internal variational approximation approach to these spaces.

Lemma 2.2 *The space V_h^2 is a subspace of $H^1[a, b]$ of dimension $2N + 1$. Every function $v_h \in V_h^2$ is uniquely defined by its values at the mesh vertices x_j, $j = 0, 1, \ldots, N$ and at the midpoints $x_{j+\frac{1}{2}} = \frac{x_j + x_{j+1}}{2} = x_j + \frac{h_{j+1}}{2}$, $j = 0, 1, \ldots, N-1$, where $h_{j+1} = x_{j+1} - x_j$:*

$$v_h(x) = \sum_{j=0}^{N} v_h(x_j) \phi_j(x) + \sum_{j=0}^{N-1} v_h\left(x_{j+\frac{1}{2}}\right) \phi_{j+\frac{1}{2}}(x), \quad \forall \; x \in [a, b],$$

where $\left\{\phi_j\right\}_{j=0}^{N}$ is the basis of the shape functions ϕ_j defined as:

$$\phi_j(x) = \phi\left(\frac{x - x_j}{h_{j+1}}\right), \quad j = 0, 1, \ldots, N, \quad \phi_{j+\frac{1}{2}}(x) = \psi\left(\frac{x - x_{j+\frac{1}{2}}}{h_{j+1}}\right),$$
$$j = 0, 1, \ldots, N-1,$$

with

$$\phi(\xi) = \begin{cases} (1+\xi)(1+2\xi), & \xi \in [-1, 0], \\ (1-\xi)(1-2\xi), & \xi \in [0, 1], \\ 0, & |\xi| > 1, \end{cases} \qquad \psi(\xi) = \begin{cases} 1 - 4\xi^2, & |\xi| \le \frac{1}{2}, \\ 0, & |\xi| > \frac{1}{2}, \end{cases} \tag{28}$$

Figure 3 shows the global shape functions for the space V_h^2 and the three quadratic Lagrange P^2 shape functions on the reference interval $[-1, 1]$.

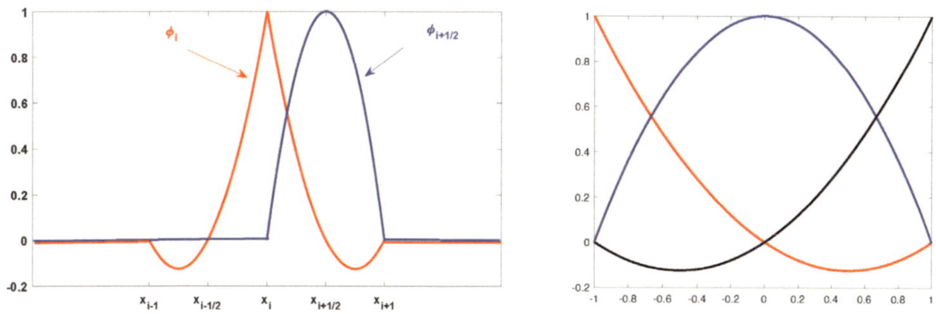

Figure 3.
(left) global shape functions for the space V_h^2. (right) the three quadratic Lagrange P^2 shape functions on the reference interval $[-1, 1]$.

Remark 2.5 *Notice that we have:*

$$\phi_j(x_j) = \delta_{ij}, \quad \phi_j\left(x_{j+\frac{1}{2}}\right) = 0, \quad \phi_{j+\frac{1}{2}}(x_j) = 0, \quad \phi_{j+\frac{1}{2}}\left(x_{j+\frac{1}{2}}\right) = \delta_{ij}.$$

Corollary 2.1 *The space $V_{0,h}^2$ is a subspace of $H_0^1[a, b]$ of dimension $2N - 1$ and every function $v_h \in V_{0,h}^2$ is uniquely defined by its values at the mesh vertices x_j, $j = 1, 2, \ldots, N - 1$ and at the midpoints $x_{j+\frac{1}{2}}$, $j = 0, 1, \ldots, N - 1$:*

$$v_h(x) = \sum_{j=1}^{N-1} v_h(x_j)\phi_j(x) + \sum_{j=0}^{N-1} v_h\left(x_{j+\frac{1}{2}}\right)\phi_{j+\frac{1}{2}}(x), \quad \forall\ x \in [a, b],$$

where $\left\{\phi_j\right\}_{j=0}^{N}$ is the basis of the shape functions ϕ_j defined as:

$$\phi_j(x) = \phi\left(\frac{x - x_j}{h_{j+1}}\right), \quad j = 0, 1, \ldots, N, \quad \phi_{j+\frac{1}{2}}(x) = \psi\left(\frac{x - x_{j+\frac{1}{2}}}{h_{j+1}}\right), \quad j = 0, 1, \ldots, N - 1,$$

with $\phi(\xi)$ and $\psi(\xi)$ are defined by (28).

2.5.1 Homogeneous boundary conditions

The variational formulation of the internal approximation of the Dirichlet BVP (3) consists now in finding $u_h \in V_{0,h}^2$, such that:

$$\int_a^b u_h' v' dx + \int_a^b q u_h v dx = \int_a^b f v dx, \quad \forall\ v \in V_{h,0}^2.$$

Here, it is convenient to introduce the notation $x_{\frac{j}{2}}$, $j = 1, \ldots, 2N - 1$ for the mesh points and $\phi_{\frac{j}{2}}$, $j = 1, \ldots, 2N - 1$ for the basis of $V_{0,h}^2$. Using these notations, we have:

$$u_h = \sum_{j=1}^{2N-1} c_{\frac{j}{2}} \phi_{\frac{j}{2}}(x),$$

where $c_{\frac{j}{2}} = u_h\left(x_{\frac{j}{2}}\right) \approx u\left(x_{\frac{j}{2}}\right)$ are the unknowns coefficients. This formulation leads to solve in $\mathbb{R}^{2N-1}$ a linear system:

$$A\mathbf{c} = \mathbf{b},$$

where $\mathbf{c} = \left[c_{\frac{1}{2}}, c_1, \dots, c_{N-\frac{1}{2}}\right]^t \in \mathbb{R}^{2N-1}$ is the unknown vector containing the coefficients $c_{\frac{j}{2}}$, $j = 1, 2, \dots, 2N-1$, A is an $(2N-1) \times (2N-1)$ matrix with entries

$$a_{ij} = \int_a^b \left(\phi_{\frac{i}{2}}' \phi_{\frac{j}{2}}' + q\phi_{\frac{i}{2}}\phi_{\frac{j}{2}}\right)dx, \quad i,j = 1, 2, \dots, 2N-1,$$

and load vector $\mathbf{b} \in \mathbb{R}^{2N-1}$ has entries

$$b_{\frac{i}{2}} = \int_a^b f\phi_{\frac{i}{2}}dx, \quad i = 1, 2, \dots, 2N-1.$$

Since the shape functions ϕ_i have a small support, the matrix A is mostly composed of zeros. However, the main difference with the Lagrange P^1 FE method, the matrix A is no longer a tridiagonal matrix.

Computer Implementation: The coefficients of the matrix A can be computed more easily by considering the following change of variables, for $\xi \in [-1, 1]$:

$$x = \frac{x_j + x_{j-1}}{2} + \frac{x_j - x_{j-1}}{2}\xi = x_{j-\frac{1}{2}} + \frac{x_j - x_{j-1}}{2}\xi, \quad \forall\ x \in \left[x_{j-1}, x_j\right],$$
$$j = 1, 2, \dots, N.$$

Hence, the shape functions can be reduced to only three basic shape functions (**Figure 3**):

$$\hat{\phi}_{-1}(\xi) = \frac{\xi(\xi-1)}{2}, \quad \hat{\phi}_0(\xi) = (1-\xi)(1+\xi), \quad \hat{\phi}_1(\xi) = \frac{\xi(\xi+1)}{2}.$$

Their respective derivatives are

$$\frac{d\hat{\phi}_{-1}(\xi)}{d\xi} = \frac{2\xi-1}{2}, \quad \frac{d\hat{\phi}_0(\xi)}{d\xi} = -2\xi, \quad \frac{d\hat{\phi}_1(\xi)}{d\xi} = \frac{2\xi+1}{2}.$$

This approach consists in considering all computations on an interval $I_i = [x_{i-1}, x_i]$ on the reference interval $[-1, 1]$. Thus, we have:

$$\frac{d\phi_i(x)}{dx} = \frac{d\phi_i\left(x_{i-1/2} + \frac{x_i - x_{i-1}}{2}\xi\right)}{d\xi}\frac{d\xi}{dx} = \frac{2}{x_i - x_{i-1}}\frac{d\hat{\phi}_k(\xi)}{d\xi} = \frac{2}{h_i}\frac{d\hat{\phi}_k(\xi)}{d\xi}.$$

In this case, the elementary contributions of the element I_i to the stiffness matrix and to the mass matrix are given by the 3×3 matrices K^{I_i} and M^{I_i}:

$$K^{I_i} = \int_{I_i} \begin{bmatrix} \phi_{i-1}'\phi_{i-1}' & \phi_{i-1}'\phi_{i-\frac{1}{2}}' & \phi_{i-1}'\phi_i' \\ \phi_{i-\frac{1}{2}}'\phi_{i-1}' & \phi_{i-\frac{1}{2}}'\phi_{i-\frac{1}{2}}' & \phi_{i-\frac{1}{2}}'\phi_i' \\ \phi_i'\phi_{i-1}' & \phi_i'\phi_{i-\frac{1}{2}}' & \phi_i'\phi_i' \end{bmatrix} dx = \frac{2}{h_i}\int_{-1}^{1} \begin{bmatrix} \hat{\phi}_{-1}'\hat{\phi}_{-1}' & \hat{\phi}_{-1}'\hat{\phi}_0' & \hat{\phi}_{-1}'\hat{\phi}_1' \\ \hat{\phi}_0'\hat{\phi}_{-1}' & \hat{\phi}_0'\hat{\phi}_0' & \hat{\phi}_0'\hat{\phi}_1' \\ \hat{\phi}_1'\hat{\phi}_{-1}' & \hat{\phi}_1'\hat{\phi}_0' & \hat{\phi}_1'\hat{\phi}_1' \end{bmatrix} d\xi$$
$$= \frac{1}{3h_i}\begin{bmatrix} 7 & -8 & 1 \\ -8 & 16 & -8 \\ 1 & -8 & 7 \end{bmatrix},$$

$$M^{I_i} = \int_{I_i} \begin{bmatrix} \phi_{i-1}\phi_{i-1} & \phi_{i-1}\phi_{i-\frac{1}{2}} & \phi_{i-1}\phi_i \\ \phi_{i-\frac{1}{2}}\phi_{i-1} & \phi_{i-\frac{1}{2}}\phi_{i-\frac{1}{2}} & \phi_{i-\frac{1}{2}}\phi_i \\ \phi_i\phi_{i-1} & \phi_i\phi_{i-\frac{1}{2}} & \phi_i\phi_i \end{bmatrix} dx = \frac{h_i}{2}\int_{-1}^{1} \begin{bmatrix} \hat{\phi}_{-1}\hat{\phi}_{-1} & \hat{\phi}_{-1}\hat{\phi}_0 & \hat{\phi}_{-1}\hat{\phi}_1 \\ \hat{\phi}_0\hat{\phi}_{-1} & \hat{\phi}_0\hat{\phi}_0 & \hat{\phi}_0\hat{\phi}_1 \\ \hat{\phi}_1\hat{\phi}_{-1} & \hat{\phi}_{-1}\hat{\phi}_0 & \hat{\phi}_1\hat{\phi}_1 \end{bmatrix} d\xi$$

$$= \frac{h_i}{30} \begin{bmatrix} 4 & 2 & -1 \\ 2 & 16 & 2 \\ -1 & 2 & 4 \end{bmatrix}.$$

Coefficients of the right-hand side b: Usually, the function f is only known by its values at the mesh points $x_{\frac{i}{2}}$, $i = 0, 1, \dots, 2N$ and thus, we use the decomposition of f in the basis of shape functions $\phi_{\frac{i}{2}}$, $i = 0, 1, \dots, 2N$ as $f(x) = \sum_{j=0}^{2N} f\left(x_{\frac{j}{2}}\right)\phi_{\frac{j}{2}}$. Each component $b_{\frac{i}{2}}$ of the right-hand side vector is obtained as $b_{\frac{i}{2}} = \sum_{k=1}^{N}\int_{x_{k-1}}^{x_k} f\phi_{\frac{i}{2}}dx$. Using the previous decomposition of f, we obtain:

$$b_{\frac{i}{2}} = \sum_{k=1}^{N}\int_{x_{k-1}}^{x_k}\sum_{j=0}^{2N} f\left(x_{\frac{j}{2}}\right)\phi_{\frac{j}{2}}\phi_{\frac{i}{2}}dx = \sum_{j=0}^{2N} f\left(x_{\frac{j}{2}}\right)\left(\sum_{k=1}^{N}\int_{x_{k-1}}^{x_k}\phi_{\frac{i}{2}}\phi_{\frac{j}{2}}dx\right).$$

Thus, the problem is reduced to computing the integrals $\int_{x_{k-1}}^{x_k}\phi_{\frac{i}{2}}\phi_{\frac{j}{2}}dx$. It is easy to see that we obtain expressions very similar to that of the mass matrix. More precisely, the element $I_i = [x_{i-1}, x_i]$ will contribute to only three components of indices $i-1$, $i-\frac{1}{2}$ and i as:

$$\mathbf{b}^{I_i} = \frac{h_i}{30}\begin{bmatrix} 4 & 2 & -1 \\ 2 & 16 & 2 \\ -1 & 2 & 4 \end{bmatrix}\begin{bmatrix} f(x_{i-1}) \\ f\left(x_{i-\frac{1}{2}}\right) \\ f(x_i). \end{bmatrix}.$$

2.5.2 Nonhomogeneous boundary conditions

Consider the following two-point BVP: find $u \in C^2(a,b)$ such that

$$-u'' + q(x)u = f(x), \quad x \in [a,b], \quad u(a) = \alpha, \quad u(b) = \beta, \tag{29}$$

where α and β are given constants and $f \in C(a,b)$ is a given function.

Multiplying (29) by a function $v \in H_0^1 = \left\{v : \|v\|^2 + \|v'\|^2 < \infty, \; v(a) = v(b) = 0\right\}$ and integrating by parts gives

$$\int_a^b fvdx = \int_a^b(-u'' + qu)vdx = -u'(b)v(b) + u'(a)v(a) + \int_a^b(u'v' + quv)dx = \int_a^b u'v'dx.$$

Hence, the weak or variational form of (29) reads: Given $u(a) = \alpha$, $u(b) = \beta$, find $u \in H^1 = \left\{v : \|v\|^2 + \|v'\|^2 < \infty\right\}$, such that

$$\int_a^b(u'v' + quv)dx = \int_a^b fvdx, \quad \forall\; v \in H_0^1.$$

Let V_h^2 and $V_{h,0}^2$, respectively, be the space of all continuous piecewise quadratic functions and the space of all continuous piecewise quadratic functions which vanish at the end points a and b, on a uniform partition $a = x_0 < x_1 < \cdots < x_N = b$ of the interval $[a, b]$.

The FE method scheme consists of finding $u_h \in V_h^2$, such that:

$$\int_a^b u_h' v' dx + \int_a^b q u_h v dx = \int_a^b f v dx, \quad \forall\ v \in V_{h,0}^2.$$

Introduce the notation $x_{\frac{j}{2}},\ j = 0, 1, \ldots, 2N-1, 2N$ for the mesh points and $\phi_{\frac{j}{2}},\ j = 0, 1, \ldots, 2N-1, 2N$ for the basis of V_h^2 and $\phi_{\frac{j}{2}},\ j = 1, \ldots, 2N-1$ for the basis of $V_{0,h}^2$. Using these notations, we have:

$$u_h = \sum_{j=0}^{2N} c_{\frac{j}{2}} \phi_{\frac{j}{2}}(x),$$

where $c_{\frac{j}{2}} = u_h\left(x_{\frac{j}{2}}\right) \approx u\left(x_{\frac{j}{2}}\right)$ are the unknowns coefficients. We note that $c_0 = u_h(x_0) = \alpha$ and $c_{2N} = u_h(x_N) = \beta$. This formulation leads to solve in $\mathbb{R}^{2N-1}$ a linear system:

$$A\mathbf{c} = \mathbf{b},$$

where $\mathbf{c} = \left[c_{\frac{1}{2}}, c_1, \ldots, c_{N-\frac{1}{2}}\right]^t \in \mathbb{R}^{2N-1}$ is the unknown vector containing the coefficients $c_{\frac{j}{2}},\ j = 1, 2, \ldots, 2N-1$, A is an $(2N-1) \times (2N-1)$ matrix with entries

$$a_{ij} = \int_a^b \left(\phi_{\frac{i}{2}}' \phi_{\frac{j}{2}}' + q \phi_{\frac{i}{2}} \phi_{\frac{j}{2}}\right) dx, \quad i,j = 1, 2, \ldots, 2N-1,$$

and the load vector $\mathbf{b} \in \mathbb{R}^{2N-1}$ has entries

$$b_{\frac{i}{2}} = \int_a^b f \phi_{\frac{i}{2}} dx - \alpha \int_a^b \left(\phi_{\frac{i}{2}}' \phi_0' + q \phi_{\frac{i}{2}} \phi_0\right) dx - \beta \int_a^b \left(\phi_{\frac{i}{2}}' \phi_N' + q \phi_{\frac{i}{2}} \phi_N\right) dx, \quad i = 1, 2, \ldots, 2N-1.$$

Clearly, the only extra terms are given in the vector with entries

$$\tilde{b}_{\frac{i}{2}} = -\alpha \int_a^b \left(\phi_{\frac{i}{2}}' \phi_0' + q \phi_{\frac{i}{2}} \phi_0\right) dx - \beta \int_a^b \left(\phi_{\frac{i}{2}}' \phi_N' + q \phi_{\frac{i}{2}} \phi_N\right) dx, \quad i = 1, 2, \ldots, 2N-1.$$

Suppose $q = 0$ then for $N \geq 2$, we have

$$\tilde{b}_{\frac{1}{2}} = -\alpha \int_a^b \phi_{\frac{1}{2}}' \phi_0' dx - \beta \int_a^b \phi_{\frac{1}{2}}' \phi_N' dx = -\alpha \int_{x_0}^{x_1} \phi_{\frac{1}{2}}' \phi_0' = \frac{8\alpha}{3h_1},$$

$$\tilde{b}_1 = -\alpha \int_a^b \phi_1' \phi_0' dx - \beta \int_a^b \phi_1' \phi_N' dx = -\alpha \int_{x_0}^{x_1} \phi_1' \phi_0' = -\frac{\alpha}{3h_1},$$

$$\tilde{b}_{\frac{i}{2}} = -\alpha \int_a^b \phi_{\frac{i}{2}}' \phi_0' dx - \beta \int_a^b \phi_{\frac{i}{2}}' \phi_N' dx = 0, \quad i = 3, \ldots, 2N-3,$$

$$\tilde{b}_{N-1} = -\alpha \int_a^b \phi_{N-1}' \phi_0' dx - \beta \int_a^b \phi_{N-1}' \phi_N' dx = -\beta \int_{x_{N-1}}^{x_N} \phi_{N-1}' \phi_N' dx = -\frac{\beta}{3h_1},$$

$$\tilde{b}_{N-\frac{1}{2}} = -\alpha \int_a^b \phi_{N-\frac{1}{2}}' \phi_0' dx - \beta \int_a^b \phi_{N-\frac{1}{2}}' \phi_N' dx = -\beta \int_{x_{N-1}}^{x_N} \phi_{N-\frac{1}{2}}' \phi_N' dx = \frac{8\beta}{3h_1}.$$

3. The FE for elliptic PDEs

Here, we apply the FE method for two-dimensional elliptic problem: Find u such that

$$-\nabla\cdot(a\nabla u)+bu=f(\mathbf{x}),\quad \mathbf{x}\in\Omega,\quad a\nabla u\cdot\mathbf{n}=\kappa(g-u),\quad \text{on }\partial\Omega, \tag{30}$$

where $a>0$, $b\geq 0$, $\kappa\geq 0$, $f\in L^2(\Omega)$ and $g\in C^0(\partial\Omega)$.

3.1 Meshes

Let $\Omega\subset\mathbb{R}^2$ bounded with $\partial\Omega$ assumed to be polygonal. A triangulation $\mathcal{T}_h$ of Ω is a set of triangles T such that $\Omega=\bigcup_{T\in\mathcal{T}_h}T$, and two triangles intersect by either a common triangle edge, or a corner, or nothing. Corners will be referred to as nodes. We let $h_T=diam(T)$ the length or the largest edge.

Let $\mathcal{T}_h$ have N nodes and M triangles. The data is stored in two matrices. The matrix $P\in\mathbb{R}^{2\times N}$ describes the nodes $((x_1,y_1),\dots,(x_N,y_N))$ and the matrix $K\in\mathbb{R}^{3\times M}$ describes the triangles, *i.e.*, it describes which nodes (numerated from 1 to N) form a triangle T and how it is orientated:

$$P=\begin{bmatrix}x_1 & x_2 & \cdots & x_N\\ y_1 & y_2 & \cdots & y_N\end{bmatrix},\quad K=\begin{bmatrix}n_1^\alpha & n_2^\alpha & \cdots & n_M^\alpha\\ n_1^\beta & n_2^\beta & \cdots & n_M^\beta\\ n_1^\gamma & n_2^\gamma & \cdots & n_M^\gamma\end{bmatrix}.$$

This means that triangle T_i is formed by the nodes n_i^α, n_i^β, and n_i^γ (enumeration in counter-clockwise direction).

The Delaunay algorithm determine a triangulation with the given points as triangle nodes. Delaunay triangulations are optimal in the sense that the angles of all triangles are maximal.

Matlab has a built in toolbox called PDE Toolbox and includes a mesh generation algorithm.

3.2 Piecewise polynomial spaces

Let T be a triangle with nodes $N_1=(x_1,y_1)$, $N_2=(x_2,y_2)$, and $N_3=(x_3,y_3)$. We define

$$P^1(T)=\{v\in C^0(T)|\ v(x,y)=c_1+c_2x+c_3y,\ \ c_1,c_2,c_3\in\mathbb{R}\}.$$

Now let $v_i=v(N_i)$ for $i=1,2,3$. Note that $v\in P^1(T)$ is determined by $\{v_i\}_{i=1}^3$. Given v_i we compute c_i by

$$\begin{bmatrix}1 & x_1 & y_1\\ 1 & x_2 & y_2\\ 1 & x_3 & y_3\end{bmatrix}\begin{bmatrix}c_1\\ c_2\\ c_3\end{bmatrix}=\begin{bmatrix}v_1\\ v_2\\ v_3\end{bmatrix}.$$

This is solvable due to

$$\det\begin{bmatrix}1 & x_1 & y_1\\ 1 & x_2 & y_2\\ 1 & x_3 & y_3\end{bmatrix} = 2|T| \neq 0, \quad \begin{bmatrix}1 & x_1 & y_1\\ 1 & x_2 & y_2\\ 1 & x_3 & y_3\end{bmatrix}^{-1}$$
$$= \frac{1}{2|T|}\begin{bmatrix}x_2y_3 - x_3y_2 & x_3y_1 - x_1y_3 & x_1y_2 - x_2y_1\\ y_2 - y_3 & y_3 - y_1 & y_1 - y_2\\ x_3 - x_2 & x_1 - x_3 & x_3 - x_1\end{bmatrix},$$

where $|T| = \frac{1}{2}\left(x_2y_3 - x_3y_2 - x_1y_3 + x_3y_1 + x_1y_2 - x_2y_1\right)$, which is $\pm$ the area of the triangle T.

Let $\lambda_j \in P^1(T)$ be given by the nodal values $\lambda_j(N_i) = \delta_{ij}$, where δ_{ij} is the Kronecker symbol. This gives us $v(x,y) = \alpha_1\lambda_1(x,y) + \alpha_2\lambda_2(x,y) + \alpha_3\lambda_3(x,y)$, where $\alpha_i = v(N_i)$ for $i = 1,2,3$. We can compute $\lambda_i(x,y)$ as follows: Let $\lambda_i(x,y) = a_i + b_ix + c_iy$. Using $\lambda_j(N_i) = \delta_{ij}$, we get

$$\begin{bmatrix}1 & x_1 & y_1\\ 1 & x_2 & y_2\\ 1 & x_3 & y_3\end{bmatrix}\begin{bmatrix}a_1\\ b_1\\ c_1\end{bmatrix} = \begin{bmatrix}1\\ 0\\ 0\end{bmatrix}, \quad \begin{bmatrix}1 & x_1 & y_1\\ 1 & x_2 & y_2\\ 1 & x_3 & y_3\end{bmatrix}\begin{bmatrix}a_2\\ b_2\\ c_2\end{bmatrix} = \begin{bmatrix}0\\ 1\\ 0\end{bmatrix}, \quad \begin{bmatrix}1 & x_1 & y_1\\ 1 & x_2 & y_2\\ 1 & x_3 & y_3\end{bmatrix}\begin{bmatrix}a_3\\ b_3\\ c_3\end{bmatrix} = \begin{bmatrix}0\\ 0\\ 1\end{bmatrix}.$$

Solving the systems, we get

$$\lambda_1(x,y) = \frac{1}{2|T|}\left(x_2y_3 - x_3y_2 + (y_2 - y_3)x + (x_3 - x_2)y\right),$$
$$\lambda_2(x,y) = \frac{1}{2|T|}\left(x_3y_1 - x_1y_3 + (y_3 - y_1)x + (x_1 - x_3)y\right),$$
$$\lambda_3(x,y) = \frac{1}{2|T|}\left(x_1y_2 - x_2y_1 + (y_1 - y_2)x + (x_3 - x_1)y\right).$$

Let $\mathcal{T}_h$ be a triangulation of Ω, then we let

$$V_h = \left\{v \in C(\Omega) |\ v|_T \in P^1(T),\ \forall\ T \in \mathcal{T}_h\right\}.$$

Functions in V_h are piecewise linear and continuous. We know that $v \in V_h$ is uniquely determined by $\{v(N_i),\ i = 1,2,\dots,N\}$. We let $\phi_j(N_i) = \delta_{ij}$ and let $\left\{\phi_j,\ j = 1,2,\dots,N\right\} \subset V_h$ be a basis for V_h (hat functions), *i.e.*,

$$v(x,y) = \sum_{i=1}^{N} \alpha_i\phi_i(x,y), \quad \alpha_i = v(N_i), \quad i = 1,2,\dots,N.$$

3.3 Interpolation

Given $u \in C(T)$ on a single triangle with nodes $N_i = \left(x_i, y_i\right)$, $i = 1,2,3$, we let

$$\pi u(x,y) = \sum_{i=1}^{3} u(N_i)\phi_i(x,y),$$

in particular $\pi u(N_i) = u(N_i)$, $i = 1,2,\dots,N$. We want to estimate the interpolation error $u - \pi u$. Let

$$\|u\|^2_{L^2(\Omega)} = \int_\Omega |u(x)|^2 dxdy, \quad \|Du\|^2_{L^2(\Omega)} = \|u_x\|^2_{L^2(\Omega)} + \|u_y\|^2_{L^2(\Omega)},$$
$$\|D^2u\|^2_{L^2(\Omega)} = \|u_{xx}\|^2_{L^2(\Omega)} + 2\|u_{xy}\|^2_{L^2(\Omega)} + \|u_{yy}\|^2_{L^2(\Omega)}.$$

Theorem 3.1 *Suppose that* $u \in C^2(T)$. *Then the following hold*

$$\|u - \pi u\|_{L^2(T)} \leq Ch_T^2 \|D^2u\|_{L^2(T)}, \quad \|D(u - \pi u)\|_{L^2(T)} \leq Ch_T \|D^2u\|_{L^2(T)},$$

where C is a generic constant independent of h_T and u, but it depends on the ratio between smallest and largest interior angle of the triangle T.

Now, we consider the piecewise continuous interpolant $\pi u = \sum_{i=1}^N u(N_i)\phi_i$.
Theorem 3.2 *Suppose that* $u \in C^2(T)$ *for all* $T \in \mathcal{T}_h$. *Then the following hold*

$$\|u - \pi u\|^2_{L^2(\Omega)} \leq C \sum_{T \in \mathcal{T}_h} h_T^4 \|D^2u\|^2_{L^2(T)}, \quad \|D(u - \pi u)\|^2_{L^2(\Omega)} \leq C \sum_{T \in \mathcal{T}_h} h_T^2 \|D^2u\|^2_{L^2(T)},$$

where C is a generic constant independent of h and u, but it depends on the ratio between smallest and largest interior angle of the triangles of $\mathcal{T}_h$. Here $\|D(u - \pi u)\|^2_{L^2(\Omega)} = \sum_{T \in \mathcal{T}_h} \|D(u - \pi u)\|^2_{L^2(T)}$.

3.4 L^2-projection

Let $\Omega \subset \mathbb{R}^2$. We consider the space $L^2(\Omega) = \{v | \int_\Omega v^2(x, y)dxdy < \infty\}$. Let $u \in L^2(\Omega)$. We define the L^2-projection $P_h : L^2(\Omega) \to V_h = \{v \in C^0(\Omega) | \ v|_T \in P^1(T), \ \forall \ T \in \mathcal{T}_h\}$ by $P_h u \in V_h$ such that

$$\int_\Omega (u - P_h u)v_h dxdy = 0, \quad \forall v_h \in V_h.$$

The problem of finding $P_h u \in V_h$ is equivalent to solve the following linear system

$$\int_\Omega (u - P_h u)\phi_i dxdy = 0, \quad i = 1, 2 \ldots, N,$$

where $\{\phi_i\}_{i=1}^N$ is a basis of V_h.

Since $P_h u \in V_h$ we can express it as $P_h u = \sum_{i=1}^N c_i \phi_i(x, y)$, where $c_i \in \mathbb{R}$. Therefore, to find $P_h u \in V_h$ we need to find $c_1, c_2, \ldots, c_N \in \mathbb{R}$ such that

$$\sum_{i=1}^N c_i \int_\Omega \phi_i \phi_j dxdy = \int_\Omega u\phi_j dxdy, \quad j = 1, 2, \ldots, N.$$

The problem can be expressed as a linear system of equations $M\mathbf{c} = \mathbf{b}$, where $\mathbf{c} = [c_1, c_2, \ldots, c_N]^t$ and the entries of the matrix $M \in \mathbb{R}^{N \times N}$ and the vector $\mathbf{b} \in \mathbb{R}^N$ are given by

$$m_{ij} = \int_\Omega \phi_i \phi_j dxdy, \quad b_j = \int_\Omega u\phi_j dxdy.$$

In general, we use a quadrature rule to approximate integrals. The general form is

$$\int_T f(x,y)dxdy \approx \sum_{j=1}^{n} \omega_j f(\overline{N}_j),$$

where the ω_j's denote the weights and the $(\overline{N}_j)$'s the quadrature points.

Lemma 3.1 *The mass matrix M with entries* $m_{ij} = \int_\Omega \phi_i \phi_j dxdy$ *is symmetric and positive definite.*

Theorem 3.3 *For any* $u \in L^2(\Omega)$ *the* L^2*-projection* $P_h u$ *exists and is unique.*

3.5 A priori error estimate

Theorem 3.4 *Let* $u \in L^2(\Omega)$ *and let* $P_h u$ *be the* L^2*-projection of* u*, then*

$$\|u - P_h u\|_{L^2(\Omega)} \le \|u - v_h\|_{L^2(\Omega)}, \quad \forall\ v_h \in V_h.$$

Theorem 3.5 *Suppose that* $u \in C^2(\Omega)$ *with* $u \in C^2(T)$ *for all* $T \in \mathcal{T}_h$*. Then there exists a constant C such that*

$$\|u - P_h u\|^2_{L^2(\Omega)} \le C \sum_{T \in \mathcal{T}_h} h_T^4 \|D^2 u\|^2_{L^2(T)}.$$

3.6 The FE method for general elliptic problem

The FE method was designed to approximate solutions to complicated equations of elasticity and structural mechanics, usually modeled by elliptic type equations, with complicated geometries. It has been developed for other applications as well.

Consider the following two-dimensional elliptic problem: Find u such that

$$-\nabla \cdot (a \nabla u) + bu = f, \quad \text{in } \Omega, \quad a\nabla u \cdot \mathbf{n} = \kappa(g - u), \quad \text{on } \partial\Omega, \tag{31}$$

where $a > 0$, $b \ge 0$, $\kappa \ge 0$, $f \in L^2(\Omega)$ and $g \in C^0(\partial\Omega)$. We seek a weak solution u in

$$V = H^1(\Omega) = \left\{ v \in L^2(\Omega) |\ v \text{ has a weak derivative and } \|v\|_{L^2(\Omega)} + \|\nabla v\|_{L^2(\Omega)} < \infty \right\}.$$

In order to derive the weak formulation, we multiply (31) with $v \in V$, integrate over Ω and use Green's formula to obtain

$$\begin{aligned}
\int_\Omega f v dxdy &= -\int_\Omega v \nabla \cdot (a \nabla u) dxdy + \int_\Omega buv dxdy \\
&= \int_\Omega a \nabla u \cdot \nabla v dxdy - \int_{\partial\Omega} v(a\nabla u) \cdot \mathbf{n} ds + \int_\Omega buv dxdy \\
&= \int_\Omega a \nabla u \cdot \nabla v dxdy + \int_\Omega buv dxdy + \int_{\partial\Omega} \kappa(u - g) v ds.
\end{aligned}$$

We obtain the weak form: Find $u \in V$ such that

$$\int_\Omega a \nabla u \cdot \nabla v dxdy + \int_\Omega buv dxdy + \int_{\partial\Omega} \kappa uv ds = \int_\Omega f v dxdy + \int_{\partial\Omega} \kappa g v ds, \quad v \in V. \tag{32}$$

We can formulate the method as in the 1D case by using the weak formulation (32). The FE method in 2D is defined as follows: Find $u_h \in V_h$ such that

$$\int_\Omega a\nabla u_h \cdot \nabla v_h dxdy + \int_\Omega b u_h v_h dxdy + \int_{\partial\Omega} \kappa u_h v_h ds = \int_\Omega f v_h dxdy + \int_{\partial\Omega} \kappa g v_h ds, \quad v_h \in V_h, \tag{33}$$

where $V_h = \{v \in V | \; v|_T \in P^1(T), \; \forall \; T \in \mathcal{T}_h\}$.

Implementation: Let $a = 1$ and $b = g = 0$. Substituting $u_h = \sum_{j=1}^N c_j \phi_j$ into (33) and picking $v_h = \phi_i$, we obtain

$$\sum_{j=1}^N c_j \left(\int_\Omega \nabla\phi_j \cdot \nabla\phi_i dxdy + \int_{\partial\Omega} \kappa\phi_j\phi_i ds \right) = \int_\Omega f\phi_i dxdy, \quad i = 1, 2, \ldots, N.$$

This gives us the system $(A + R)\mathbf{c} = \mathbf{b}$, where $\mathbf{c} = [c_1, c_2, \ldots, c_N]^t \in \mathbb{R}^N$ is the unknown vector and the entries of $A \in \mathbb{R}^{N\times N}$, $R \in \mathbb{R}^{N\times N}$, and $\mathbf{b} \in \mathbb{R}^N$ are given by

$$a_{ij} = \int_\Omega \nabla\phi_j \cdot \nabla\phi_i dxdy, \quad r_{ij} = \int_{\partial\Omega} \kappa\phi_j\phi_i ds, \quad b_i = \int_\Omega f\phi_i dxdy, \quad i,j = 1, 2, \ldots, N.$$

Assembly of the stiffness matrix A: We can again identify the local contributions that come form a particular triangle T

$$a_{ij}^T = \int_\Omega \nabla\phi_j \cdot \nabla\phi_i dxdy, \quad i,j = 1, 2, 3.$$

where T is an arbitrary triangle with vertices $N_i = (x_i, y_i)$ and ϕ_i are the hat functions *i.e.*, $\phi_j(N_i) = \delta_{ij}$. Let $\phi_i(x,y) = \alpha_i + \beta_i x + \gamma_i y$, for $i = 1, 2, 3$. Then, we compute $\alpha_i, \; \beta_i, \; \gamma_i$ by

$$\begin{bmatrix} 1 & x_1 & y_1 \\ 1 & x_2 & y_2 \\ 1 & x_3 & y_3 \end{bmatrix} \begin{bmatrix} \alpha_1 \\ \beta_1 \\ \gamma_1 \end{bmatrix} = \begin{bmatrix} 1 \\ 0 \\ 0 \end{bmatrix}, \quad \begin{bmatrix} 1 & x_1 & y_1 \\ 1 & x_2 & y_2 \\ 1 & x_3 & y_3 \end{bmatrix} \begin{bmatrix} \alpha_2 \\ \beta_2 \\ \gamma_2 \end{bmatrix} = \begin{bmatrix} 0 \\ 1 \\ 0 \end{bmatrix}, \quad \begin{bmatrix} 1 & x_1 & y_1 \\ 1 & x_2 & y_2 \\ 1 & x_3 & y_3 \end{bmatrix} \begin{bmatrix} \alpha_3 \\ \beta_3 \\ \gamma_3 \end{bmatrix} = \begin{bmatrix} 0 \\ 0 \\ 1 \end{bmatrix}.$$

In general we have $B\alpha_i = \mathbf{e}_i$ for $i = 1, 2, 3$, where

$$B = \begin{bmatrix} 1 & x_1 & y_1 \\ 1 & x_2 & y_2 \\ 1 & x_3 & y_3 \end{bmatrix}, \quad \alpha_i = \begin{bmatrix} \alpha_i \\ \beta_i \\ \gamma_i \end{bmatrix}, \quad \mathbf{e}_1 = \begin{bmatrix} 1 \\ 0 \\ 0 \end{bmatrix}, \quad \mathbf{e}_2 = \begin{bmatrix} 0 \\ 1 \\ 0 \end{bmatrix}, \quad \mathbf{e}_3 = \begin{bmatrix} 0 \\ 0 \\ 1 \end{bmatrix}.$$

Furthermore, we obviously have $\nabla\phi_i = [\beta_i, \gamma_i]^t$, which gives

$$a_{ij}^T = \int_\Omega \left(\beta_i\beta_j + \gamma_i\gamma_j\right) dx = \left(\beta_i\beta_j + \gamma_i\gamma_j\right)|T|, \quad i,j = 1, 2, 3.$$

Assembly of boundary matrix R: Let Γ_h^{out} denote the set of boundary edges of the triangulation, *i.e.* $\Gamma_h^{out} = \{E | \; E = T \cap \partial\Omega, \; for \; T \in \mathcal{T}_h\}$. Assume that κ is constant on E. For an edge $E \in \Gamma_h^{out}$, we define $R^E \in \mathbb{R}^{2\times 2}$ by the entries

$$r_{ij}^{E} = \int_{E} \kappa \phi_j \phi_i ds = \frac{\kappa}{6}(1+\delta_{ij})|E|, \quad i,j = 1,2,$$

where $|E|$ is the length of E and δ_{ij} is 1 for $i = j$ and 0 else.

Assembly of load vector: We use a corner quadrature rule for approximating the integral. We obtain for $T \in \mathcal{T}_h$

$$b_i^T = \int_T f\phi_i dxdy \approx \frac{|T|}{3} f(N_i), \quad i = 1, 2, \ldots, N.$$

Given A, R and $\mathbf{b}$, we can solve $(A+R)\mathbf{c} = \mathbf{b}$ and write $u_h = \sum_{j=1}^{N} c_j \phi_j$.

3.7 The Dirichlet problem

Consider the following Dirichlet Problem: Find u such that

$$-\Delta u = f, \quad \text{in } \Omega, \quad u = g, \quad \text{on } \partial\Omega, \tag{34}$$

where $f \in L^2(\Omega)$ and $g \in C^0(\partial\Omega)$. We seek a weak solution u in $V_g = \{v \in V |\ v|_{\partial\Omega} = g\}$. Multiplying (34) by a test function $v \in V_0$ and integrating over Ω, we get

$$\int_\Omega fvdxdy = -\int_\Omega v\Delta udxdy = \int_\Omega \nabla u \cdot \nabla vdxdy - \int_{\partial\Omega} v\nabla u \cdot \mathbf{n}ds = \int_\Omega \nabla u \cdot \nabla vdxdy.$$

So the weak problem reads: Find $u \in V_g$ such that

$$\int_\Omega \nabla u \cdot \nabla vdxdy = \int_\Omega fvdxdy, \quad v \in V_0.$$

Assume that g is piecewise linear on $\partial\Omega$ with respect to the triangulation. Then the FE method in 2D is defined as follows: Find $u_h \in V_{h,g} = \{v \in V_h |\ v|_{\partial\Omega} = g\}$ such that

$$\int_\Omega \nabla u_h \cdot \nabla v_h dxdy = \int_\Omega fv_h dxdy, \quad v_h \in V_{h,0}.$$

Assume that we have N nodes and J boundary nodes, then the matrix form of the FE method problem reads:

$$\begin{bmatrix} A_{0,0} & A_{0,g} \\ A_{g,0} & A_{0,g} \end{bmatrix} \begin{bmatrix} \mathbf{c}_0 \\ \mathbf{c}_1 \end{bmatrix} = \begin{bmatrix} \mathbf{b}_0 \\ \mathbf{b}_1 \end{bmatrix},$$

where $A_{0,0} \in \mathbb{R}^{(N-J)\times(N-J)}$, $A_{g,g} \in \mathbb{R}^{J\times J}$, $A_{0,g} \in \mathbb{R}^{(N-J)\times J}$, $A_{g,0} \in \mathbb{R}^{J\times(N-J)}$. Note that $\mathbf{c}_1 \in \mathbb{R}^J$ is known (it contains the values of g in the boundary nodes). We can therefore solve the simplified problem reading: find $\mathbf{c}_0 \in \mathbb{R}^{N-J}$ with $A_{0,0}\mathbf{c}_0 = \mathbf{b}_0 - A_{0,g}\mathbf{c}_1$.

3.8 The Neumann problem

Consider the following Neumann Problem: Find u such that

$$-\Delta u = f, \quad \text{in } \Omega, \quad \nabla u \cdot \mathbf{n} = g, \quad \text{on } \partial\Omega, \tag{35}$$

where $f \in L^2(\Omega)$ and $g \in C^0(\partial\Omega)$. Let us try to seek a solution to this problem in the space $V = \left\{ v | \ \|v\|_{L^2(\Omega)} + \|\nabla v\|_{L^2(\Omega)} < \infty \right\}$. Multiplying (35) by a test function $v \in V$, integrating over Ω, and using Green's formula, we get

$$\begin{aligned}\int_\Omega f v dxdy &= -\int_\Omega v \Delta u dxdy = \int_\Omega \nabla u \cdot \nabla v dxdy - \int_{\partial\Omega} v \nabla u \cdot \mathbf{n} ds \\ &= \int_\Omega \nabla u \cdot \nabla v dxdy - \int_{\partial\Omega} v g ds.\end{aligned}$$

Thus, the variational formulation reads: find $u \in V$ such that

$$\int_\Omega \nabla u \cdot \nabla v dxdy - \int_{\partial\Omega} v g ds = \int_\Omega f v dxdy, \quad \forall \ v \in V.$$

In order to guarantee solvability, we note that if $v = 1$ then we have

$$0 = \int_\Omega \nabla u \cdot \nabla 1 dxdy = \int_\Omega f dxdy + \int_{\partial\Omega} g ds.$$

Therefore we need to assume the following compatibility condition

$$\int_\Omega f dxdy + \int_{\partial\Omega} g ds = 0,$$

to ensure that a solution can exist. Note that if u exists, it is only determined up to a constant, since $u + c$ is a solution if u is a solution and $c \in \mathbb{R}$. To fix this constant and obtain a unique solution a common trick is to impose the additional constraint $\int_\Omega u dxdy = 0$. We therefore define the weak solution space

$$\hat{V} = \left\{ v \in V | \int_\Omega v dxdy = 0 \right\},$$

which contains only functions with a zero mean value. This is a called a quotient space. This space guarantees a unique weak solution (with weak formulation as usual with test functions in V). So the weak problem reads: Find $u \in \hat{V}$ such that

$$\int_\Omega \nabla u \cdot \nabla v dxdy - \int_{\partial\Omega} v g ds = \int_\Omega f v dxdy, \quad \forall \ v \in V.$$

Now, the FE method takes the form: find $u_h \in \hat{V}_h \subset \hat{V}$ such that

$$\int_\Omega \nabla u_h \cdot \nabla v_h dxdy - \int_{\partial\Omega} v_h g ds = \int_\Omega f v_h dxdy, \quad \forall \ v_h \in \hat{V}_h,$$

where $\hat{V}_h$ is the space of all continuous piecewise linear functions with a zero mean.

3.9 Finite elements for mixed Dirichlet-Neumann conditions

Here we describe briefly how Neumann conditions are handled in two-dimensional finite elements. Suppose Ω is a domain in either $\mathbb{R}^2$ or $\mathbb{R}^3$ and assume that $\partial\Omega$ has been partitioned into two disjoint sets: $\partial\Omega = \Gamma_1 \cup \Gamma_2$. We consider the following BVP:

$$-\nabla\cdot(\kappa(\mathbf{x})\nabla u)=f(\mathbf{x}),\quad \mathbf{x}\in\Omega,\quad u=0,\quad \mathbf{x}\in\Gamma_1,\quad \nabla u\cdot\mathbf{n}=0,\quad \mathbf{x}\in\Gamma_2, \tag{36}$$

where $f\in L^2(\Omega)$. As for the 1-D case, Dirichlet conditions are termed essential boundary conditions because they must be explicitly imposed in the FE method, while Neumann conditions are called natural and need not be mentioned. We therefore define the space of test functions by

$$\hat{V}=\{v\in C^2(\overline{\Omega}):\ v(\mathbf{x})=0,\ \mathbf{x}\in\Gamma_1\}.$$

Multiplying (36) by a test function $v\in\hat{V}$ and integrating over Ω, we get

$$\begin{aligned}\int_\Omega fv\,dxdy&=-\int_\Omega v\nabla\cdot(\kappa(\mathbf{x})\nabla u)dxdy=\int_\Omega \kappa(\mathbf{x})\nabla u\cdot\nabla v\,dxdy-\int_{\partial\Omega}\kappa(\mathbf{x})v\nabla u\cdot\mathbf{n}ds\\&=\int_\Omega \kappa(\mathbf{x})\nabla u\cdot\nabla v\,dxdy-\int_{\Gamma_1}\kappa(\mathbf{x})v\nabla u\cdot\mathbf{n}ds-\int_{\Gamma_2}\kappa(\mathbf{x})v\nabla u\cdot\mathbf{n}ds\\&=\int_\Omega \kappa(\mathbf{x})\nabla u\cdot\nabla v\,dxdy,\end{aligned}$$

since $v=0$ on Γ_1 and $\nabla u\cdot\mathbf{n}$ on Γ_1. Thus the weak form of (36) is: Find $u\in\hat{V}$ such that

$$\int_\Omega \kappa(\mathbf{x})\nabla u\cdot\nabla v\,dxdy=\int_\Omega fv\,dxdy,\quad v\in\hat{V}. \tag{37}$$

We now restrict our discussion once more to two-dimensional polygonal domains. To apply the FE method, we must choose an approximating subspace of $\hat{V}$. Since the boundary conditions are mixed, there are at least two points where the boundary conditions change from Dirichlet to Neumann. We will make the assumption that the mesh is chosen so that all such points are nodes (and that all such nodes belong to Γ_1, that is, that Γ_1 includes its "endpoints"). We can then choose the approximating subspace of $\hat{V}$ as follows:

$$V_h=\{v\in C(\overline{\Omega}):\ \text{v is linear on } \mathcal{T}_\text{h},\ v(\mathbf{z})=0\ \text{ for all nodes } \mathbf{z}\in\Gamma_1\}.$$

A basis for V_h is formed by including all basis functions corresponding to interior boundary nodes that do not belong to Γ_1. If the BVP includes only Neumann conditions, then the stiffness matrix will be singular, reflecting the fact that BVP either does not have a solution or has infinitely many solutions. Special care must be taken to compute a meaningful solution to the resulting linear system.

3.10 The method of shifting the data

3.10.1 Inhomogeneous Dirichlet conditions on a rectangle

In a two-dimensional problem, inhomogeneous boundary conditions are handled just as in one dimension. Inhomogeneous Dirichlet conditions are addressed via the method of shifting the data (with a specially chosen piecewise linear function), while inhomogeneous Neumann conditions are taken into account directly when deriving the weak form. Both types of boundary conditions lead to a change in the load vector.

The method of shifting the data can be used to transform an inhomogeneous Dirichlet problem to a homogeneous Dirichlet problem. This technique works just as it did for a one-dimensional problem, although in two dimensions it is more difficult to find a function satisfying the boundary conditions. We consider the BVP

$$-\Delta u = f(\mathbf{x}), \quad \mathbf{x} \in \Omega = (0,a) \times (0,b), \quad u(\mathbf{x}) = g(\mathbf{x}) = \begin{cases} g_1(\mathbf{x}), & \mathbf{x} \in \Gamma_1, \\ g_2(\mathbf{x}), & \mathbf{x} \in \Gamma_2, \\ g_3(\mathbf{x}), & \mathbf{x} \in \Gamma_3, \\ g_4(\mathbf{x}), & \mathbf{x} \in \Gamma_4, \end{cases} \tag{38}$$

where Γ_1, Γ_2, Γ_3, and Γ_4 are, respectively, the bottom, right, top, and left boundary edges of the rectangular domain $\Omega = (0,a) \times (0,b)$. We will assume that the boundary data are continuous, so

$$g_1(0) = g_4(0), \quad g_1(a) = g_2(0), \quad g_2(b) = g_3(a), \quad g_3(0) = g_4(b).$$

Suppose we find a function w defined on $\overline{\Omega}$ and satisfying $w(\mathbf{x}) = g(\mathbf{x})$ for all $\mathbf{x} \in \partial\Omega$. We then define $v = u - w$ and note that

$$-\Delta v = -\Delta u + \Delta w = f(\mathbf{x}) + \Delta w = \hat{f}(\mathbf{x}),$$

and $v(\mathbf{x}) = u(\mathbf{x}) - w(\mathbf{x}) = 0$ for all $\mathbf{x} \in \partial\Omega$. We can then solve

$$-\Delta v = \hat{f}(\mathbf{x}), \quad \mathbf{x} \in \Omega, \quad v(\mathbf{x}) = 0, \quad \mathbf{x} \in \partial\Omega. \tag{39}$$

Finally, the solution u will be given by $u = v + w$.

We now describe a method for computing a function w that satisfies the given Dirichlet conditions. We first note that there is a polynomial of the form $q(x,y) = c_0 + c_1 x + c_2 y + c_3 xy$, which assumes the desired boundary values at the corners:

$$q(0,0) = g_1(0) = g_4(0), \quad q(a,0) = g_1(a) = g_2(0), \quad q(a,b) = g_2(b) = g_3(a), \quad q(0,b) = g_3(0) = g_4(b).$$

A direct calculation shows that

$$c_0 = g_1(0), \quad c_1 = \frac{g_1(a) - g_1(0)}{a}, \quad c_2 = \frac{g_4(b) - g_4(0)}{b},$$
$$c_3 = \frac{g_2(b) + g_1(0) - g_1(a) - g_4(b)}{ab}.$$

We then define

$$h(\mathbf{x}) = \begin{cases} h_1(x) = g_1(x) - \left(g_1(0) + \dfrac{g_1(a) - g_1(0)}{a} x\right), & \mathbf{x} \in \Gamma_1, \\ h_2(y) = g_2(y) - \left(g_2(0) + \dfrac{g_2(b) - g_2(0)}{b} y\right), & \mathbf{x} \in \Gamma_2, \\ h_3(x) = g_3(x) - \left(g_3(0) + \dfrac{g_3(a) - g_3(0)}{a} x\right), & \mathbf{x} \in \Gamma_3, \\ h_4(y) = g_4(y) - \left(g_4(0) + \dfrac{g_4(b) - g_4(0)}{b} y\right), & \mathbf{x} \in \Gamma_4. \end{cases}$$

We have thus replaced each g_i by a function h_i which differs from g_i by a linear function, and which has value zero at the two endpoints:

$$h_1(0) = h_1(a) = h_2(0) = h_2(b) = h_3(0) = h_3(a) = h_4(0) = h_4(b) = 0.$$

Finally, we define

$$w(x,y) = (c_0 + c_1x + c_2y + c_3xy) + \left(h_1(x) + \frac{h_3(x) - h_1(x)}{b}y\right) + \left(h_4(y) + \frac{h_2(y) - h_4(y)}{a}x\right).$$

The reader should notice how the second term interpolates between the boundary values on Γ_1 and Γ_3, while the third term interpolates between the boundary values on Γ_2 and Γ_4. In order for these two terms not to interfere with each other, it is necessary that the boundary data be zero at the corners. It was for this reason that we transformed the g_i's into the h_i's. The first term in the formula for w undoes this transformation. It is straightforward to verify that w satisfies the desired boundary conditions.

3.10.2 Inhomogeneous Neumann conditions on a rectangle

We can also apply the technique of shifting the data to transform a BVP with inhomogeneous Neumann conditions to a related BVP with homogeneous Neumann conditions. However, the details are somewhat more involved than in the Dirichlet case. Consider the following BVP with the Neumann conditions

$$-\Delta u = f(\mathbf{x}), \quad \mathbf{x} \in \Omega = (0,a) \times (0,b), \quad \mathbf{n} \cdot \nabla u(\mathbf{x}) = g(\mathbf{x}) = \begin{cases} g_1(\mathbf{x}), & \mathbf{x} \in \Gamma_1, \\ g_2(\mathbf{x}), & \mathbf{x} \in \Gamma_2, \\ g_3(\mathbf{x}), & \mathbf{x} \in \Gamma_3, \\ g_4(\mathbf{x}), & \mathbf{x} \in \Gamma_4, \end{cases} \tag{40}$$

where Γ_1, Γ_2, Γ_3, and Γ_4 are, respectively, the bottom, right, top, and left boundary edges of the rectangular domain $\Omega = (0,a) \times (0,b)$. We first note that this is equivalent to

$$-u_y(\mathbf{x}) = g_1(x), \ \mathbf{x} \in \Gamma_1, \quad u_x(\mathbf{x}) = g_2(y), \ \mathbf{x} \in \Gamma_2, \quad u_y(\mathbf{x}) = g_3(x), \ \mathbf{x} \in \Gamma_3, \quad -u_x(\mathbf{x}) = g_4(y), \ \mathbf{x} \in \Gamma_4.$$

We make the following observation: If there is a twice-continuously differentiable function u satisfying the given Neumann conditions, then, since $u_{xy} = u_{yx}$, we have

$$-u_{xy}(x,0) = g_1'(x), \quad -u_{yx}(0,y) = g_{4'}(y),$$

which together imply that $g_1'(0) = g_{4'}(0)$. By similar reasoning, we have all of the following conditions:

$$g_1'(0) = g_{4'}(0), \quad g_1'(0) = g_{4'}(0), \quad -g_1'(a) = g_2'(0), \quad g_2'(b) = g_3'(a). \tag{41}$$

We will assume that (41) holds.

We now explain how to compute a function that satisfies the desired Neumann conditions. The method is similar to that used to shift the data in a Dirichlet problem: we will "interpolate" between the Neumann conditions in each dimension and arrange things so that the two interpolations do not interfere with each other. We use the fact that

$$\psi(x) = -\alpha x + \frac{\alpha + \beta}{2a} x^2 \quad \text{satisfies} \quad \psi'(0) = -\alpha, \quad \psi'(a) = \beta. \tag{42}$$

The first step is to transform the boundary data $g_l(x)$ to a function $h_1(x)$ satisfying $h_{1'}(0) = h_{1'}(a) = 0$, and similarly for g_2, g_3, g_4 and h_2, h_3, h_4. Since these derivatives of the boundary data at the corners are (plus or minus) the mixed partial derivatives of the desired function at the corners, it suffices to find a function $q(x,y)$ satisfying the conditions

$$u_{xy}(0,0) = -g_1'(0), \quad u_{xy}(a,0) = -g_1'(a), \quad u_{xy}(0,b) = -g_3'(0), \quad u_{xy}(a,b) = -g_2'(b).$$

We can satisfy these conditions with a function of the form $q(x,y) = c_0 xy + c_1 x^2 y + c_2 x y^2 + c_3 x^2 y^2$. The reader can verify that the necessary coefficients are

$$c_0 = -g_1'(0), \quad c_1 = \frac{g_1'(0) - g_1'(a)}{2a}, \quad c_2 = \frac{g_3'(0) + g_1'(0)}{2b},$$

$$c_3 = \frac{g_2'(b) + g_1'(a) - g_3'(0) - g_1'(0)}{4ab}.$$

If w is to satisfy the desired Neumann conditions, then $w - q = h_i$ on Γ_i, $i = 1 - 4$, where

$$h_1(x) = g_1(x) + c_0 x + c_1 x^2, \quad h_2(y) = g_2(y) - (c_0 + 2ac_1)y - (c_2 + 2ac_3)y^2,$$

$$h_3(x) = g_3(x) - (c_0 + 2bc_2)x - (c_1 + 2bc_3)x^2, \quad h_4(y) = g_4(y) + c_0 y + c_2 y^2.$$

We can now define $w - q$ by the interpolation described by (42):

$$w(x,y) = q(x,y) - h_1(x)y + \frac{h_3(x) + h_1(x)}{2b} y^2 - h_4(y)x + \frac{h_2(y) + h_4(y)}{2a} y x^2.$$

Then w satisfies the original Neumann conditions, as the interested reader can verify directly.

3.11 Eigenvalue problem

Consider the following Eigenvalue Problem: Find $\lambda \in \mathbb{R}$ and u such that

$$-\Delta u = \lambda u, \quad \text{in } \Omega, \quad \nabla u \cdot \mathbf{n} = 0, \quad \text{on } \partial\Omega. \tag{43}$$

In order to derive the weak formulation, we multiply (43) with $v \in V$, integrate over Ω and use Green's formula to obtain

$$\lambda \int_\Omega uv dxdy = -\int_\Omega v \Delta u dxdy = \int_\Omega \nabla u \cdot \nabla v dxdy - \int_{\partial\Omega} v \nabla u \cdot \mathbf{n} ds = \int_\Omega \nabla u \cdot \nabla v dxdy.$$

We obtain the weak form: Find $u \in V$ such that

$$\int_\Omega \nabla u \cdot \nabla v dxdy = \lambda \int_\Omega uvdxdy, \quad v \in V. \tag{44}$$

The FE method in 2D is defined as follows: Find $\lambda_h \in \mathbb{R}$ and $u_h \in V_h$ such that

$$\int_\Omega \nabla u_h \cdot \nabla v_h dxdy = \lambda_h \int_\Omega u_h v_h dxdy, \quad v_h \in V_h, \tag{45}$$

where $V_h = \{v \in V | \ v|_T \in P^1(T), \ \forall \ T \in \mathcal{T}_h\}$.

Implementation: Substituting $u_h = \sum_{j=1}^N c_j \phi_j$ into (45) and picking $v_h = \phi_i$, we obtain

$$\sum_{j=1}^N c_j \left(\int_\Omega \nabla \phi_j \cdot \nabla \phi_i dxdy - \lambda_h \int_\Omega \phi_i \phi_j dxdy \right) = 0, \quad i = 1, 2, \ldots, N.$$

This leads to an algebraic system of the form $A\mathbf{c} = \lambda_h M \mathbf{c}$, *i.e.* an algebraic eigenvalue problem.

3.12 Error analysis

Consider the following model Problem: Find u such that

$$-\Delta u = f, \quad \text{in } \Omega, \qquad u = 0, \quad \text{on } \partial\Omega.$$

The weak form: Find $u \in V_0$ such that

$$\int_\Omega \nabla u \cdot \nabla v dxdy = \int_\Omega fv dxdy, \quad v \in V_0.$$

The FE approximation is defined as follows: Find $u_h \in V_{h,0}$ such that

$$\int_\Omega \nabla u_h \cdot \nabla v_h dxdy = \int_\Omega f v_h dxdy, \quad v_h \in V_{h,0},$$

where $V_h = \{v \in V | \ v|_T \in P^1(T), \ \forall \ T \in \mathcal{T}_h\}$. Expressing $u_h = \sum_{j=1}^N c_j \phi_j$ and picking $v_h = \phi_i$, we obtain

$$\sum_{j=1}^N c_j \left(\int_\Omega \nabla \phi_j \cdot \nabla \phi_i dxdy \right) = \int_\Omega f \phi_i dxdy, \quad i = 1, 2, \ldots, N.$$

This leads to system of the form $A\mathbf{c} = \mathbf{b}$, where the entries of $A \in \mathbb{R}^{N \times N}$ and $\mathbf{b} \in \mathbb{R}^N$ are

$$a_{ij} = \int_\Omega \nabla \phi_j \cdot \nabla \phi_i dxdy, \quad b_i = \int_\Omega f \phi_i dxdy, \quad i,j = 1, 2, \ldots, N.$$

Theorem 3.6 *The stiffness matrix A is symmetric and positive definite.*

Theorem 3.7 (Galerkin orthogonality) *Let $u \in V_0$ denote the weak solution and $u_h \in V_{h,0}$ the corresponding FE method approximation. Then*

$$\int_\Omega \nabla (u - u_h) \cdot \nabla v_h dxdy = 0, \quad v_h \in V_{h,0}.$$

Now, let $|||v|||^2 = \int_\Omega \nabla v \cdot \nabla v dxdy = \int_\Omega |\nabla v|^2 dxdy$ be the energy norm on V_0.

There are two different kinds of error estimates, *a priori* estimates, where the error is bounded in terms of the exact solution, and *a posteriori* error estimates, where the error is bounded in terms of the computed solution.

Theorem 3.8 (A priori error bound) *Let $u \in V_0$ denote the weak solution and $u_h \in V_{h,0}$ the corresponding FE method approximation. Then*

$$|||u - u_h||| \leq |||u - v_h|||, \quad v_h \in V_{h,0}.$$

Theorem 3.9 *Let $u \in V_0$ denote the weak solution and $u_h \in V_{h,0}$ the corresponding FE method approximation. If $u \in C^2(\Omega)$, then there exists C independent of h_T and u such that*

$$|||u - u_h|||^2_{L^2(\Omega)} \leq C \sum_{T \in \mathcal{T}_h} h_T^2 ||D^2 u||^2_{L^2(T)}.$$

3.13 The FE method for elliptic problems with a convection term

Consider the following convection-diffusion problem: Find u such that

$$-\nabla \cdot (a \nabla u) + \mathbf{b} \cdot \nabla u + cu = f, \quad \text{in } \Omega, \quad u = 0, \quad \text{on } \partial\Omega. \tag{46}$$

We seek a weak solution u in $V_0 = \{v \in V | \ v|_{\partial\Omega} = 0\}$. In order to derive the weak formulation, we multiply (46) with $v \in V_0$, integrate over Ω and use Green's formula to obtain

$$\begin{aligned}\int_\Omega fv dxdy &= -\int_\Omega v \nabla \cdot (a \nabla u) dxdy + \int_\Omega v \mathbf{b} \cdot \nabla u dxdy + \int_\Omega cuv dxdy \\ &= \int_\Omega a \nabla u \cdot \nabla v dxdy - \int_{\partial\Omega} v \nabla u \cdot \mathbf{n} ds + \int_\Omega v \mathbf{b} \cdot \nabla u dxdy + \int_\Omega cuv dxdy \\ &= \int_\Omega a \nabla u \cdot \nabla v dxdy + \int_\Omega v \mathbf{b} \cdot \nabla u dxdy + \int_\Omega cuv dxdy.\end{aligned}$$

Note that there is no need to apply Green's formula to $\int_\Omega v \mathbf{b} \cdot \nabla u dxdy$. We obtain the weak form: Find $u \in V_0$ such that

$$\int_\Omega a \nabla u \cdot \nabla v dxdy + \int_\Omega v \mathbf{b} \cdot \nabla u dxdy + \int_\Omega cuv dxdy = \int_\Omega fv dxdy, \quad v \in V_0.$$

The FE method in 2D is defined as follows: Find $u_h \in V_{h,0} = \{v \in V_h | \ v|_{\partial\Omega} = 0\}$ such that

$$\int_\Omega a \nabla u_h \cdot \nabla v_h dxdy + \int_\Omega v_h \mathbf{b} \cdot \nabla u_h dxdy + \int_\Omega cu_h v_h dxdy = \int_\Omega f v_h dxdy, \quad v_h \in V_{h,0}, \tag{47}$$

where $V_h = \{v \in V | \ v|_T \in P^1(T), \ \forall \ T \in \mathcal{T}_h\}$.

Implementation: Substituting $u_h = \sum_{j=1}^N c_j \phi_j$ into (47) and picking $v_h = \phi_i$, we obtain

$$\sum_{j=1}^N c_j \left(\int_\Omega a \nabla \phi_j \cdot \nabla \phi_i dxdy + \int_\Omega \phi_i \mathbf{b} \cdot \nabla \phi_j dxdy + \int_\Omega c \phi_i \phi_j dxdy \right) = \int_\Omega f \phi_i dxdy, \quad i = 1, 2, \ldots, N.$$

This gives us the system $(A+B+C)\mathbf{c}=\mathbf{d}$, where $\mathbf{c}=[c_1,\dots,c_N]^t\in\mathbb{R}^N$ is the unknown vector and the entries of $A,B,C\in\mathbb{R}^{N\times N}$ and $\mathbf{d}\in\mathbb{R}^N$ are given by

$$a_{ij}=\int_\Omega a\nabla\phi_j\cdot\nabla\phi_i dxdy,\quad b_{ij}=\int_\Omega \phi_i\mathbf{b}\cdot\nabla\phi_j dxdy,\quad c_{ij}=\int_\Omega c\phi_i\phi_j dxdy,\quad d_i=\int_\Omega f\phi_i dxdy,$$

for $i,j=1,2,\dots,N$. Note that B is not symmetric, *i.e.* $b_{ij}\neq b_{ji}$.

4. The FE method for the heat equation

Consider the following heat/diffusion problem: Find $u(\mathbf{x},t)$ such that

$$\dot{u}-\Delta u=f,\quad \text{in } \Omega\subset\mathbb{R}^2,\quad t\in[0,T], \tag{48}$$

$$u(\cdot,t)=0,\quad \text{on } \partial\Omega \text{ and } t\in[0,T], \tag{49}$$

$$u(\mathbf{x},0)=u_0(\mathbf{x}),\quad \text{for } \mathbf{x}\in\Omega \text{ and } t=0. \tag{50}$$

We seek a weak solution u in $V_0=\{v|\ \|v\|+\|\nabla v\|<\infty,\ v|_{\partial\Omega}=0\}$. In order to derive the weak formulation, we multiply (48) with $v\in V_0$, integrate over Ω and use Green's formula to obtain, for $t\in[0,T]$,

$$\int_\Omega fv d\mathbf{x}=\int_\Omega \dot{u}v d\mathbf{x}+\int_\Omega \nabla u\cdot\nabla v d\mathbf{x}-\int_{\partial\Omega} v\nabla u\cdot\mathbf{n} ds=\int_\Omega \dot{u}v d\mathbf{x}+\int_\Omega \nabla u\cdot\nabla v d\mathbf{x}.$$

The weak form therefore reads: Find $u(\cdot,t)\in V_0$ such that for $t>0$

$$\int_\Omega \dot{u}v d\mathbf{x}+\int_\Omega \nabla u\cdot\nabla v d\mathbf{x}=\int_\Omega fv d\mathbf{x},\quad v\in V_0. \tag{51}$$

The semi-discrete FE method in 2D is defined as follows: Find $u_h(\cdot,t)\in V_{h,0}=\{v\in V_h|\ v|_{\partial\Omega}=0\}$ such that

$$\int_\Omega \dot{u}_h v_h d\mathbf{x}+\int_\Omega \nabla u_h\cdot\nabla v_h d\mathbf{x}=\int_\Omega f v_h d\mathbf{x},\quad v_h\in V_{h,0}, \tag{52}$$

where $V_h=\{v\in V|\ v|_T\in P^1(T),\ \forall\ T\in\mathcal{T}_h\}$.

Implementation: Substituting $u_h(\mathbf{x},t)=\sum_{j=1}^N c_j(t)\phi_j(\mathbf{x})$ into (52) and choosing $v_h=\phi_i$, we obtain

$$\sum_{j=1}^N \dot{c}_j\int_\Omega \phi_j\phi_i d\mathbf{x}+\sum_{j=1}^N c_j\int_\Omega \nabla\phi_j\cdot\nabla\phi_i d\mathbf{x}=\int_\Omega f\phi_i d\mathbf{x},\quad i=1,2,\dots,N.$$

This gives us the system of ODEs

$$M\dot{\mathbf{c}}(t)+A(t)\mathbf{c}(t)=\mathbf{b}(t),\quad t\in(0,T],\quad \mathbf{c}(0)=\mathbf{c}_0,$$

where $\mathbf{c}=[c_1,c_2,\dots,c_N]^t=[u_h(N_1,t),\dots,u_h(N_N,t)]^t\in\mathbb{R}^N$ (here N_i denotes the node that belongs to the basis function ϕ_i) is the unknown vector and the entries of $M,\ A\in\mathbb{R}^{N\times N}$ and $\mathbf{b}\in\mathbb{R}^N$ are given by

$$m_{ij}=\int_\Omega \phi_i\phi_j d\mathbf{x},\quad a_{ij}=\int_\Omega \nabla\phi_j\cdot\nabla\phi_i d\mathbf{x},\quad b_i=\int_\Omega f\phi_i d\mathbf{x},\quad i,j=1,2,\dots,N.$$

Finally, the system of ODEs can be solved with *e.g.*, the backward Euler method as follows: Let $0 = t_0 < t_1 < \cdots < t_M = T$ be a discretization, let $k_m = t_m - t_{m-1}$ for $m = 1, 2, \ldots, M$ be the time step size and let $\mathbf{c}^m \approx \mathbf{c}(t_m)$ for $m = 1, 2, \ldots, M$ denote corresponding approximations. Then, we can compute $\mathbf{c}^m$ using

$$(M + k_m A_m)\mathbf{c}^m = M\mathbf{c}^{m-1} + k_m \mathbf{b}_m, \quad m = 1, 2, \ldots, M,$$

where $\mathbf{c}^0$ is obtained from $u_0(\mathbf{x})$. We can either use $\mathbf{c}^0 = [c_1(0), \ldots, c_N(0)]^t = [u_0(N_1), \ldots, u_0(N_N)]^t$, or we can let $\mathbf{c}^0$ to be the L^2-projection of u_0. We set $u_h^0 = \sum_{j=1}^N c_j^0 \phi_j(\mathbf{x})$ and solve for c_j^0 using

$$\sum_{j=1}^N c_j^0 \int_\Omega \phi_j \phi_i d\mathbf{x} = \int_\Omega u_0 \phi_i d\mathbf{x}, \quad i = 1, 2, \ldots, N.$$

Theorem 4.1 (Stability) *There hold continuous and discrete stability estimates*

$$\|u(\cdot,t)\| \le \|u(\cdot,0)\| + \int_0^t \|f(\cdot,s)\| ds, \quad \|u_h^m\| \le \|u_h^{m-1}\| + k_m \|f_m\| \le \|u_h^0\| + \sum_{i=1}^m k_i \|f_i\|.$$

5. The FE method for the wave equation

Many physical phenomena exhibit wave characteristics. For instance light which is an electromagnetic wave have the ability to disperse and create diffraction patterns, which is typical of waves.

Consider the following wave problem: Find $u(\mathbf{x}, t)$ such that

$$\ddot{u} - \nabla \cdot (\varepsilon \nabla u) = f, \quad \text{in } \Omega \subset \mathbb{R}^2, \quad t \in [0, T], \tag{53}$$

$$\mathbf{n} \cdot \nabla u(\cdot, t) = 0, \quad \text{on } \partial\Omega \text{ and } t \in [0, T], \tag{54}$$

$$u(\mathbf{x}, 0) = u_0(\mathbf{x}), \quad \dot{u}(\mathbf{x}, 0) = v_0(\mathbf{x}), \quad \text{for } \mathbf{x} \in \Omega \text{ and } t = 0, \tag{55}$$

where f is a given load, $\varepsilon = \varepsilon(\mathbf{x}, t)$ is a positive parameter, u_0 and v_0 are a prescribed initial conditions, and Ω is a bounded domain with boundary $\partial\Omega$ and unit outward normal $\mathbf{n}$.

We seek a weak solution u in $V = H^1(\Omega) = \{v| \ \|v\| + \|\nabla v\| < \infty\}$. Multiplying the wave Eq. (53) with $v \in V$, integrating over Ω, and using Green's formula, we obtain, for $t \in [0, T]$,

$$\int_\Omega f v d\mathbf{x} = \int_\Omega \ddot{u} v d\mathbf{x} - \int_\Omega v \nabla \cdot (\varepsilon \nabla u) d\mathbf{x} = \int_\Omega \ddot{u} v d\mathbf{x} + \int_\Omega \varepsilon \nabla u \cdot \nabla v d\mathbf{x} - \int_{\partial\Omega} v \varepsilon \nabla u \cdot \mathbf{n} ds$$
$$= \int_\Omega \ddot{u} v d\mathbf{x} + \int_\Omega \varepsilon \nabla u \cdot \nabla v d\mathbf{x}.$$

The weak form (variational formulation) therefore reads: Find $u(\cdot, t) \in V = H^1(\Omega)$ such that for all $t > 0$

$$\int_\Omega \ddot{u} v d\mathbf{x} + \int_\Omega \varepsilon \nabla u \cdot \nabla v d\mathbf{x} = \int_\Omega f v d\mathbf{x}, \quad v \in V. \tag{56}$$

Let $V_h = \{v \in V| \ v|_T \in P^1(T), \ \forall \ T \in \mathcal{T}_h\} \subset V$ be the space of all continuous piecewise linear functions on a triangle mesh of Ω. The semi-discrete FE method in 2D is defined as follows: Find $u_h(\cdot, t) \in V_h$ such that

$$\int_\Omega \ddot{u}_h v_h d\mathbf{x} + \int_\Omega \varepsilon \nabla u_h \cdot \nabla v_h d\mathbf{x} = \int_\Omega f v_h d\mathbf{x}, \quad v_h \in V_h. \tag{57}$$

Implementation: Substituting $u_h(\mathbf{x},t) = \sum_{j=1}^N c_j(t)\phi_j(\mathbf{x})$ into (57) and choosing $v_h = \phi_i$, we obtain

$$\sum_{j=1}^N \ddot{c}_j \int_\Omega \phi_j \phi_i d\mathbf{x} + \sum_{j=1}^N c_j \int_\Omega \varepsilon \nabla \phi_j \cdot \nabla \phi_i d\mathbf{x} = \int_\Omega f \phi_i d\mathbf{x}, \quad i = 1, 2, \dots, N.$$

This gives us the system

$$M\ddot{\mathbf{c}}(t) + A(t)\mathbf{c}(t) = \mathbf{b}(t), \quad t \in (0, T], \tag{58}$$

where $\mathbf{c} = [c_1, \dots, c_N]^t = [u_h(N_1, t), \dots, u_h(N_N, t)]^t \in \mathbb{R}^N$ (here N_i denotes the node that belongs to the basis function ϕ_i) is the unknown vector and the entries of the mass and stiffness matrices $M,\ A \in \mathbb{R}^{N \times N}$ and the load vector $\mathbf{b} \in \mathbb{R}^N$ are given by

$$m_{ij} = \int_\Omega \phi_i \phi_j d\mathbf{x}, \quad a_{ij} = \int_\Omega \varepsilon \nabla \phi_j \cdot \nabla \phi_i d\mathbf{x}, \quad b_i = \int_\Omega f \phi_i d\mathbf{x}, \quad i,j = 1, 2, \dots, N.$$

Eq. (58) is a semi-discretization of the wave equation in the sense that it does not contain any unknowns with spatial derivatives.

Time discretization: We first transform the system of ODEs into a first-order system. Let $\mathbf{d}(t) = \dot{\mathbf{c}}(t)$, we get the new coupled system

$$M\dot{\mathbf{c}}(t) - M\mathbf{d}(t) = 0, \quad M\dot{\mathbf{d}}(t) + A(t)\mathbf{c}(t) = \mathbf{b}(t), \quad t \in (0, T].$$

Let $\mathbf{w} = [\mathbf{c}, \mathbf{d}]^t$ then the system is equivalent to $\hat{M}\dot{\mathbf{w}}(t) + \hat{A}(t)\mathbf{w}(t) = \hat{\mathbf{b}}(t)$, $t \in (0, T]$, where

$$\hat{M} = \begin{bmatrix} M & 0 \\ 0 & M \end{bmatrix}, \quad \hat{A} = \begin{bmatrix} 0 & -M \\ A & 0 \end{bmatrix}, \quad \hat{\mathbf{b}} = \begin{bmatrix} 0 \\ \mathbf{b} \end{bmatrix}.$$

Finally, the system of ODEs can be solved with *e.g.*, the backward Euler method as follows: Let $0 = t_0 < t_1 < \cdots < t_M = T$ be a discretization, let $k_m = t_m - t_{m-1}$ for $m = 1, 2, \dots, M$ be the time step size and let $\mathbf{w}^m \approx \mathbf{w}(t_m)$ for $m = 1, 2, \dots, M$ denote corresponding approximations. Then, we can compute $\mathbf{w}^m$ using

$$\left(\hat{M} + k_m \hat{A}_m\right)\mathbf{w}^m = \hat{M}\mathbf{w}^{m-1} + k_m \hat{\mathbf{b}}_m, \quad m = 1, 2, \dots, M,$$

where $\mathbf{w}^0$ is obtained from $u_0(\mathbf{x})$ and $v_0(\mathbf{x})$.

There are several possible choices of initial data. We can either use $\mathbf{w}^0 = [w_1(0), \dots, c_{2N}(0)]^t = [u_0(N_1), \dots, u_0(N_N), v_0(N_1), \dots, v_0(N_N)]^t$, or we can let $\mathbf{w}^0 = [\mathbf{w}_1^0, \mathbf{w}_2^0]^t$, where $\mathbf{w}_1^0$ and $\mathbf{w}_2^0$ are the L^2-projection of u_0 and v_0, respectively. We set $w_{h,1}^0 = \sum_{j=1}^N w^0_{j,1}\phi_j(\mathbf{x})$ and $w_{h,2}^0 = \sum_{j=1}^N w^0_{j,2}\phi_j(\mathbf{x})$ and solve for $w^0_{j,1}, w^0_{j,2}$ using

$$\sum_{j=1}^N w^0_{j,1} \int_\Omega \phi_j \phi_i d\mathbf{x} = \int_\Omega u_0 \phi_i d\mathbf{x}, \quad \sum_{j=1}^N w^0_{j,2} \int_\Omega \phi_j \phi_i d\mathbf{x} = \int_\Omega v_0 \phi_i d\mathbf{x}, \quad i = 1, 2, \dots, N.$$

We can also use Crank–Nicolson scheme

$$\left(\hat{M} + \frac{k_m}{2}\hat{A}_m\right)\mathbf{w}^m = \left(\hat{M} - \frac{k_m}{2}\hat{A}_{m-1}\right)\mathbf{w}^{m-1} + \frac{k_m}{2}\left(\hat{\mathbf{b}}_{m-1} + \hat{\mathbf{b}}_m\right) \equiv \mathbf{g}_m.$$

Theorem 5.1 (Conservation of energy) *If* $f = 0$, *then*

$$\|\dot{u}_h(\cdot,t)\|^2_{L^2(\Omega)} + \varepsilon\|\nabla u_h(\cdot,t)\|^2_{L^2(\Omega)} = \|\dot{u}(\cdot,0)\|^2_{L^2(\Omega)} + \varepsilon\|\nabla u(\cdot,0)\|^2_{L^2(\Omega)}.$$

6. Conclusion

In this chapter, we introduced the finite element (FE) method for approximation the solutions to ODEs and PDEs. More specifically, the FE method is presented for first-order initial-value problems for OEDs, second-order boundary-value problems for ODEs, second-order elliptic PDEs, second-order heat and wave equations. The remaining chapters of this textbook are based on the FE method. The derivation of the FE method for other problems is straightforward. In the remaining chapters, the FE method will developed to solve complicated problems in engineering, notably in elasticity and structural mechanics modeling involving elliptic partial differential equations and complicated geometries. For more details, we refer the reader to [1–4, 6–9] and the references therein.

Author details

Mahboub Baccouch
Department of Mathematics, University of Nebraska at Omaha, Omaha, NE, USA

*Address all correspondence to: mbaccouch@unomaha.edu

References

[1] Ainsworth M. and Oden J. T. *A posteriori Error Estimation in Finite Element Analysis*. John Wiley, New York, 2000.

[2] Brenner S. C. and Scott L. R. *The Mathematical Theory of Finite Element Methods, second edition*. Springer-Verlag, New York, 2002.

[3] Ciarlet P. G. *The finite element method for elliptic problems*. North-Holland Pub. Co., Amsterdam-New York-Oxford, 1978.

[4] Johnson C. *Numerical Solution of Partial Differential Equations by the Finite Element Method*. Cambridge University Press, New York, 1987.

[5] Kaltenbacher M. *Numerical simulation of mechatronic sensors and actuators: finite elements for computational multiphysics*. Springer, Heidelberg, 2015.

[6] Larson M. *The finite element method: theory, implementation, and applications*. Springer, Berlin New York, 2013.

[7] Oden J. T. and Carey G. F. *Finite Elements, Mathematical Aspects*. Prentice Hall, Englewood Cliffs, 1983.

[8] Schwab C. *p– and hp– Finite Element Methods*. Oxford University Press, New York, 1998.

[9] Szabo B. and Babu I. š ka. *Finite element analysis*. Wiley, New York, 1991.

[10] Hrennikoff A. Solution of problems of elasticity by the framework method. *Journal of Applied Mechanics*, 8(4):169–175, 1941.

[11] Courant R. Variational methods for the solution of problems of equilibrium and vibrations. bulletin of the american mathematical society. 49: 1–23. doi: 10.1090. Technical report, 1943.

[12] Strang G. and Fix G. J. An analysis of the finite element method. 1973.

Chapter 2

Fluid Structure Interaction Study of Stenosed Carotid Artery Considering the Effects of Blood Pressure and Altered Gravity

S.M. Abdul Khader, Nitesh Kumar and Raghuvir Pai

Abstract

Atherosclerosis is a very common cardiovascular disease (CVD) causing increased morbidity. Atherosclerosis is a disease that involves several factors and usually affects the wall of the arterial bifurcations. Advanced Computational Fluid Dynamics (CFD) techniques has the potential to shed more light in understanding of the causes of atherosclerosis and perhaps in its early diagnosis. Fluid Structure Interaction (FSI) study was carried out on two different three dimensional patient specific cases (a) Normal carotid bifurcation and (b) Stenosed carotid bifurcation. Physiological conditions were considered to evaluate hemodynamic parameters and understand the origin and progression of atherosclerosis in the carotid artery bifurcation, first for the normal and then with hypertension disease. Commercial software ANSYS and ANSYS CFX (version 19.0) was used to perform a two-way FSI using a fully implicit second-order backward Euler differencing scheme. Arterial response was calculated by employing an Arbitrary Lagrangian–Eulerian (ALE) formulation and using the temporal blood response. The carotid artery bifurcation caused a velocity reduction and backflow was observed causing a reduction in the shear stress. A low shear stress resulted due to an oscillatory behavior at the start point of the internal carotid artery near the carotid sinus. Shear stresses are obtained by using anatomically realistic 3D geometry and representative physiological conditions. Results of this study agree with those in the literature showing that the regions with low wall shear stress. Geometry and flow conditions greatly affected the hemodynamics of the carotid artery. Furthermore, regions of relatively low wall shear stress were observed post stenosis, which is a known cause of plaque development and progression. Under altered gravity conditions the same artery was studied to determine the flow conditions and predict the progression of plague.

Keywords: fluid structure interaction, stenosis, carotid artery, blood pressure, altered gravity

1. Introduction

The most important and essential system in human body is the cardiovascular system, also known as circulatory system. In the circulatory system, the heart acts as a pump, supplying the blood to different tissues, organs and muscles of the body through the dense network of ducts: arteries and veins. The normal blood flow through arteries can be altered significantly by arterial diseases such as atherosclerosis [1]. The atherosclerosis is characterized by the thickening, narrowing and stiffening of the arterial walls. The hardened substance along the walls of the arteries is called plaque and the plaque deposit gradually narrows the artery. The artery, hereby loses its flexibility, which ultimately leads to the blockage of the artery [2]. The narrowing of artery will obstruct and severely reduce the blood flow leading to the organ disfunction [3]. The detailed study of the gradual narrowing or bulging of the artery will help in understanding the underlying mechanisms for unusual behavior of blood flow [4]. The fluid mechanical forces due to the interaction of the blood flow and the arterial wall have a strong influence on the initiation and progression of narrowing or bulging of the artery [5]. Detailed study of hemodynamics in stenosis will be useful in the diagnosis and treatment of vascular diseases [6]. Clinically analyzing the hemodynamics will not yield the detailed investigation using current diagnostic imaging options such as angiography, CTA, MRA or duplex scanning [7]. The detailed information about the hemodynamics in diseased vessels can be obtained from the numerical simulations, and such simulations will help in obtaining a better insight in predicting the hemodynamics. It was observed that hemodynamics of the carotid artery was very much affected by the geometry and flow conditions. Furthermore, regions of relatively low wall shear stress were observed post stenosis, which is a known cause of plaque development and progression [8].

Another major cause of stroke is hypertension. Some of the studies have investigated the effect of hypertension on aneurysms and stenosed arteries. Researchers observed that hypertension increases the WSS and deformation in aneurysm regions, which were initiated by extreme stress - strain conditions [9], whereas Milad et al. [10] experimentally elucidated that the increase in stenosis severity in the carotid artery. This observation is found to fluctuate the hemodynamic parameters especially at throat of the stenosis. All the parameters help predict the locations of potential plaque growth and these results helped in studying the plaque growth and arterial remodeling [10]. It is also found that, the effect of hypertension on the atherosclerotic arteries was studied as it poses a major risk in the rupture of the plaque [11]. A significant correlation between carotid strain parameters and peak and mean WSS in hypertension was also observed [11].

Moreover, the numerical study focusing on flow variation due to gravitational effect during change of different postures such as sitting, sleeping and standing have been discussed with more focus on clinical aspects [12]. Some of the studies have justified the variation in flow behavior during change of postures as observed clinically [13]. The analytical models representing the vascular network in order to predict the variation in flow and pressure during change of postures were also developed. [14]. However, most of these studies are investigated for space application with different conditions such zero gravity and hyper gravity [15]. In another attempt, patient specific numerical study is simulated to demonstrate the gravitational effects on the brain circulation under auto-regulation for change of postures [16]. Hence, the influence of gravity also plays a very vital role in studying the detailed blood flow analysis because of constant change of postures.

Overall, these numerical simulations will aid in interpreting the existing *in-vivo* data, and eventually lead to the development of improved imaging techniques [17]. It is also recognized that narrowing or bulging of arteries are closely related with blood flow characteristics, such as areas of flow reversal or low and oscillatory shear stress [18]. Therefore, a detailed understanding of the local hemodynamics can have useful applications, for instance in predicting potential regions for the formation and development of atherosclerosis, or the consequences of surgical intervention [19].

However, the effect of variation in blood pressure on atherosclerosis has been limited and few studies have attempted to predict that hypertension increases the WSS and arterial deformation initiated by extreme stress – strain conditions [20–22]. The effects of increase blood pressure or hypertension with more focus on stenosis related to patient specific cases is one of the potential area for numerical investigation. Also, there are very limited studies with clinical relevance which supports that a change of posture will certainly cause symptoms/stroke in patients altered cerebral auto regulation. Due to the risks involved in stroke (plaque rupture) in patients, studies supporting the change of posture in such patients are not possible to investigate clinically. Hence, this kind of observation of minor flow changes in healthy individuals and significant variation in patients with different postures under the influence of altered gravity are demonstrated in this chapter/section using numerical simulation approach. This chapter/section, therefore summarizes the investigation on effects of hypertension in comparison with normal blood pressure on normal and stenosed carotid artery bifurcation. In addition, effects of altered gravity is also discussed during change of posture from sleeping to standing under normal blood pressure condition.

2. Methodology

2.1 FSI theory

The blood flow behavior in both cases (a) and (b) of this study is assumed to be governed by the Navier–Stokes equations of incompressible flows. The fluid domain in FSI simulation is solved using modified momentum equation adopting moving velocity concept along with continuity equation as given in Eq. (1) [2, 11, 12].

$$\frac{\partial}{\partial t}\int_{\Omega}\rho\partial\Omega + \int_{S}\rho(v - v_b)n\partial S = \int_{S}(\tau_{ij}i_j - Pi_i)n\partial S + \int_{\Omega}b_i\partial\Omega \tag{1}$$

The artery wall is assumed to be elastic, isotropic, incompressible and homogeneous and the transient dynamic structural solution is given by Eq. (2) [4]. The stiffness matrix is updated in each time step and the Newmark method is adopted in updating the displacement terms at each time interval and further the stiffness matrix is solved using direct solver in particular sparse solver for each time step.

$$[M]\{\ddot{U}\} + [C]\{\dot{U}\} + [K]\{U\} = \{F^a\} \tag{2}$$

FSI Algorithm: Based on the Newtonian assumption with incompressible flow for blood and linear elastic property of arterial wall, the two-way transient FSI analysis is performed using system coupling FSI solver in ANSYS-18.0.

This coupling solver solves fluid and solid domain separately using ANSYS CFX and ANSYS STRUCTURAL respectively as shown in the **Figure 1**. The pressure loads obtained from initial ANSYS CFX solution is transferred to the structure through FSI interface and later ANSYS structural domain is solved. Further details of FSI solver are described in [2, 19].

2.2 Modeling

The present study discuss two different patient specific case, (a) healthy and normal carotid bifurcation without any symptoms of stroke and (b) stenosis of 75% at ECA root, while ICA and CCA appears to be normal. The required geometric

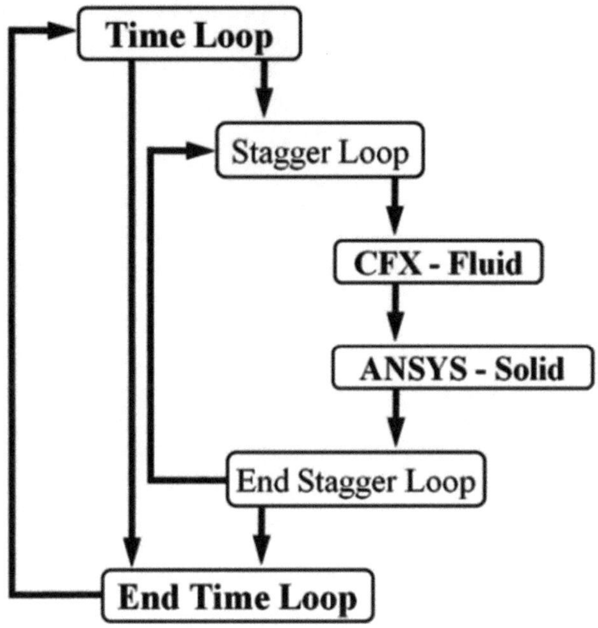

Figure 1.
Fluid structure interaction algorithm in ANSYS.

model is generated based on data obtained from CT angio scan. **Figure 2** shows the different views of CT scan data and the encircled area highlights the location of carotid bifurcation on both left and right side in addition to the 3D geometric model generated in MIMICS. The 3D fluid and solid models of both these cases left and right carotid system are generated using MIMICS-16 based on CT angio data. The solid model is generated using CATIAV5R20.0, versatile geometric modeling software and further transferred to ANSYS 18.0 for the meshing. Case-(a) carotid system consisting of fluid and structural model is meshed with 233,750 and 43,455 hexahedral elements respectively. Similarly, fluid and solid models of case-(b) carotid system is meshed with 254,220 and 41,760 hexahedral elements as shown in the **Figure 3**.

2.3 Analysis

Generally blood is known to be non-Newtonian physiologically, however in the present study, since the focus in on large arteries, Newtonian assumption is acceptable as relatively high shear rate occurs [23]. In medium and smaller arteries, non-Newtonian assumption is valid as shear rate is lower than 100 s^{-1} and shear stresses depend non-linearly on the deformation rate [24]. A time varying velocity pulse is applied at inlet of both the carotid cases based on available literature [8]. A typical inlet velocity profile as shown in the **Figure 4** is applied for both cases (a) and (b) without altered gravity behavior which contributes to sleeping posture. However, under altered gravity condition, such as standing posture, the inlet velocity profile will be as shown in the **Figure 4**. Under the standing posture condition, the inlet velocity will change considering the hydrostatic pressure, which is related to gravity and referred as ρgh, where ρ is fluid density, g is the acceleration due gravity and h is the height of the hydrostatic column [21]. The height of the column of blood is always referred at the level of the heart.

Also, to include the peripheral resistance, a time varying pressure wave form is applied at the outlet as shown in the **Figure 4** [5]. The range of pulse pressure is

DOI: http://dx.doi.org/10.5772/intechopen.93908

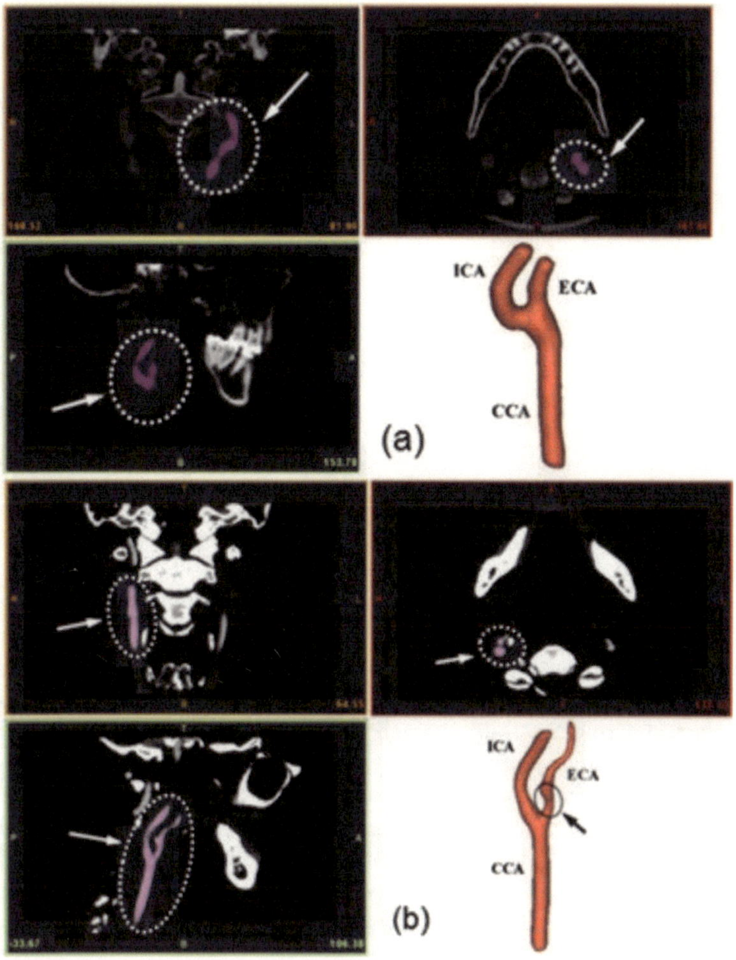

Figure 2.
Different views of CT scan of carotid bifurcation, case-(a) Normal carotid artery model and case-(b) Stenosed carotid artery model.

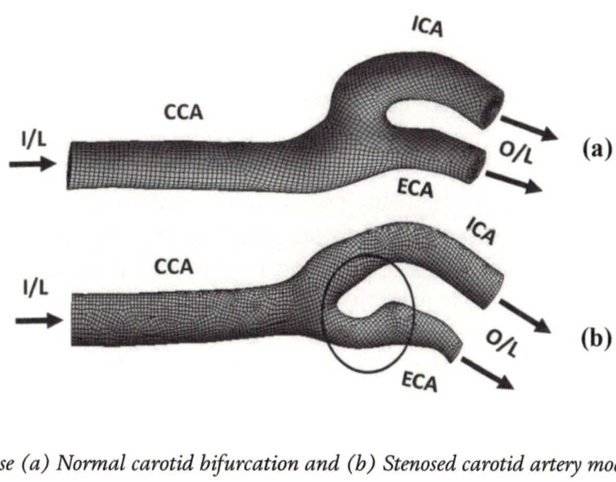

Figure 3.
Meshed model of case (a) Normal carotid bifurcation and (b) Stenosed carotid artery model.

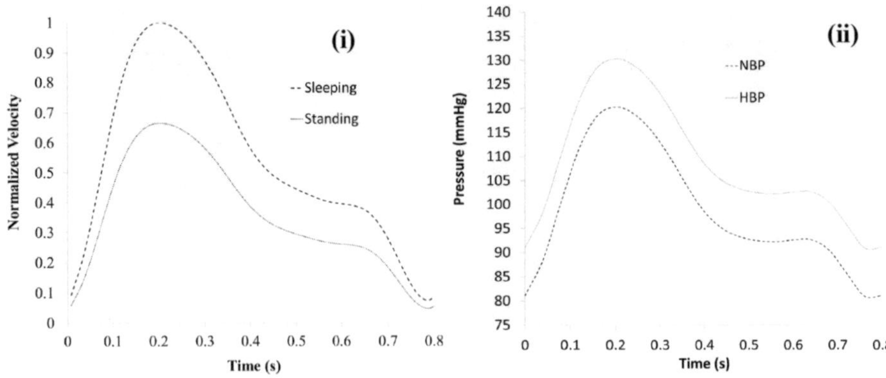

Figure 4.
(i) Normalized inlet velocity waveform and (ii) outlet pressure waveform with NBP and HBP.

different as the simulation is carried out for both NBP and HBP having 80–125 mm Hg and 100–170 mm Hg, respectively. Each pulse cycle for a time period of 0.8 s is discretized into 180 time steps to simulate the flow behavior more accurately. In this study, blood flow properties in form of density and dynamic viscosity are considered to be 1050 kg/m^3 and 0.004 $N\text{-}sec/m^2$ respectively [15]. The arterial wall is assumed to behave linearly-elastic having elastic modulus is 0.9 MPa and Poisson's ratio of 0.40 with density of 1120 kg/m^3 [20, 25]. The convergence criteria of fluid flow and across the fluid-surface interface is set at 10^{-4} and 10^{-3} respectively and low Reynolds k-ω model is used to model the turbulence behavior [22]. In this study, with sleeping position as reference, effects of NBP and HBP on flow behavior are investigated and further altered gravity evaluation is performed for change of posture from sleeping to standing under NBP condition. These simulation results provide useful data in quantifying the hemodynamic changes during different blood pressures (NBP and HBP) and also during change of posture from sleeping to standing.

3. Results and discussion

Numerical simulation in this study of both the cases (a) and (b) is carried out for 3 pulse cycle and results obtained in the last cycle is considered for the investigation. The hemodynamic parameters like velocity, WSS and arterial wall deformation are studied at specific instants of pulse cycle like peak systole (i), early systole (ii) and late diastole (iii). Inlet velocity is considered with reference to sleeping position and effects of variation in pressure parameter is investigated under NBP and HBP conditions. Under altered gravity assumption, inlet velocity is considered for change of position from sleeping to standing under NBP condition only. Flow behavior will be less intense during standing posture in contrast to sleeping condition. WSS is considered to be the most crucial and interesting hemodynamic parameters related to the atherosclerotic progression. It varies with time due to the pulsatility of the flow waveform and the maximum value generally occurs at the peak systole when the inflow is maximum.

Velocity: The velocity streamlines of the case (a) normal carotid bifurcation subjected to NBP & HBP is shown in **Figure 5**. In this case, higher velocity magnitude is observed at peak systole for all the blood pressure models. Flow recirculation region is almost similar at peak systole and early diastole, however, in late diastole, it is more chaotic at the carotid bulb.

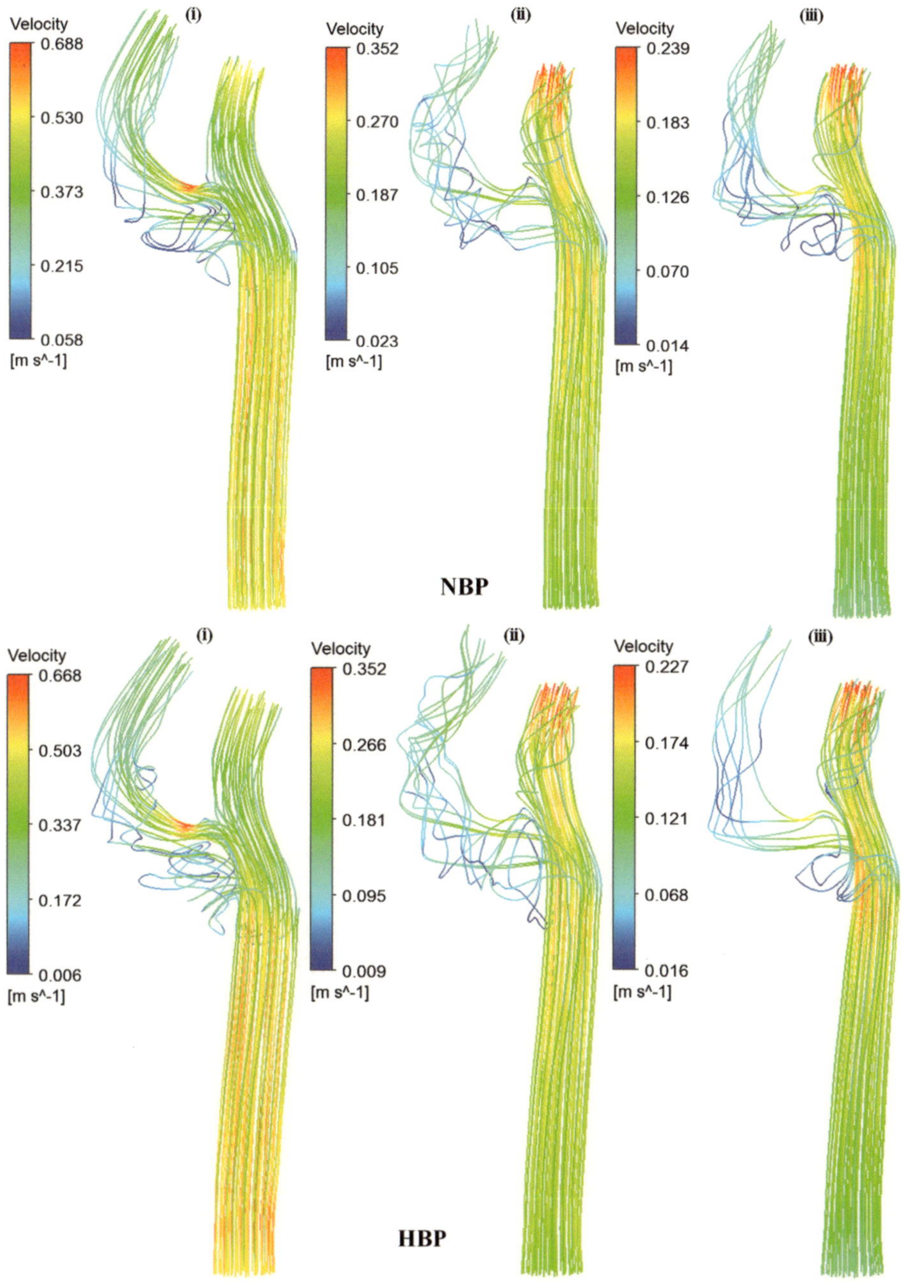

Figure 5.
Velocity streamline plot for case (a): Normal carotid artery model under NBP & HBP during (i) peak systole, (ii) early systole (iii) late diastole.

High-velocity gradients is seen at the bifurcation and flow reversals along the outer wall of the ICA due to bifurcation of the arterial geometry and carotid bulb is located at the outer wall region of the ICA leading to flow reversals. Similar behavior is observed with higher blood pressure. As compared to NBP, streamlines tends to be more laminar at higher blood pressure. The flow separation was observed to be leading to vortices and the vortex shedding was observed at elevated flow rates due to the increased momentum of flow. At peak systole, higher velocity was observed, and flow separation was occurring at the upper portion of the CCA due to bifurcation and due to sudden increase in diameter at carotid sinus. The velocity magnitude was inversely proportional to the variation in blood pressure. This was due to higher peripheral resistance due to blood pressure, which also leads to an increase in arterial deformation. The velocity is higher in the ECA as compared to the ICA due to the geometry, where the centerline of is almost in line with the CCA. In case (b) of stenosed carotid artery model, the flow is almost similar at peak systole, except minor flow recirculation areas at the carotid sinus and post stenotic region in in ECA which gets magnified with the increase in the blood pressure as shown in the **Figure 6**. However, at early diastole and late diastole, the flow turns chaotic post stenosis and at the carotid sinus. The stenosis further induces abrupt flow disturbance creating complex vortex formation in the downstream of the narrowed ECA.

The vortex induced in the downstream is highly complex and extends till the distal end of the ECA and more prominent in the later part of the cardiac cycle. The magnitude of the velocity is reduced with an increase in blood pressure at peak systole and it tends to increase at late diastole at ICA. The flow recirculation region

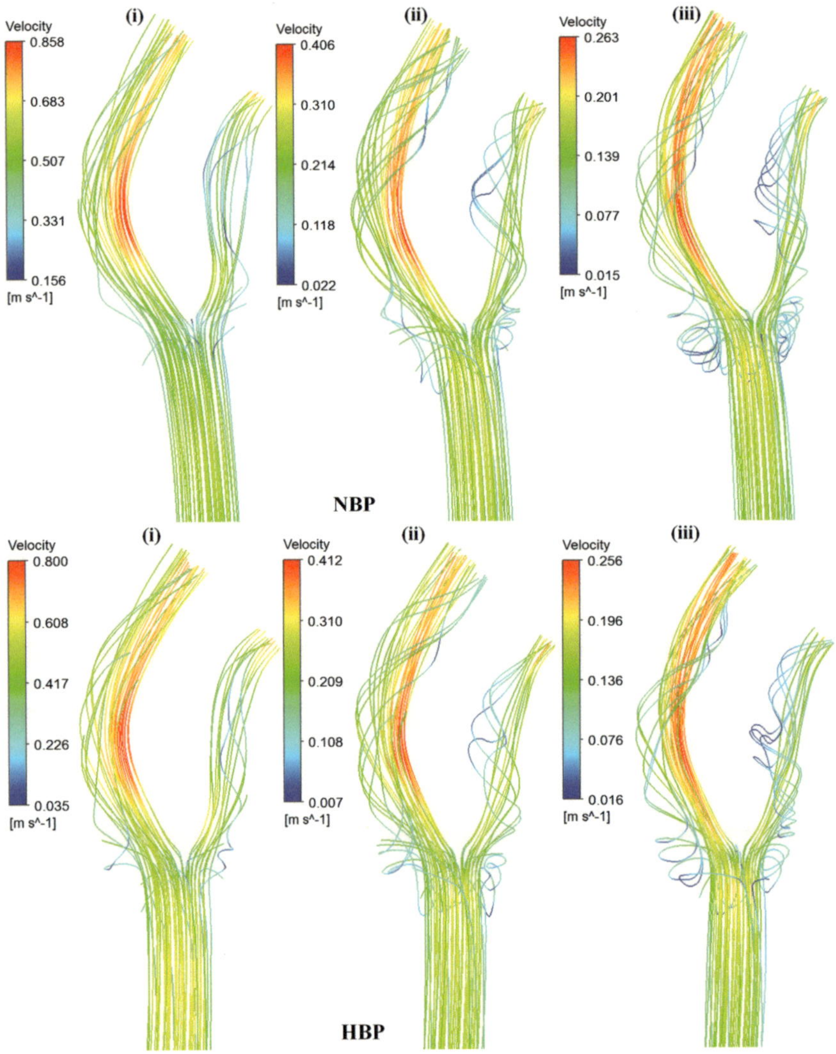

Figure 6.
Velocity streamline plot for case (b): Stenosed carotid artery model under NBP & HBP during (i) peak systole, (ii) early systole (iii) late diastole.

is increased with higher blood pressure at late diastole. Under the altered gravity, the hemodynamic characteristics in both cases (a) and (b) will be similar to that as observed during sleeping posture. However, the flow will be less intense during standing posture in contrast to sleeping condition. The flow changes observed during sleeping and standing in case-(a) and (b) is plotted in the **Figure** 7. With the change of posture from sleeping to standing, the velocity changes abruptly in both these cases as shown in the **Figure 7** with similar pattern. In case (a), the velocity in standing position decreases 25 to 30% with mild variation in flow separation. However, in case (b), the variation in magnitude of flow velocity shows the elevated velocities as compared to case (a). The percentage variation of flow velocity during the change of posture from sleeping to standing indicates a drop in 30–35%, as observed in the **Figure** 7.

Wall Shear Stress: Wall shear stress is a significant parameter as it is related to degeneration of the arterial wall. **Figure 8** shows the WSS contours at peak systole, early and late diastole phase of the cardiac cycle in case (a) normal carotid artery. Maximum WSS is observed at NBP and varies inversely with blood pressure and the lowest WSS magnitude is observed at HBP. In both NBP and HBP cases, the WSS is concentrated at the bifurcation point towards the inner wall at the stagnation point due to high velocity gradient. The low WSS at the outer wall of the ICA at the bifurcation decreases with the increase in blood pressure. At peak systole the flow separation occurs at the base of the bifurcation near the carotid sinus leading to lower WSS and leading along the outer wall of the ICA at late diastole.

Significantly lower WSS is observed at late diastole where the flow recirculation is maximum. The decrease in WSS is due to reduced flow velocity and enlargement of the arterial wall due to increased blood pressure. **Figure 8** shows the WSS contours for NBP and HBP at different phases of the cardiac cycle in case (b) of stenosed carotid artery model. Maximum WSS magnitude at peak systole for NBP, HBP are 10.147 Pa, 10.176 Pa respectively. The WSS tends to concentrate mainly near the stenosis region at the bifurcation and at the inner wall of the ICA.

At peak systole, the WSS is concentrated mainly at the inner wall of the ICA at the curved region, whereas the low WSS region is predominant immediately after the stenosis for both BP cases. The intensity of low shear region increases at early diastole and it spreads all over the inner wall of ECA at late diastole. The inner wall of the ECA at the bifurcation zone have traces of low WSS (>2 Pa) which have slightly reduced influence of progression of atherosclerosis [19]. In addition, at this instant WSS increases pre stenosis at the neck of the stenosis. Low WSS is also observed at the carotid sinus and post stenosis regions in the ECA due to flow recirculation caused by sudden increase in diameter. This low WSS (>0.4 Pa)

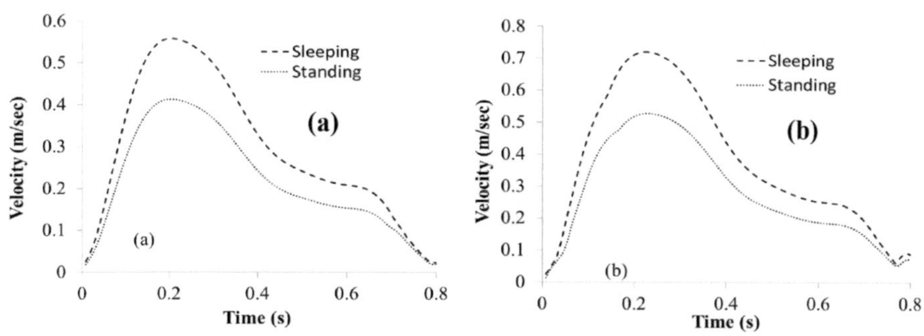

Figure 7.
Comparison of velocity during of change of posture from sleeping to standing in case (a) and case (b) in NBP condition.

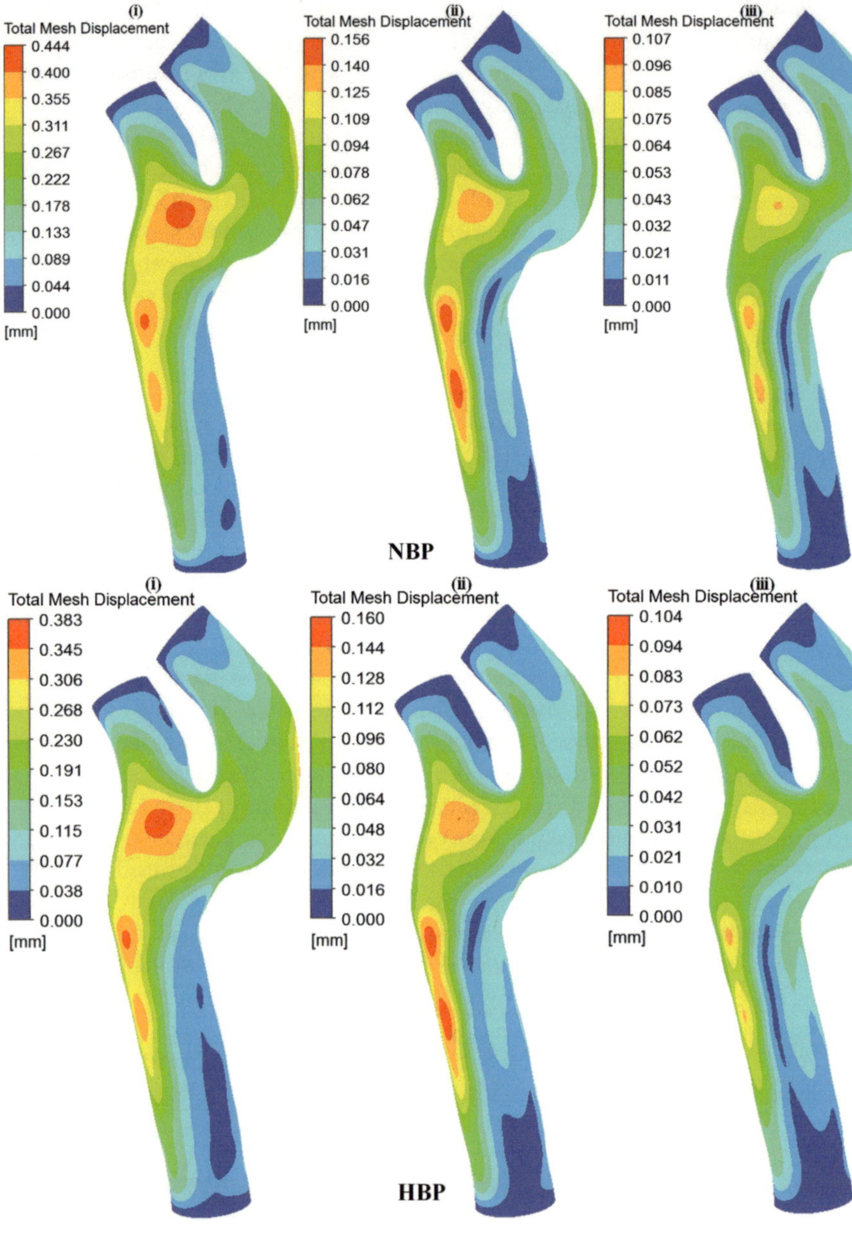

Figure 8.
WSS contour plot for case (a): Normal carotid artery model under NBP & HBP during (i) peak systole, (ii) early systole (iii) late diastole.

encourage the progression of atherosclerosis. The low WSS region is at the neck of the bifurcation below the carotid sinus where there is maximum flow recirculation. The low WSS at higher BP will certainly trigger atherosclerosis progression and endothelial cell disorientation [22] (**Figure 9**).

Even though the WSS pattern across the carotid models in both cases (a) and (b) will be similar to that of sleeping condition, however, during the change of posture from sleeping to standing, due to less flow, the maximum variation in WSS is observed during peak systole as compared to rest of the pulse cycle. WSS changes observed in both these cases (a) and (b) are compared in the **Figure 10** through the entire pulse cycle. The change of posture from sleeping to standing drops the WSS

DOI: http://dx.doi.org/10.5772/intechopen.93908

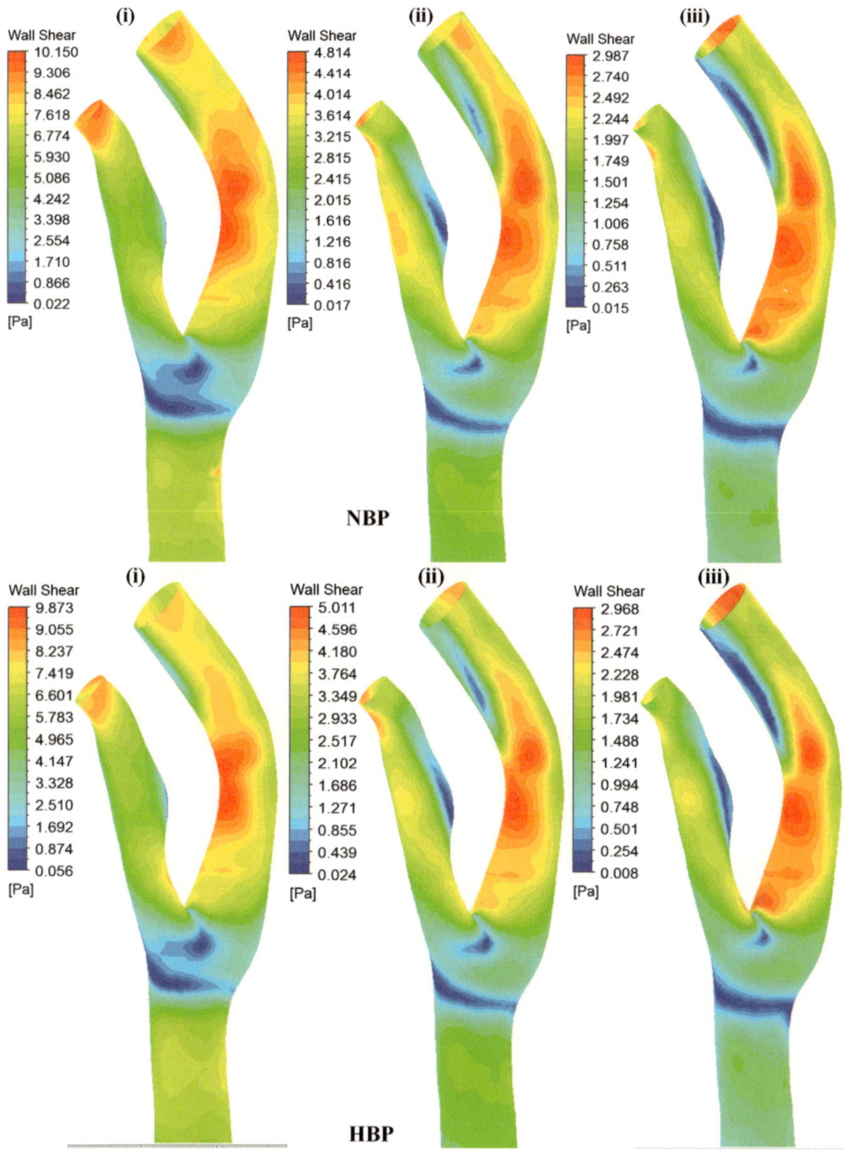

Figure 9.
WSS contour plot for case (b): Stenosed carotid artery model under NBP & HBP during (i) peak systole, (ii) early systole (iii) late diastole.

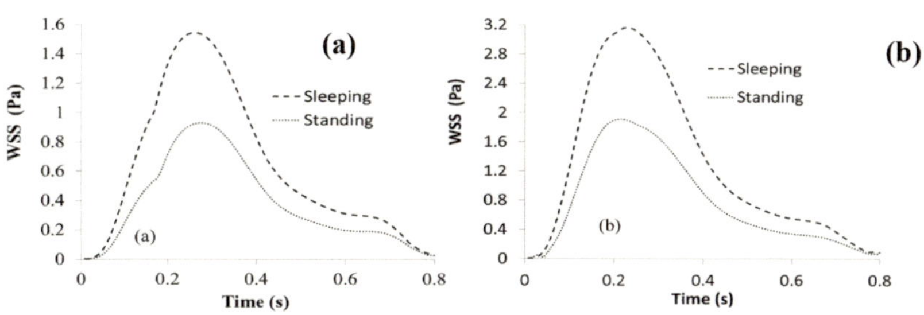

Figure 10.
Comparison of WSS during of change of posture from sleeping to standing in case (a) and case (b) in NBP condition.

by 40–45%, as observed in case (a). The WSS is found to be less disturbed without any major complexity due to normal flow behavior. In case (b), the WSS variation as compared for case (a) substantially drop by 50% during the change of posture from sleeping to standing. Significant WSS variation during the change of postures will certainly trigger the damage to arterial wall and induce the plaque rupture [26].

Wall Deformation: Figure 11 shows the arterial wall deformation contours at normal carotid artery of case (a) at different phases of the cardiac cycles. The maximum deformation is at the bifurcation region, mainly at the base of the branching of ICA and ECA due to reduced arterial stiffness because of the curvature. Generally, maximum deformation is at the location where the pressure is maximum, especially at the apex of the bifurcation. The curvature of the

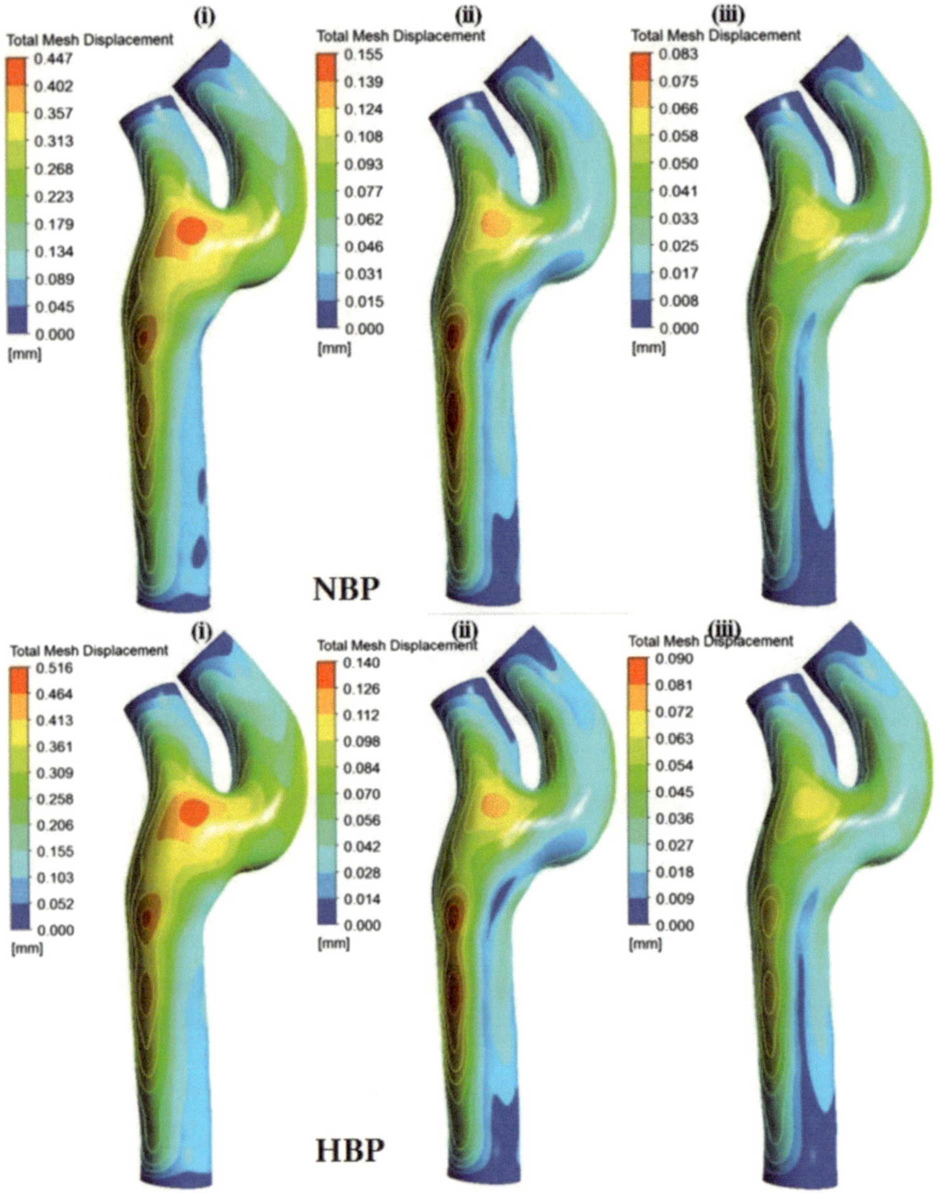

Figure 11.
WSS contour plot for case (a): Normal carotid artery model under NBP & HBP during (i) peak systole, (ii) early systole (iii) late diastole.

bifurcation reduces the stiffness of the wall, and therefore have high wall deformation [27]. The outer wall of the ICA is also subjected to moderate deformations along with the bifurcation region. Low WSS along with higher wall deformation is one of the possible causes of atherosclerosis development. Maximum deformation of 0.447 mm, 0.516 mm, is observed at peak systole for NBP, and HBP conditions respectively.

The maximum arterial deformation occurs at peak systole as observed in the stenosed carotid bifurcation of case (b) as shown in **Figure 10** for both the NBP and HBP conditions. The location of the maximum deformation is observed to be in the bifurcation region. The plaque at the root of the ECA has higher stiffness resulting in reduced elastic deformation. The reduced wall stiffness is localized around the ECA resulted in reduced wall deformation across the stenosed location. Another observation is post stenotic deformation in distal side of the ECA because of eccentric stenosis with a lower profile due to increased stiffness of the plaque. The partial restriction offered for the flow in the ECA diverts the flow through the ICA. However, the increased pressure in the upstream of the narrowed region to compensate the flow has increased the deformation distribution. Therefore, the

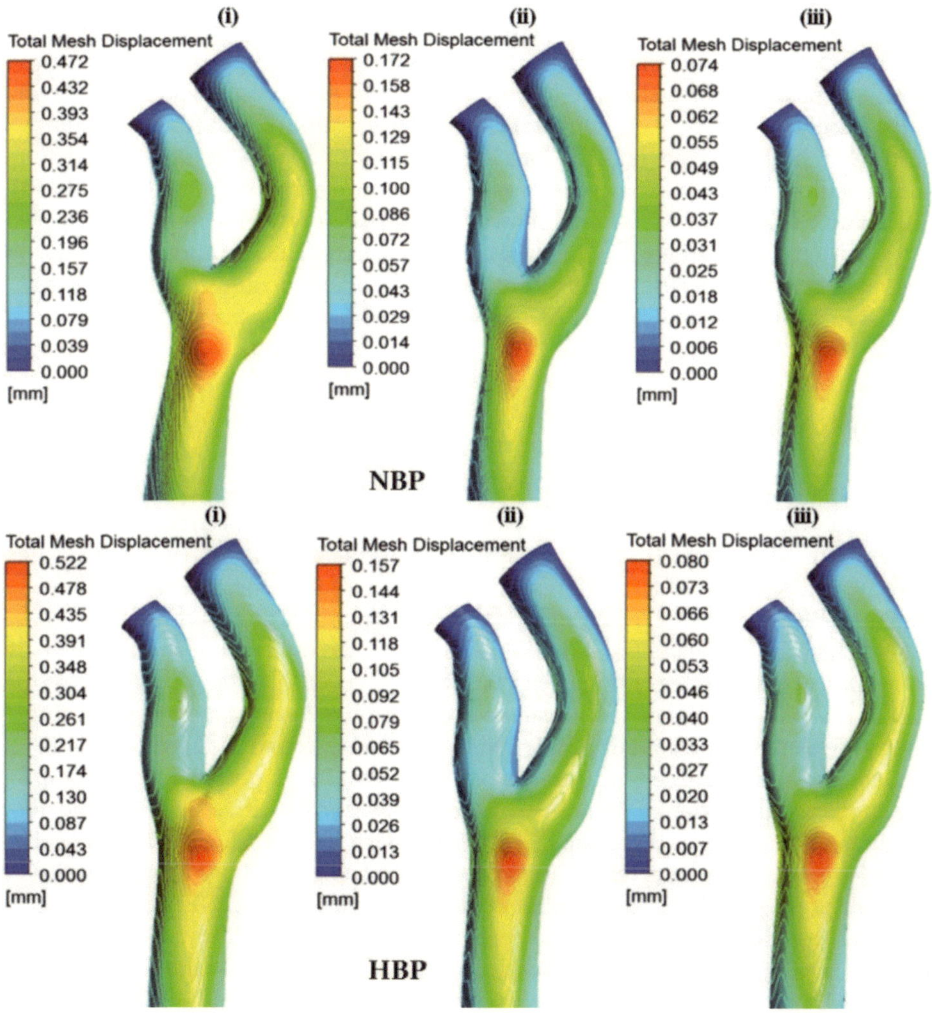

Figure 12.
WSS contour plot for case (b): Stenosed carotid artery model under NBP & HBP during (i) peak systole, (ii) early systole (iii) late diastole.

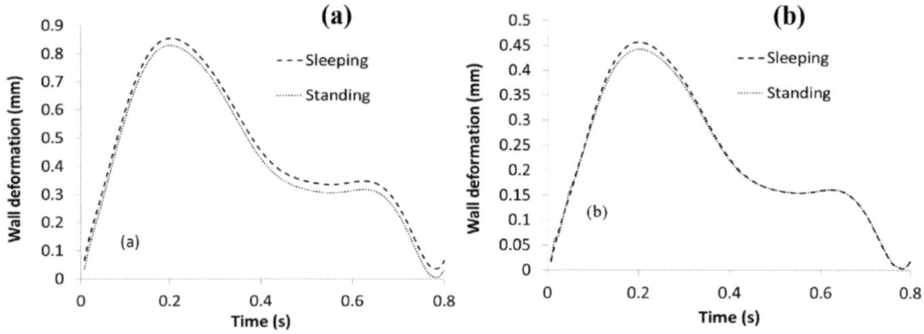

Figure 13.
Comparison of WSS during of change of posture from sleeping to standing in case (a) and case (b) in NBP condition.

maximum deformation is during the peak systole at the entrance of the ICA in the bifurcation region. The deformation profile is typical of normal carotid bifurcation as observed in published literature [28]. There is no significant difference in arterial deformation between the rheological models considered in the study (**Figure 12**).

The variation in wall deformation behavior throughout the pulse cycle during change of postures from sleeping to standing in both the case-(a) and (b) is shown in the **Figure 13**. There is no remarkable difference among sleeping and standing postures in both these cases [29]. The change of position from sleeping to standing notices a drop of less than 5% and 8% as observed for both case (a) and (b) [13, 30]. The obtained deformation pattern shows considerable change during different postures and agrees well with the clinical observation [31]. The high pressure in upstream of stenosis causes the maximum wall deformation and the intense pressure drop at the throat region will result in wall collapse.

4. Conclusion

In this study, the pulsatile flow of blood with different physiological pressure conditions and altered gravity was studied. In normal carotid bifurcation case (a) during both NBP and HBP cases, the curvature of the bifurcation has influenced in reducing the stiffness of the wall resulting in higher wall deformation. The outer wall of the ICA is also subjected to moderate deformations along with the bifurcation region. The WSS is found to be concentrated at the bifurcation and intense flow separation at this zone resulting in lower WSS. However, in case (b), stenosis at the root of the ECA has higher stiffness resulting in reduced elastic deformation. The reduced wall stiffness is localized around the ECA resulted in reduced wall deformation across the stenosed location. Another observation is post stenotic deformation in distal side of the ECA because of eccentric stenosis with a lower profile due to increased stiffness of the plaque. The WSS tends to concentrate mainly near the stenosis region at the bifurcation and at the inner wall of the ICA. The low WSS region is predominant in post stenotic region for both NBP and HBP cases. In both the cases (a) and (b), low WSS at different regions of carotid bifurcation shall significantly influences the progression of atherosclerosis. Under the altered gravity, case-1, demonstrated typical flow behavior as that of normal carotid bifurcation, but minor variations are present due to tortuous bifurcation. It is clear from the results that the wall deformation has dropped by less than 5% during standing posture. However, velocity and WSS show a considerable drop of 25% and 45%

respectively during standing. In case (b), the velocity and WSS drops by 30–35% and 50% while the arterial wall deformation reduces by 8% in standing posture. The moderately higher WSS at the level of maximum stenosis has slightly reduced the arterial wall stiffness resulting in low risk factor of disease progression during standing posture. It can be concluded that risk factors are quite low and the flow behavior is also within the physiological limits, during the change of postures from sleeping to standing. Further, the results of this study demonstrate the potential of numerical simulation in understanding of the causes of atherosclerosis and pave the way in developing innovative computational solutions to aid the early diagnosis of atherosclerosis.

Acknowledgements

Authors thanks Department of Radio-Diagnosis and Imaging, Kasturba Hospital, Manipal for providing the patient specific data to carry out the numerical simulation.

Conflict of interest

The authors declare no conflict of interest.

Abbreviations and nomenclature

CCA	Common Carotid Artery
ICA	Internal Carotid Artery
ECA	External Carotid Artery
NBP	Normal Blood Pressure
HBP	High Blood Pressure
WSS	Wall Shear Stress
FSI	Fluid Structure Interaction
ρ	Density
τ	Stress tensor
υ	Velocity vector
υ_b	Grid velocity
P	Pressure
b_i	Body force at time t
M	Structural mass matrix
C	Structural damping matrix
K	Structural stiffness matrix
F^a	Applied load vector
$\ddot{U}$	Acceleration component
$\dot{U}$	Velocity
U	Displacement vector

Author details

S.M. Abdul Khader, Nitesh Kumar and Raghuvir Pai*
Department of Mechanical and Manufacturing Engineering, Manipal Institute of Technology, Manipal Academy of Higher Education, Manipal, India

*Address all correspondence to: raghuvir.pai@manipal.edu

References

[1] Hosseini V, Mallone A, Mirkhani N, Noir J, Salek M, Pasqualini FS, et al. A Pulsatile Flow System to Engineer Aneurysm and Atherosclerosis Mimetic Extracellular Matrix. Adv Sci. 2020; 2000173.

[2] Jiang F, Zhu Y, Gong C, Wei X. Atherosclerosis and Nanomedicine Potential: Current Advance and Future Opportunities. Curr Med Chem. 2020;

[3] Michail M, Davies JE, Cameron JD, Parker KH, Brown AJ. Pathophysiological coronary and microcirculatory flow alterations in aortic stenosis. Nat Rev Cardiol. 2018;15 (7):420–31.

[4] Yang H, Yu PK, Cringle SJ, Sun X, Yu D-Y. Microvascular network and its endothelial cells in the human iris. Curr Eye Res. 2018;43(1):67–76.

[5] Wang R, Yu X, Zhang Y. Mechanical and structural contributions of elastin and collagen fibers to interlamellar bonding in the arterial wall. Biomech Model Mechanobiol. 2020;1–14.

[6] Pai R, Khader SM, Ayachit A, Ahmad KA, Zubair M, Rao VRK, et al. Fluid-Structure Interaction Study of Stenotic Flow in Subject Specific Carotid Bifurcation—A Case Study. J Med Imaging Heal Informatics. 2016;6(6):1494–9.

[7] Rayz VL, Cohen-Gadol AA. Hemodynamics of Cerebral Aneurysms: Connecting Medical Imaging and Biomechanical Analysis. Annu Rev Biomed Eng. 2020;22.

[8] Kumar N, Khader SMA, Pai R, Khan SH, Kyriacou PA. Fluid structure interaction study of stenosed carotid artery considering the effects of blood pressure. Int J Eng Sci [Internet]. 2020; 154:103341. Available from: http://www.sciencedirect.com/science/article/pii/S0020722520301294

[9] Yang JW, Im Cho K, Kim JH, Kim SY, Kim CS, You GI, et al. Wall shear stress in hypertensive patients is associated with carotid vascular deformation assessed by speckle tracking strain imaging. Clin Hypertens. 2014;20(1):10.

[10] Samaee M, Tafazzoli-Shadpour M, Alavi H. Coupling of shear–circumferential stress pulses investigation through stress phase angle in FSI models of stenotic artery using experimental data. Med Biol Eng Comput. 2017;55(8):1147–62.

[11] Rabby MG, Razzak A, Molla MM. Pulsatile non-Newtonian blood flow through a model of arterial stenosis. Procedia Eng [Internet]. 2013;56:225–31. Available from: http://dx.doi.org/10.1016/j.proeng.2013.03.111

[12] Abdul Khader SM, Ayachit A, Pai R, Zubair M, Ahmed KA, Rao VR. Study of the influence of Normal and High Blood pressure on normal and stenosed Carotid Bifurcation using Fluid-Structure Interaction. In: Applied Mechanics and Materials. 2013. p. 982–6.

[13] Azran A, Hirao Y, Kinouchi Y, Yamaguchi H, Yoshizaki K. Variations of the maximum blood flow velocity in the carotid, brachial and femoral arteries in a passive postural changes by a Doppler ultrasound method. In: The 26th Annual International Conference of the IEEE Engineering in Medicine and Biology Society. 2004. p. 3708–11.

[14] Alirezaye-Davatgar MT. Numerical simulation of blood flow in the systemic vasculature incorporating gravitational force with application to the cerebral circulation. University of New South Wales; 2006.

[15] Gisolf J, others. Postural changes in humans: effects of gravity on the circulation. 2005.

[16] Kim CS, Kiris C, Kwak D, David T. Numerical simulation of local blood flow in the carotid and cerebral arteries under altered gravity. 2006;

[17] Mittal R, Seo JH, Vedula V, Choi YJ, Liu H, Huang HH, et al. Computational modeling of cardiac hemodynamics: current status and future outlook. J Comput Phys. 2016;305:1065–82.

[18] Carpenter HJ, Gholipour A, Ghayesh MH, Zander AC, Psaltis PJ. A review on the biomechanics of coronary arteries. Int J Eng Sci. 2020;147:103201.

[19] Wong KKL, Wang D, Ko JKL, Mazumdar J, Le T-T, Ghista D. Computational medical imaging and hemodynamics framework for functional analysis and assessment of cardiovascular structures. Biomed Eng Online. 2017;16(1):35.

[20] Hirschhorn M, Tchantchaleishvili V, Stevens R, Rossano J, Throckmorton A. Fluid–structure interaction modeling in cardiovascular medicine–A systematic review 2017–2019. Med Eng Phys. 2020;

[21] Al-Sharea A, Lee MKS, Whillas A, Michell DL, Shihata WA, Nicholls AJ, et al. Chronic sympathetic driven hypertension promotes atherosclerosis by enhancing hematopoiesis. Haematologica. 2019;104(3):456–67.

[22] Sun HT, Sze KY, Tang AYS, Tsang ACO, Yu ACH, Chow KW. Effects of aspect ratio, wall thickness and hypertension in the patient-specific computational modeling of cerebral aneurysms using fluid-structure interaction analysis. Eng Appl Comput Fluid Mech. 2019;13(1): 229–44.

[23] Perktold K, Peter R, Resch M. Pulsatile non-Newtonian blood flow simulation through a bifurcation with an aneurysm. Biorheology. 1989;26(6): 1011–30.

[24] Khader SMA, Azriff A, Johny C, Pai R, Zuber M, Ahmad KA, et al. Haemodynamics Behaviour in Normal and Stenosed Renal Artery using Computational Fluid Dynamics. J Adv Res Fluid Mech Therm Sci. 2018;51(1): 80–90.

[25] Dhawan SS, Nanjundappa RPA, Branch JR, Taylor WR, Quyyumi AA, Jo H, et al. Shear stress and plaque development. Expert Rev Cardiovasc Ther [Internet]. 2010;8(4):545–56. Available from: https://www.ncbi.nlm.nih.gov/pmc/articles/PMC5467309/pdf/nihms213759.pdf

[26] Ku DN. Blood flow in arteries. Annu Rev Fluid Mech. 1997;29(1): 399–434.

[27] Zhao SZ, Xu XY, Hughes AD, Thom SA, Stanton A V, Ariff B, et al. Blood flow and vessel mechanics in a physiologically realistic model of a human carotid arterial bifurcation. J Biomech. 2000;33(8):975–84.

[28] Toloui M, Firoozabadi B, Saidi MS. A numerical study of the effects of blood rheology and vessel deformability on the hemodynamics of carotid bifurcation. Sci Iran [Internet]. 2011; 19(1):119–25. Available from: http://dx.doi.org/10.1016/j.scient.2011.12.008

[29] Toloui M, Firoozabadi B, Saidi MS. A numerical study of the effects of blood rheology and vessel deformability on the hemodynamics of carotid bifurcation. Sci Iran [Internet]. 2012;19(1):119–26. Available from: http://dx.doi.org/10.1016/j.scient.2011.12.008

[30] Kim CS, Kiris C, Kwak D, David T. Numerical Simulation of Local Blood Flow in the Carotid and Cerebral Arteries Under Altered Gravity. J Biomech Eng [Internet]. 2006;128(2): 194. Available from: http://biomechanical.asmedigitalcollection.asme.org/article.aspx?articleid=1415462

[31] Savin E, Bailliart O, Checoury A, Bonnin P, Grossin C, Martineaud J-P. Influence of posture on middle cerebral artery mean flow velocity in humans. Eur J Appl Physiol Occup Physiol. 1995; 71(2–3):161–5.

Chapter 3

The Finite Element Method Applied to the Magnetostatic and Magnetodynamic Problems

Dang Quoc Vuong and Bui Minh Dinh

Abstract

Modelling of realistic electromagnetic problems is presented by partial differential equations (FDEs) that link the magnetic and electric fields and their sources. Thus, the direct application of the analytic method to realistic electromagnetic problems is challenging, especially when modeling structures with complex geometry and/or magnetic parts. In order to overcome this drawback, there are a lot of numerical techniques available (e.g. the finite element method or the finite difference method) for the resolution of these PDEs. Amongst these methods, the finite element method has become the most common technique for magnetostatic and magnetodynamic problems.

Keywords: finite element method, magnetostastics, magnetodynamics, Maxwell's equations, weak formulations

1. Introduction

Mathematical modeling of realistic problems in the framework of electromagnetics leads to a set of partial derivates equations that have to be solved on a domain with complex geometry associated with boundary conditions and initial conditions. This complexity makes any analytical approach unpracticable. In the past (until 1960), people used experimentation (very expensive, sometimes destructive) or analogic simulation (lack of generality) to solve these problems. Since 1970, the growth of computer capabilities makes the numerical simulation a tool that is more and more used by the people interested in solving these complex problems. When using the computer, the continuous problem is represented with a finite number of degrees of freedom (d.o.f.). The continuous problem is then replaced by a discrete problem. There are a lot of numerical techniques available. We will see that the most common ones can be derived from the same general principle of weighted residuals.

A continuous formulation of a problem cannot generally be solved analytically and some numerical methods have to be used in order to obtain quantitative information about the solution. The unknown functions of a continuous problem belong to continuous function spaces which are usually of infinite dimensions, that is, those functions are usually described by an infinite number of parameters. The basis of any numerical method is to discretize such a problem in order to obtain a similar discrete problem, characterized by a finite number of unknowns which are called degrees of freedom. This discretization process consists of replacing the considered continuous function spaces by some discrete function spaces, whose dimensions are finite, and which are usually subspaces of them. Those spaces are also called approximation spaces and their elements are called approximation functions.

The function spaces are defined in a particular studied domain. If this one is discretized, that is, if it is defined as the union of geometric elements of simple shapes, and if the discrete function spaces are built in such a way that their functions are piecewise defined, then the approximation numerical method is called the finite element method (FEM). It is this kind of method we are interested in. We can thus see that the finite element method necessitates a double discretization: a discretization of some function spaces and a discretization of the studied geometric domain, which leads to a mesh.

Weak formulations are well adapted to the finite element method, which will appear in the following. Such formulations make use of several kinds of Green formulas.

2. Numerical technique

2.1 The Laplacian problem

The formalism used in the case of a Laplacian problem is sufficiently simple to be very understandable without lack of generality. The description of a Laplacian problem is presented now. Let us consider a bounded domain Ω and its boundary $\Gamma = \Gamma_h \cup \Gamma_e$ (**Figure 1**).

The Laplace equation has to be solved in Ω [1–3]:

$$\Delta u(\mathbf{x}) = \frac{\partial^2 u}{\partial x^2} + \frac{\partial^2 u}{\partial y^2} + \frac{\partial^2 u}{\partial z^2} = 0, \tag{1}$$

where u is the unknown field defined at each point x (x, y, z) of the studied domain. The associated boundary conditions are respectively Dirichlet and Neumann conditions, that is

$$u(\mathbf{x}) = \overline{u}(\mathbf{x}), \quad \mathbf{x} \in \Gamma_h, \tag{2}$$

$$v(\mathbf{x}) = \frac{\partial u(\mathbf{x})}{\partial n} = \overline{v}(\mathbf{x}), \quad \mathbf{x} \in \Gamma_e. \tag{3}$$

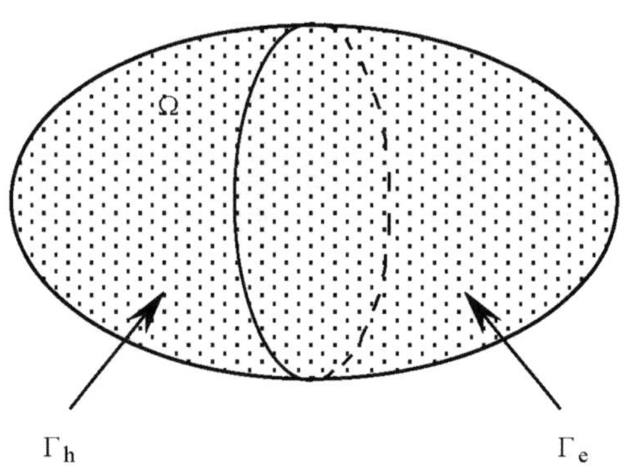

Figure 1.
Studied domain Ω and its boundary $\Gamma = \Gamma_h \cup \Gamma_e$.

This diffusion equation describes a wide range of physical phenomena. The next table shows some of these phenomena.

Problems	u	on Γ_h	on Γ_e
Thermostatics	T (temperature)	T fixed	thermal flux fixed
Electrostatics	V (voltage)	V fixed (electrode)	fixed flux of electrical displacement
Perfect fluids	ψ (flow function) I (current function)		
Magnetostatics	B_r (reduced potential)	fixed magnetic flux density	fixed tangential magnetic field

A natural way to discretize the problem is to impose the error on the equation and on the boundary conditions weighted by a trial function w to be equal to zero, that is

$$\int_{\Omega} \Delta u\, w\, d\Omega + \int_{\Gamma_h} (\overline{u} - u) w\, d\Gamma_e + \int_{\Gamma_e} (\overline{v} - v) w\, d\Gamma_h = 0. \tag{4}$$

Equations (1–3) are then meanly solved, the sense of the mean being the principle of the numerical method. In fact, the numerical method used (F.D.M, F.E.M or B.E.M) are directly related to the chosen trial functions.

2.2 Green formulas

The following notations are used for integration of products of scalar or vector fields over a volume Ω or on a surface Γ, where L^2 and $\mathbf{L}^2$ are the spaces of square-summable scalar and vector functions [2, 3]:

$$(\mathrm{u}, \mathrm{v}) = \int_{\Omega} \mathrm{u}(\boldsymbol{x}) \mathrm{v}(\boldsymbol{x}) \mathrm{d}\boldsymbol{x}, \quad \mathrm{u}, \mathrm{v} \in L^2(\Omega)$$

$$(\boldsymbol{u}, \boldsymbol{v}) = \int_{\Omega} \boldsymbol{u}(\boldsymbol{x}) \cdot \boldsymbol{v}(\boldsymbol{x}) \mathrm{d}x, \quad \boldsymbol{u}, \boldsymbol{v} \in \boldsymbol{L}^2(\Omega),$$

$$\langle \mathrm{u}, \mathrm{v} \rangle_{\Gamma} = \int_{\Gamma} \mathrm{u}(\mathbf{x}) \cdot \mathrm{v}(\mathbf{x}) \mathrm{ds}, \quad \langle \mathbf{u}, \mathbf{v} \rangle_{\Gamma} = \int_{\Gamma} \mathrm{u}(\mathbf{x}) \cdot \mathrm{v}(\mathbf{x}) \mathrm{ds},$$

A first relation of vectorial analysis

$$\mathbf{u} \bullet \mathrm{grad} \mathrm{v} + \mathrm{v} \bullet \mathrm{div}\, \mathbf{u} = \mathrm{div}\, (\mathrm{v}\, \mathbf{u}),$$

integrated in the domain Ω, after applying the divergence theorem, gives the Green formula said of kind grad-div in Ω, that is

$$(\boldsymbol{u}, \mathrm{grad}\, v) + (\mathrm{div}\, \boldsymbol{u}, v) = <v, \boldsymbol{n} \cdot \boldsymbol{u}>_{\Gamma}, \forall \boldsymbol{u} \in \boldsymbol{H}^1(\Omega), \forall v \in H^1(\Omega) \tag{5}$$

where $\boldsymbol{H}^1(\Omega)$ and $v \in H^1(\Omega)$ are function spaces built for scalar and vector fields, respectively.

Another relation of vectorial analysis

$$\mathbf{u}.\mathrm{curl}\, v - \mathrm{curl}\, \boldsymbol{u} \cdot \boldsymbol{v} = \mathrm{div}\, (\boldsymbol{v} \times \boldsymbol{u}) \tag{6}$$

integrated in the domain Ω, after applying the divergence theorem, gives the Green formula said of kind curl-curl in Ω, that is

$$(\mathbf{u}, \text{curl}\, \boldsymbol{v}) - (\text{curl}\, \boldsymbol{u}, \boldsymbol{v}) = <\boldsymbol{u} \times \boldsymbol{n}, \boldsymbol{v}>_{\Gamma},. \quad \forall \boldsymbol{u}, \boldsymbol{v} \in \boldsymbol{H}^1(\Omega) \tag{7}$$

Note that the surface integral term appearing in this last formula can take the following similar forms:

$$\langle \boldsymbol{u} \times \boldsymbol{n}, \boldsymbol{v} >_{\Gamma} = \langle \boldsymbol{v} \times \boldsymbol{u}, \boldsymbol{n} \rangle_{\Gamma} = -\langle \boldsymbol{v} \times \boldsymbol{n}, \boldsymbol{u} >_{\Gamma}$$

It is possible to define a **generalized Green formula** by

$$(\mathrm{Lu}, \mathrm{v}) - (\mathrm{u}, \mathrm{L}^* \mathrm{v}) = \int_{\Gamma} \mathrm{Q}(\mathrm{u}, \mathrm{v}) \mathrm{ds}, \forall \mathrm{u} \in \mathrm{dom}(\mathrm{L}) \text{ and} \forall \mathrm{v} \in \mathrm{dom}\ (\mathrm{L}^*), \tag{8}$$

where L and L* are differential operators of order n which act respectively on functions u and v defined in $\bar{\Omega}$, with $\bar{\Omega} = \Omega \cup \Gamma$; Q is a bi-linear function of u and v. The operator L* is called the **dual operator** of L. It can easily be seen that formulas (6) and (7) are particular cases of (8).

2.3 Weak formulations

Consider a partial differential problem of the form [4].

$$\mathrm{L\,u} = \mathrm{f} \text{ in } \Omega, \tag{9}$$

$$\mathrm{B\,u} = \mathrm{g} \text{ on } \Gamma = \partial\Omega, \tag{10}$$

where L is a differential operator of order n, B is an operator which defines a boundary condition, f and g are functions respectively defined in Ω and on its boundary Γ, and u is an unknown function from a function space U and defined in $\bar{\Omega}$, that is, $u \in U(\bar{\Omega})$. Note that f can eventually depend on u.

Problems (9 and 10) constitute what is called a **classical formulation**, or strong formulation. A function $u \in U(\bar{\Omega})$ which verifies this problem is called a **classical solution**, or strong solution. Particularly, as L is of order n, the function u has to be n–1 times continuously differentiable, that is, u ∈ Cn–1(Ω).

A **weak formulation** of problem (9) is defined as having the generalized form.

$$(\mathrm{u}, \mathrm{L}^* \mathrm{v}) - (\mathrm{f}, \mathrm{v}) + \int_{\Gamma} \mathrm{Q_g}(\mathrm{v}) \mathrm{ds} = 0, \quad \forall \mathrm{v} \in \mathrm{V}(\Omega) \tag{11}$$

where L* is the dual operator of L, defined by the generalized Green formula (8), Qg is a linear form in v which depend on g, and the space V (Ω) is a space of **test functions** which has to be defined according to the operator L* and particularly according to the boundary condition (9 and 10). A function u which satisfies this equation for any test function v ∈ V (Ω) is called a **weak solution**.

The generalized Green formula (8) can be applied to formulation (11) in order to get L instead of L*, which usually consists of performing an integration by parts. It is then possible to find again, thanks to a judicious choice of test functions, the equations and relations of the classical formulation of the problem, that is, Eq. (9) and boundary condition (11).

It is often easy to check that a classical solution is also a weak solution. Nevertheless, it is not always straightforward that a weak solution is also a classical solution because it has to be regular enough in order to be defined at the classic sense.

One mathematical advantage of weak formulations is that they usually enable to prove the existence of a solution easier than classical formulations do. The solution has then to be proved to be regular enough to be also a classical solution. Another advantage of weak formulations is that they are well adapted to a discretization using finite elements and then to a numerical solution, which is not the case with classical formulations.

In some cases, it is possible to define a minimization problem similar to the weak formulation (11).

2.4 A weak formulation for the magnetodynamic problem

In order to illustrate the notion of weak formulation, consider the magnetodynamic problem, limited to the domain Ω, with boundary $\partial\Omega = \Gamma = \Gamma_h \cup \Gamma_e$ (**Figure 2**), whose equations and material relations are written in Euclidean space $\mathbb{R}^3$ [5, 6].

$$\text{curl}\,\mathbf{h} = \mathbf{j}_s \tag{12}$$

$$\text{curl}\,\mathbf{e} = -j\omega\,\boldsymbol{b} \tag{13}$$

$$\text{div}\,\boldsymbol{b} = 0 \tag{14}$$

with behavior relations of materials.

$$\boldsymbol{b} = \mu\boldsymbol{h} \tag{15}$$

$$\boldsymbol{j} = \sigma\boldsymbol{e}, \tag{16}$$

where j is called the imaginary unit, $\boldsymbol{b}$ is the magnetic induction (T), $\boldsymbol{e}$ is the electric field (V/m), $\boldsymbol{j}_s$ is the current density (A/m2), $\boldsymbol{h}$ is the magnetic field (A/m), μ is the relative permeability and σ is the electric conductivity (S/m). From the Eq. (13), the field $\boldsymbol{b}$ can be obtained from a magnetic vector potential $\boldsymbol{a}_i$ via the term:

$$\boldsymbol{b} = \text{curl}\,\boldsymbol{a}. \tag{17}$$

Combining (15 and 16) into (14), one has curl $(\boldsymbol{e} + \partial_t\boldsymbol{a}) = 0$, that leads to the presentation of an electric scalar potential ν through $\boldsymbol{e} = -\partial_t a - \text{grad}\,\upsilon$.

By starting from the Ampere's law (12), the weak form of magnetic vector potential is written as [4, 6].

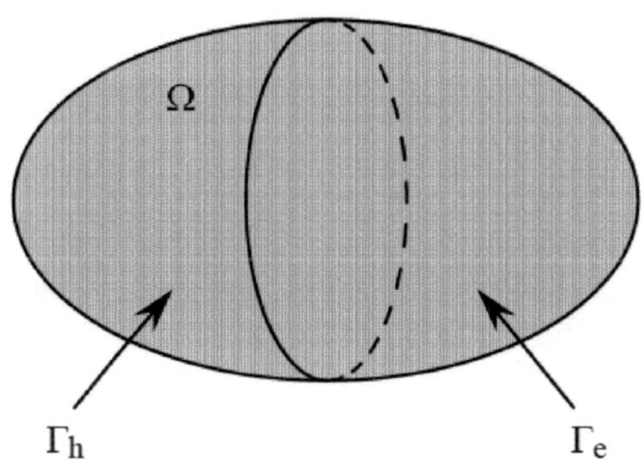

Figure 2.
Studied domain Ω and its boundary.

$$\left(\mu^{-1}\mathrm{b}, \mathrm{curl\ a}'\right)_{\Omega} - (\sigma\mathrm{e}, \mathrm{a}')_{\Omega_c} + \langle \mathrm{n} \times \mathrm{b}, \mathrm{a}' \rangle_{\Gamma} = \left(\mathrm{j_s}, \mathrm{a}'\right)_{\Omega_s} \forall \mathrm{a}' \in \mathrm{H}_{\mathrm{e}}^{0}(\mathrm{curl}, \Omega) \quad (18)$$

Combining the magnetic vector potential ***a*** and the electrical field ***e***, one has

$$\begin{aligned}&\left(\mu^{-1}\mathrm{curl}\, \boldsymbol{a}, \mathrm{curl}\, \boldsymbol{a}'\right)_{\Omega} + (\sigma\partial_t \boldsymbol{a}, \boldsymbol{a}')_{\Omega_c} + (\sigma \mathrm{grad}\, v, \mathrm{curl}\, \boldsymbol{a}')_{\Omega_c} + \langle \boldsymbol{n} \times \boldsymbol{h}, \boldsymbol{a}' \rangle_{\Gamma_h} \\ &\quad = \left(\boldsymbol{j}_s, \boldsymbol{a}'\right)_{\Omega_s}, \forall \boldsymbol{a}_i' \in F_e^0(\mathrm{curl}, \Omega_i), \end{aligned} \quad (19)$$

where $H_e^0(\mathrm{curl}, \Omega)$ is a function space defined on Ω containing the basis functions for ***a*** as well as for the test function ***a***′ (at the discrete level, this space is defined by edge FEs; notations (·, ·) and < ·, · > are respectively a volume integral in and a surface integral of the product of their vector field arguments.

2.5 A weak formulation for the magnetostatic problem

The magnetostatic problem is considered as a simplification of the magnetodynamic formulation where all time dependent phenomena are neglected. In a same way, by starting from the Ampere's law (12), this initial form of the problem is its **classical formulation.**

Consider the Green formula of type grad-div in Ω (5) applied to the field ***b*** and to a scalar field ϕ' to be defined, that is [6–8]

$$(\boldsymbol{b}, \mathrm{grad}\ \phi') + (\mathrm{div}\ \boldsymbol{b}, \phi') = \langle \boldsymbol{n}\ \ \boldsymbol{b}, \phi' >_{\Gamma}, \forall \phi' \in \Phi\ (\Omega). \quad (20)$$

If the space Φ(Ω) is defined such as

$$\Phi(\Omega) = \{\phi \in \mathrm{H1}(\Omega); \phi | \Gamma_{\mathrm{h}} = 0\}, \quad (21)$$

then the last term of Eq. (20) is reduced to < ***n.b***, ϕ' > Γ_e and is equal to zero if condition (15) is taken into account. Moreover, the second term of this equation is equal to zero because of Eq. (16). Eq. (20) can then be reduced to.

$$(\boldsymbol{b}, \mathrm{grad}\ \phi') = 0, \forall \phi' \in \Phi(\Omega). \quad (22)$$

This last form is called a **weak formulation** of the problem. It has been established starting from a Green formula but it can be considered now as an a priori posed form whose enclosed information can be deduced.

In fact, weak formulation (22) contains both **Eq.** (12) and **boundary condition** (15). Indeed, by applying the Green formula of type grad-div to it, we get.

$$(\mathrm{div}\ \boldsymbol{b}, \phi') = \langle \boldsymbol{n} \bullet \boldsymbol{b}, \phi' >_{\Gamma}, \forall \phi' \in \Phi(\Omega). \quad (23)$$

This equation is verified for any test function ϕ' ∈ Φ(Ω) and thus, particularly, for any function ϕ' whose value is equal to zero on Γ, that is, ϕ' ∈ Φ_0 (Ω) because $\Phi_0(\Omega) \subset \Phi(\Omega)$. Therefore, it comes that (div ***b***, ϕ') = 0 for any function ϕ' of this kind and, consequently, that div **b** = 0 in Ω, that is, Eq. (12) is satisfied. Then, Eq. (23) is reduced to $\langle \boldsymbol{n} \bullet \boldsymbol{b}, \phi' >_{\Gamma} = 0$, and by considering now all the functions ϕ' ∈ Φ(Ω) without any restriction, that is, which can vary freely on Γ_e, it comes that $\boldsymbol{n} \bullet \boldsymbol{b}|_{\Gamma_e} = 0$, that is, that condition (13) is satisfied.

It is possible to obtain more information from the weak formulation, particularly as far as the **transmission conditions** on surfaces inside Ω are concerned. Consider for that two subdomains Ω_1 and Ω_2 of Ω separated by an interface Σ (**Figure 3**) [7].

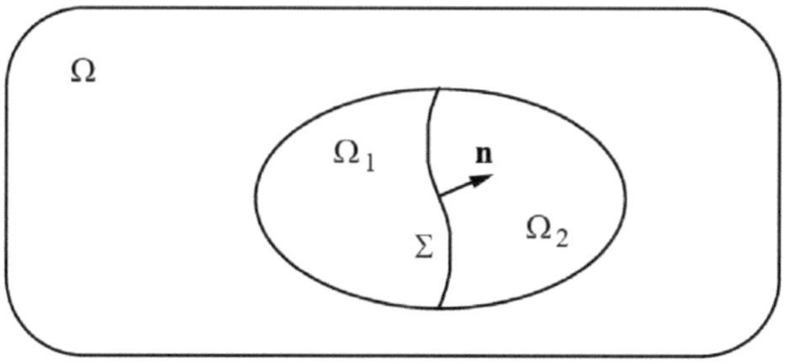

Figure 3.
Interface Σ between two subdomains Ω_1 and Ω_2.

Let us apply the Green formula of type grad-div (5) to the fields $\boldsymbol{b}$ and ϕ' successively in both subdomains Ω_1 and Ω_2. By taking into account that div $\boldsymbol{b}$ = 0 in Ω, and thus particularly in Ω_1 and Ω_2, then by summing the obtained relations, we get the relation [6, 7].

$$(\boldsymbol{b}, \text{grad } \phi')\Omega_1 \approx \Omega_2 = <\boldsymbol{n}.(\boldsymbol{b}_1 – \boldsymbol{b}_2), \phi'>\Sigma + <\boldsymbol{n}.\boldsymbol{b}, \phi'> \ (\Omega_1 \approx \Omega_2), \forall \phi' \in \Phi(\Omega), \tag{24}$$

where $\boldsymbol{b}_1$ and $\boldsymbol{b}_2$ represent the field $\boldsymbol{b}$ on both sides of Σ in the respective domains Ω1 and Ω2. Considering the test functions ϕ' whose support is $\Omega_1 \cup \Omega_2$ and which are equal to zero on _$(\Omega_1 \cup \Omega_2)$, it remains from (24) the well known transmission condition $\boldsymbol{n}.(\boldsymbol{b}_1 – \boldsymbol{b}_2)|\Sigma = 0$. Note that the first term of (24) vanishes thanks to Eq. (22) indeed, the domain of integration $\Omega_1 \cup \Omega_2$ can be extended to Ω thanks to the chosen test functions.

The way to establish a weak formulation of Eq. (13) has been described and the richness of the information enclosed in such a formulation has been shown up. Using a similar procedure, a weak formulation associated with Eq. (12) can be established, but we will proceed differently in order to keep some classical equations.

If the field $\boldsymbol{h}$ is decomposed into a given source component $\boldsymbol{h}_s$, such as curl $\boldsymbol{h}_s = \boldsymbol{j}$, and a reaction component $\boldsymbol{h}_r$, then curl $\boldsymbol{h}_r = 0$ and $\boldsymbol{h}_r$ is therefore of the form $\boldsymbol{h}_r = -$ grad ϕ (if Ω is simply connected). This consists of satisfying Eq. (15) classically. Taking into account the behavior law (15), we can write $\boldsymbol{b} = \mu\ (\boldsymbol{hs} - \text{grad } \phi)$ and put this last expression in (24) to obtain.

$$(\mu\ (\boldsymbol{h}_s - \text{grad } \phi), \text{grad } \phi') = 0, \forall \phi' \in \Phi(\Omega). \tag{25}$$

This formulation contains the whole problem (12 and 13). The potential ϕ is the unknown and all the other fields can be deduced from ϕ thanks to the equations which have been kept on a classical form. It appears that the potential ϕ belongs to the same space of the test functions or at least to a space Φ_r (Ω) which is parallel to it, that is, where the boundary condition relative to ϕ on Γh is not necessarily homogeneous, that is, $\phi|\Gamma_h$ = constant. Note that this boundary condition on Γ_h is implicitly taken into account in the space Φ(Ω).

Weak formulation (25) can be considered as **a system of an infinite number of equations with an infinite number of unknowns**. It will be seen in the following how such a problem can be approximated to lead to a numerical solution. This approximation will constitute the phase of discretization.

A similar minimization problem

It is possible to define a **minimization problem** associated with (25). For that, let us define the functional [2, 3].

$$W(\phi) = (\mu(\mathbf{hs} - \text{grad}\ \phi), \boldsymbol{h}_s - \text{grad}\ \phi), \tag{26}$$

and let us pose the following minimization problem:

$$\text{find}\ \phi \in \Phi_r(\Omega)\ \text{such as}\ W(\phi) \le W(\phi'), \forall \phi' \in \Phi r(\Omega). \tag{27}$$

The physical materials are considered having linear magnetic behavior, but the following can be generalized easily for nonlinear materials.

By stationarizing functional (25) in relation to ϕ, it can be verified that (25) is obtained. It can also be verified that the solution ϕ of (25) minimizes this functional. Indeed, let us suppose that $\phi \in \Phi_r(\Omega)$ is solution of (25) and let us consider any function $\psi \in \Phi_r(\Omega)$; then let us pose $\eta = \psi - \phi$, which implies $\eta \in \Phi(\Omega)$; we have

$$W(\psi) = W(\phi + \eta) = (\mu\ (\boldsymbol{h}_s - \text{grad}\ (\phi + \eta)), \boldsymbol{h}_s - \text{grad}\ (\phi + \eta)).$$

and thus.

$$W(\psi) = W(\phi) + (\mu\ \text{grad}\ \eta, \text{grad}\ \eta) + (\mu(\boldsymbol{h}_s - \text{grad}\ \phi), -\text{grad}\ \eta).$$

As the second term of this sum is positive or equal to zero and the third term is equal to zero, because of (25), it comes that $W(\psi)$ and $W(\phi)$.

Formulations (25) and (27) are then similar. Note that $W(\phi)$ is the magnetic coenergy and that the problem actually consists of minimizing this coenergy.

If the continuous function spaces are replaced by **discrete spaces**, and if the considered test functions are limited to these spaces, **then the information inside a weak formulation will only be satisfied approximately**, or weakly.

The basis of the discretization of weak formulations can be illustrated for the above magnetostatic problem, whose weak formulation is (25), that is

$$(\mu(\boldsymbol{h}_s - \text{grad}\ \phi), \text{grad}\ \phi') = 0, \forall \phi' \in \Phi(\Omega), \tag{28}$$

with $\phi \in \Phi(\Omega)$. The space $\Phi(\Omega)$ has to be replaced by a discrete space $\Phi_h(\Omega)$ which is a subset of it, that is, $\Phi_h(\Omega) \subset \Phi(\Omega)$. This space has a finite dimension, denoted N, and can then be defined by N linearly independent base functions. The principle is then to look for the function ϕ in $\Phi_h\ (\Omega)$, which consists of determining N unknown parameters. This function will be only an approximation of the exact solution $\phi \in \Phi\ (\Omega)$. The more the functions of $\Phi_h\ (\Omega)$ approximate well those of $\Phi\ (\Omega)$, the higher the quality of the approximation is. Each test function ϕ' will lead to an equation of the form (28) and, as the number of equations and unknowns has to be the same, N linearly independent test functions have to be chosen. This choice can be made on the base functions of $\Phi_h\ (\Omega)$ and the method is called the **Galerkine method**. Such base functions are defined thanks to finite elements.

3. Finite elements

3.1 Definition of a finite element

A **finite element** is defined by the **three element set (K, PK, ΣK)** where [2, 6, 7]:

- K is a domain of space called a **geometric element** (usually of simple shape, that is, a tetrahedron, a hexahedron or a prism);
- PK is a **function space** of dimension nK, defined in K with **base functions**;

- ΣK is a set of nK **degrees of freedom** represented by nK linear functionals ϕi, $1 \leq i \leq nK$, defined in space PK;

moreover, any function $u \in PK$ must be defined uniquely by the degrees of freedom of ΣK, which defines the unisolvance of the finite element (K, PK, ΣK).

The role of a finite element is to interpolate a field in a function space of finite dimension. Several finite elements can be defined on the same geometric element and then, under certain conditions, can form mixed finite elements. **Figure 4** shows the various spaces which occur in the definition of a finite element; the definition of the subspace of points $\kappa \subset K$ is actually associated with the definition of the functionals.

For the most commonly used finite elements, the degrees of freedom are associated with nodes of K and the functionals ϕi are reduced to functions of the coordinates in K; these elements are called **nodal finite elements**. Nevertheless, the above definition is more general thanks to the freedom let in the choice of the functionals. It will be shown that these can be, in addition to nodal values, integrals along segments, on surfaces or in volumes; the subspace of points $\kappa \subset K$ (**Figure 4**) is then respectively a point, a segment, a surface or a volume.

3.2 Unisolvant finite element

The finite element (K, PK, ΣK) is **unisolvant** if [6].

$$\forall p \in P_K, \phi i(p) = 0; \forall \phi i \in \Sigma K \Rightarrow p \equiv 0.$$

In this case, for any function u regular enough, one can define a **unique interpolation** uK, called **P_K-interpolant**, such as.

$$\phi i(u - u_K) = 0, \forall \phi i \in \Sigma_K; u_K \in P_K. \tag{29}$$

The set Σ_K is said P_K - unisolvant.

Proof:

Each function $p \in P_K$ can be written as a linear combination of functions of a base of P_K, denoted $\{pi, 1 \leq i \leq n_K\}$, that is

$$p = \sum_{i=1}^{n_K} a_i p_i,$$

where the pi, $1 \leq i \leq nK$, are called **base functions**.

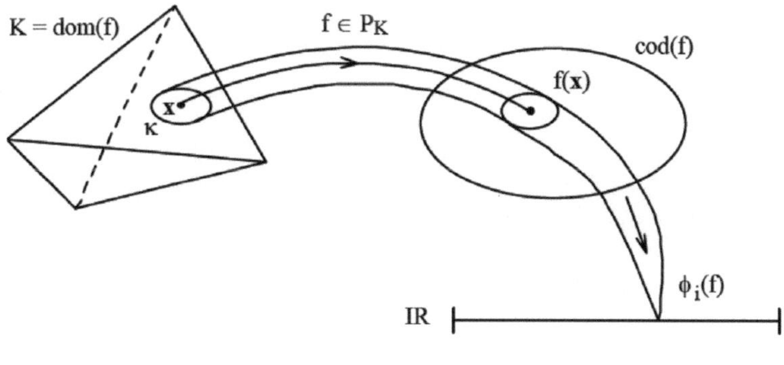

Figure 4.
Spaces associated with a finite element (K, P_K, Σ_K) [6].

As the functionals ϕj, $1 \le j \le n_K$, are linear, we have

$$\phi_j(p) = \sum_{i=1}^{n_K} a_i \phi_j(p_i), 1 \le j \le n_K.$$

And, as $\phi j(p) = 0, 1 \le j \le n_K$, leads to $p \equiv 0$, the determinant of the matrix Φ ($\Phi_{ji} = \phi_j(p_j), 1 \le i, j \le n_K$) is not equal to zero; indeed the solution of the corresponding system must be identically equal to zero (i.e., $ai = 0, 1 \le i \le n_K$). Consequently, the system

$$\phi_j(u) = \phi_j(u_K) \Leftrightarrow \phi_j(u) = \sum_{i=1}^{n_K} a_i \phi_j(p_i), 1 \le j \le n_K.$$

has a unique solution (a_j, $1 \le i \le n_K$).

3.3 Degrees of freedom

The interpolation of a function u, in the space P_K and in K, is given by the expression [4]

$$u_K = \sum_{i=1}^{n_K} a_i p_i, u_K \in P_K,$$

where the n_K coefficients a_j associated with the base functions $p_j \in P_K$ can be determined thanks to relations (26), that is, thanks to the solution of the linear system.

$$\phi_j(u) = \sum_{i=1}^{n_K} a_i \phi_j(p_i), 1 \le j \le n_K,$$

provided that the function u is sufficiently regular for the $\phi j(u)$, $1 \le j \le n_K$, to exist.

This solution is simplified to the maximum if we define the functionals so that

$$\phi_i\left(p_j\right) = \delta_{ij}, 1 \le i, j \le n_K \tag{30}$$

where $\delta_{i,j}$ is the Kronecker symbol, that is

$$\delta_{ij} = \begin{cases} 1\, s_i\, i = j \\ 0\, s_i\, i \neq j \end{cases}$$

The matrix of the system is then the unit matrix and the solution is

$$a_j = \phi_j(u), 1 \le j \le n_K$$

In this case, the interpolation $u_K \in P_K$ is expressed by

$$u_K = \sum_{j=1}^{n_K} \phi_j(u) p_j, \tag{31}$$

where the coefficients $\phi j(u) = \phi j(u_K)$, $1 \le j \le n_K$, are called **degrees of freedom**.

3.4 Finite element spaces

A **finite element space** X_h can be built on a set of geometric elements and associated finite elements. Its definition depends on the **mesh** M_h of the domain Ω as well as the knowledge of the finite element (K, P_K, Σ_K) associated with each domain $K \in M_h$ [6]

Given a function u defined in Ω, regular enough, its interpolant $u_h \in X_h$ is uniquely defined such as [6]:

- The restriction $u_h|_K$, that is, the form of u_h in the geometric element K, belongs to the space P_K;

- The restriction Finite element spaces

A **finite element space** X_h can be built on a set of geometric elements and associated finite elements. Its definition depends on the **mesh** M_h of the domain Ω as well as the knowledge of the finite element (K, P_K, Σ_K) associated with each domain $K \in M_h$

Given a function u defined in Ω, regular enough, its interpolant $u_h \in X_h$ is uniquely defined such as [6]:

- The restriction $u_h|_K$, that is, the form of uh in the geometric element K, belongs to the space P_K;

- The restriction $u_h|_K$ is entirely determined by the knowledge of the set of values $\Sigma_K(u)$ of the degrees of freedom of the function u - this is a consequence of the unisolvance;

Some continuous conditions have to be ensured across the interfaces between geometric elements, which is the property of conformity.

A **mesh** M_h of the studied domain Ω is defined as a collection of geometric elements which have in common either a facet, or an edge, or a node, or nothing (**Figure 5**). The elements cannot overlap each other.

The finite element space X_h has a finite dimension, denoted Dh. It can be characterized by a set of degrees of freedom Σ_h linked up to the sets $\Sigma_K, \forall K \in M_h$, that is

$$\Sigma_h = \left\{\phi_{h,j},\quad 1 \le j \le D_h\right\}.$$

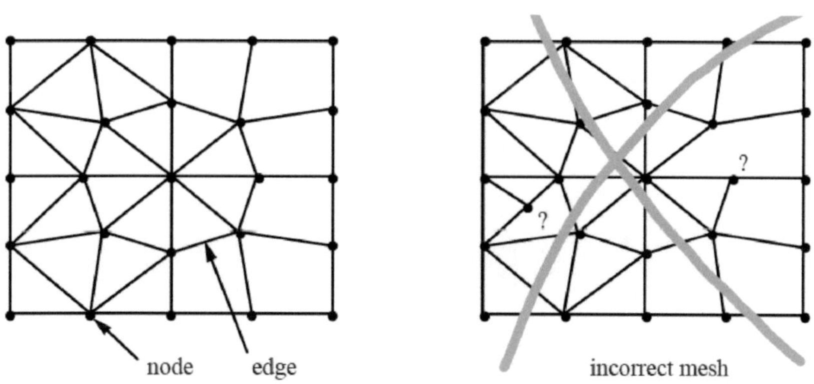

Figure 5.
Mesh of a part of a two-dimensional domain Ω.

It is also possible to define the base functions $p_{h,j}$, $1 \leq i \leq Dh$, of the space Xh from the base functions of the spaces P_K, $\forall$ K$\in$ Mh. Those have to verify the relations

$$\phi_j\left(p_j\right) = \delta_{ij}, 1 \leq i, j \leq n_K \tag{32}$$

similar to relations (32). They are actually piecewise defined and their supports are as "small" as possible, that is, are constituted by a limited number of geometric elements.

Then, with any function u regular enough so that the degrees of freedom $\phi_{h,j}(u)$, $1 \leq j \leq D_h$, are well defined, it can be associated a function u_h, called **X_h-interpolant**, defined by

$$u_h = \sum_{j=1}^{D_h} \phi_{h,j}(u) p_{h,j} \tag{33}$$

4. Construction of a sequence of finite element spaces

4.1 Geometric elements

A mesh of a domain is considered which is built with a collection of geometric elements which can be tetrahedra (4 nodes), hexahedra (8 nodes) and prisms (6 nodes) (**Figure 6**) [4, 6, 9].

These elements are called volumes and their vertices represent nodes. The sets of nodes, edges, facets and volumes of this mesh are denoted by N, E, F and V, respectively. Their sizes are #N, #E, #F and #V.

The i-th node of the mesh is denoted by ni or {i}. The edges and facets can be defined with ordered sets of nodes. An edge is denoted by eij or {i, j}, a triangular facet by fijk or {i, j, k}, and a quadrangular facet by fijkl or {i, j, k, l}. These geometric entities are shown in **Figure 7**.

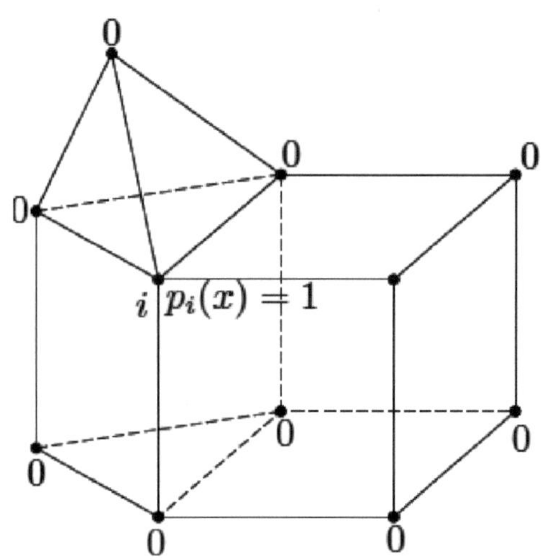

Figure 6.
Collection of different geometric elements [6].

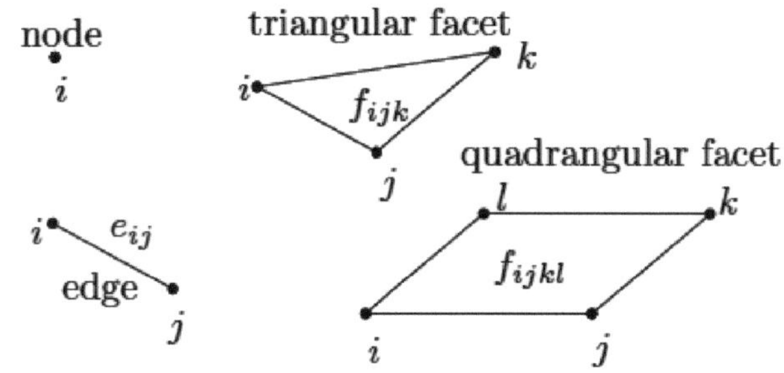

Figure 7.
Geometric entities: Node, edge and facets (i, j, k, l N) [6].

4.2 Base functions of spaces S^i

Consider the function pi(**x**) of coordinates of point **x** and relative to node ni, which is equal to 1 at this node, varies continuously in geometric elements having this node in common, and becomes equal to 0 in other elements without discontinuity (**Figure 6**). This function is nothing else than the base function, relative to node ni, of the function space of nodal finite elements built on the considered geometric elements. The function subspaces associated with each of the finite elements have respective dimensions 4, 8 or 6, for tetrahedra, hexahedra and prisms [4, 6, 9].

With **node** $n_i = \{i\}$, is associated the function

$$s_{n_i}(x) = p_i(x). \tag{34}$$

The finite dimensional space generated by all sn_i's is denoted by S^0.

With **edge** $e_{ij} = \{i, j\}$, is associated the vector field

$$s_{e_{ij}} = p_j \text{grad} \sum_{r \in N_{F,j\bar{i}}} p_r - p_i \text{ grad} \sum_{r \in N_{F,i\bar{j}}} p_r, \tag{35}$$

where $N_{F,m\bar{n}}$ is the set of nodes which belong to the facet of the geometrical element including evaluation point **x**, and including node m but not node n; such a facet is uniquely defined for three-edge-per-node elements. Its determination is shown in **Figure 8**, where either a triangular or a quadrangular facet is involved, and where shown edges belong to the geometric element including point **x**. Directions of dotted edges can be modified in order to schematize either a tetrahedron, a hexahedron or a prism. The defined set of nodes comes into view as being either {{m}, {o}, {p}} or {{m}, {o}, {p}, {q}}, respectively. The set $N_{F,m\bar{n}}$ depends on point **x**, thus on elements. Particularly, it is empty (no node) in elements which have not

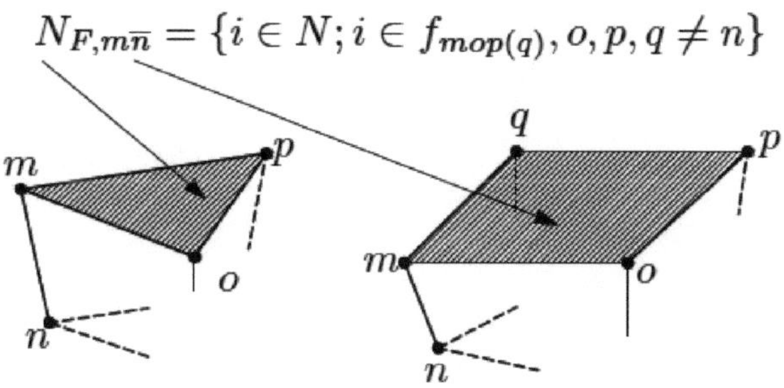

Figure 8.
Determination of the facet associated with $N_{F,m\bar{n}}$ *[6].*

edge {m, n} in common. Consequently, field $\mathbf{s}_{eij}$ is zero in all the elements non adjacent to edge e_{ij}.

The vector field space generated by all $\mathbf{s}_e$ is denoted by S^1.

With **facet** f = f_{ijk} = {i, j, k} = {q_1, q_2, q_3} or f = f_{ijkl} = {i, j, k, l} = {q_1, q_2, q_3, q_4}, is associated the vector field

$$s_f = a_f \sum_{c=1}^{\#N_f} p_{q_c} \text{grad}\left(\sum_{r \in N_{F,q_c\overline{q_c}+1}} p_r \right) \times \text{grad}\left(\sum_{r \in N_{F,q_c\overline{q_c}-1}} p_r \right) \tag{36}$$

where #Nf is the number of nodes of facet f, a_f = 2 if $\#N_f$ = 3, a_f = 1 if $\#N_f$ = 4, and the list of q_i's is made circular by setting $q_0 \equiv q\#N_f$ and $q\#N_f + 1 \equiv q_1$. Field $\mathbf{s}_f$ is zero in all the elements non adjacent to facet f.

Vector fields $s_{fijk(l)}$'s generate the space S^2.

With **volume** v, is associated the function sv, equal to 1/vol(v) on v and 0 elsewhere. The space S3 is generated by these functions.

Some developments give the following results: sn_i is equal to 1 at node ni, and to 0 at other nodes; the circulation of $\mathbf{s}_{eij}$ is equal to 1 along edge eij, and to 0 along other edges; the flux of $s_{fijk(l)}$ is equal to 1 across facet $s_{fijk(l)}$, and to 0 across other facets; and the volume integration of sv is equal to 1 over volume v, and to 0 over other volumes; that is

$$s_i(x_j) = \delta_{ij}, \quad \forall i, j \in N \tag{37}$$

$$\int_j s_i \cdot dl = \delta_{ij}, \quad \forall i, j \in E \tag{38}$$

$$\int_j s_i \cdot \mathbf{n} ds = \delta_{ij}, \quad \forall i, j \in F \tag{39}$$

$$\int_j s_i dv = \delta_{ij}, \quad \forall i, j \in V \tag{40}$$

where δ_{ij} = 1 if i = j and δ_{ij} = 0 if i ≠ j.

These properties show up various kinds of functionals and involve that functions s_n, $\mathbf{s}_e$, $\mathbf{s}_f$, s_v form bases for the spaces they generate. They are then called nodal, edge, facet and volume **base functions**. The associated finite elements are called **nodal**, **edge**, **facet** and **volume finite elements**.

4.3 Geometric interpretation of edge and facet functions

A geometric interpretation of edge and facet functions may be helpful to verify some of their properties. The vector field [4, 6, 9]

$$\text{grad}P_{F,m\bar{n}} = \text{grad} \sum_{r \in N_{F,m\bar{n}}} p_r, \tag{41}$$

involved in both expressions (31) and (32), should be analyzed at first.

The continuous scalar field,

$$P_{F,m\bar{n}} = \sum_{r \in N_{F,m\bar{n}}} p_r, \tag{42}$$

has the characteristic of being equal to 1 at every point on the facet associated with $N_{F,m\bar{n}}$. This is a property of the nodal base functions. Therefore, vector field (42) is orthogonal to this facet at every point on it (**Figure 9**).

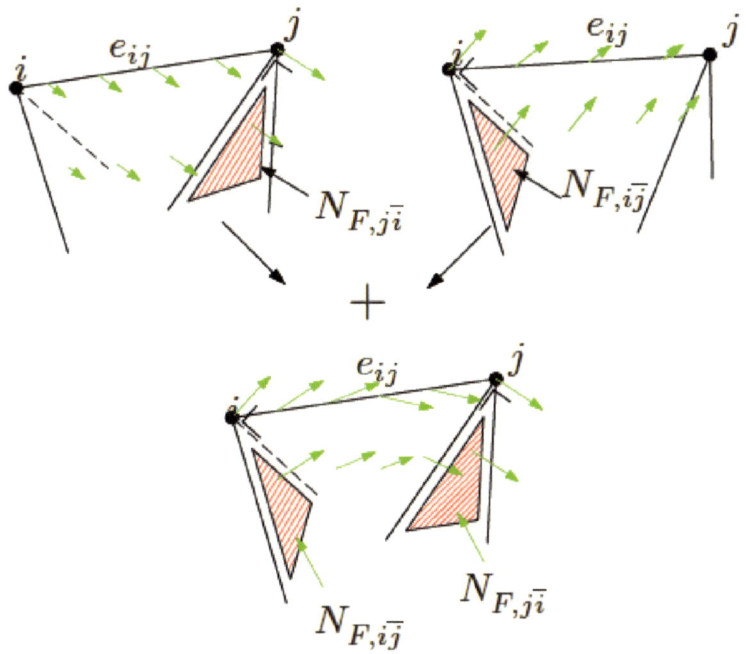

Figure 9.
Geometric interpretation of the edge function s_e (35) [6].

The vector field which is the product of pm and (35),

$$p_m \operatorname{grad} \sum_{r \in N_{F,m\bar{n}}} p_r, \tag{43}$$

is considered now. This field is said to be associated with edge {m, n}.
As far as the function pm is concerned, it is equal to 0 on all the edges of the geometric element including point **x**, except those which are incident to node {m}. Therefore, the circulation of (43) is equal to 0 along all the edges except emn; field (43) is either simply equal to zero on them, or orthogonal to them (**Figure 9**). The combination of two fields of form (43) associated with edges {j, i} and {i, j}, as in (35), leads to a vector field which has the same properties as (43) (**Figure 9**), and

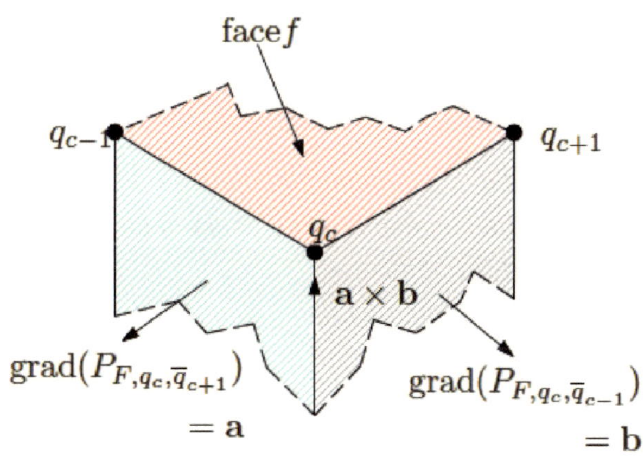

Figure 10.
Vector field $\boldsymbol{a} \times \boldsymbol{b}$ involved in s_f (33) [6].

has consequently the announced properties of $\mathbf{s}_{eij}$. The fact that its circulation along edge eij is equal to 1 needs some calculation to be proved.

The vector field

$$p_{q_c} \operatorname{grad} P_{F,q_c\bar{q}_{c+1}} \times \operatorname{grad} P_{F,q_c\bar{q}_{c-1}} \tag{44}$$

which appears in expression (35) of $\mathbf{s}_f$, is considered now. Both gradients in (44) are shown in **Figure 10**. Each one is orthogonal to its associated facet and, therefore, their cross product (i.e., $\mathbf{a} \times \mathbf{b}$ in **Figure 10**) is parallel to both these facets.

The flux of this cross product, and in consequence the one of (44), is then equal to 0 across these facets. The term pq_c in (44) enables the flux of (44) to be equal to zero across all other facets except facet f. The summation in (44) keeps the same property. The flux of $\mathbf{s}_f$ across facet f is then the only one to differ from zero (**Figure 11**).

4.4 Degrees of freedom

The expression of a field in the base of a space S^i –S^0 or S^3 for a scalar field, S^1 or S^2 for a vector field– gives scalar coefficients, called **degrees of freedom**. Fields $\phi \in S^0$, $\mathbf{h} \in S^1$, $\mathbf{j} \in S^2$ and $\rho \in S^3$ can be expressed as [4, 6, 9]

$$\Phi = \sum_{n \in N} \phi_n s_n, \phi \in S^0, \phi_n = \phi(x_n), n \in N, \tag{45}$$

$$\mathbf{h} = \sum_{e \in E} h_e s_e, \mathbf{h} \in S^1, h_e = \int_e \mathbf{h} \cdot \mathbf{dl}, e \in E \tag{46}$$

$$\mathbf{j} = \sum_{f \in F} j_f s_f, \mathbf{j} \in S^2, \quad j_f = \int_f \mathbf{j} \cdot \mathbf{n} ds, f \in F \tag{47}$$

$$\sigma = \sum_{v \in V} \sigma_v s_v, \rho \in S^3, \sigma_v = \int_v \sigma dv, v \in V \tag{48}$$

The degrees of freedom ϕ_n, h_e, j_f and ρ_v are thus, respectively, values at nodes, circulations along edges, fluxes across facets or volume integrals, of the associated fields. This is a consequence of the base functions. The associated linear functionals,

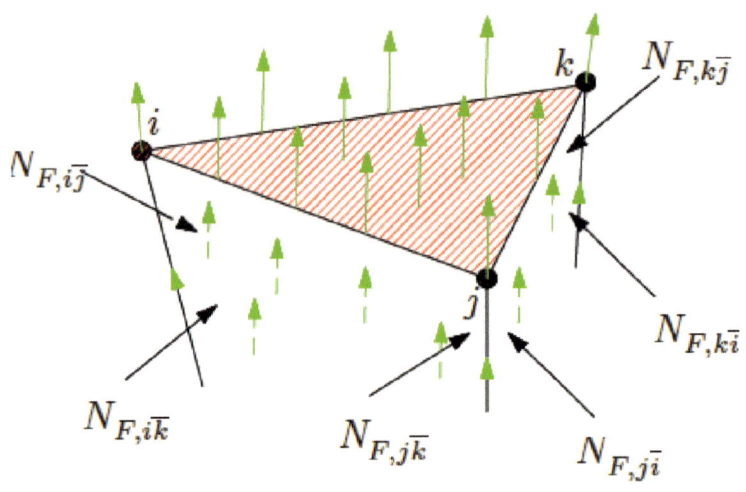

Figure 11.
Geometric interpretation of the facet function s_f (36) [6].

as mentioned in the definition of finite elements, are thus respectively pointwise evaluations, line, surface and volume integrals.

4.5 Continuity of base functions across facets

It can be proved that the function sn is continuous across facets. The same holds true for the tangential component of $\mathbf{s}_e$ and for the normal component of $\mathbf{s}_f$. As for function sv, it is discontinuous. This property, called **conformity**, allows to take exactly into account interface conditions for fields used in the modeling of physical problems. For example, in electromagnetic problems, vector fields of S^1 can represent vector fields like magnetic field **h** or electric field **e** whose tangential components are continuous across interfaces between materials, and those of S^2 can represent fields like induction field **b** or current density field **j** whose normal components are continuous across interfaces between these materials.

4.6 Spaces S^i form a sequence

The notion of **incidence** is first defined [4, 6, 9]:

The incidence of node n in edge e, denoted by i (n, e), is equal to 1 if n is the extremity of e, −1 if n is the origin of e, and 0 if n does not belong to e.

Next, the incidence of edge e in facet f is denoted by i(e, f). If e belongs to f, and if the ordered set of nodes of e appears as a direct subset in the circular set of nodes of f, then it is equal to 1. It is equal to −1 in the case of an inverse subset. If e does not belong to f, it is equal to 0.

Finally, the incidence of facet f in volume v is denoted by i(f, v). If f belongs to v, and if the normal to f, whose direction is given by the ordered set of nodes of f (right-hand rule), is outer to v, then it is equal to 1. It is equal to −1 in the case of an inner normal. If f does not belong to v, it is equal to 0.

Thanks to this notion, the following equalities can be proved,

$$\sum_{e \in E} i(n, e) s_e = \mathrm{grad}\, s_n \tag{49}$$

$$\sum_{f \in F} i(e, f) s_f = \mathrm{curl}\; s_e, \tag{50}$$

$$\sum_{v \in V} i(f, v) s_v = \mathrm{div}\; s_f. \tag{51}$$

The following inclusions are then verified,

$$\mathrm{grad}\left(S^0\right) \subset S^1, \mathrm{curl}\left(S^1\right) \subset S^2, \mathrm{div}\left(S^2\right) \subset S^3. \tag{52}$$

Therefore, the spaces Si, i = 0 to 3, form a **sequence**, that can be schematized by the diagram in **Figure 12**.

These spaces can then constitute approximation spaces for some continuous spaces Fi, i = 0 to 3, which contain scalar and vector fields associated with electromagnetic fields. The associated finite elements can then be called **mixed elements**.

$$S^0 \xrightarrow{\mathrm{grad}} S^1 \xrightarrow{\mathrm{curl}} S^2 \xrightarrow{\mathrm{div}} S^3$$

Figure 12.
The sequence of spaces S^i.

All the established properties of base functions are **valid for any collection of considered geometric elements**, that is, for any mixing of tetrahedra, hexahedra and prisms.

5. Practical information about finite elements

5.1 Isoparametric elements

An **isoparametric element** is a finite element whose nodal base functions, which enable the interpolation of scalar fields, are also used to parametrize the associated geometric element. The base functions are usually piecewise defined, in each of the geometric elements which cover the studied domain, and some continuity conditions have to be satisfied at the interfaces between elements. Then, there will be no discontinuity of the interpolated scalar fields, nor of the coordinates after transformation from the reference elements towards the real ones. Such base functions are said to be **conformal.**

Consider a nodal finite element (K, PK, ΣK). If NK is the set of nodes of K, whose coordinates are **xi**, i ∈ N, and if the pi(**u**), i ∈ NK, are its base functions expressed in the coordinates **u** of the reference element Kr associated with K, then the parametrization of K (i.e., **x** = **x**(**u**)) is given by [6]

$$\mathbf{x} = \sum_{i \in N_K} \mathbf{x}_i p_i(\mathbf{u}) \tag{53}$$

where **x**∈K, **u**∈Kr; this element is isoparametric.

5.2 Reference elements

We define here the reference elements which are associated with the considered geometric elements, that is, with tetrahedra, hexahedra and prisms. Nodal, edge, facet and volume finite elements are defined in these geometric elements.

5.2.1 Reference tetrahedron of type I

The reference tetrahedron of type I is an element with 4 nodes whose coordinates are given in **Figure 13**. The associated geometric entities, as well as their notation, are shown in **Figure 13**. The nodal and edge base functions of this element are given in **Tables 1** and **2**. **Table 3** shows the notation of facets. The incidence matrices are given by (53), (54) and (55) (**Figure 14**).

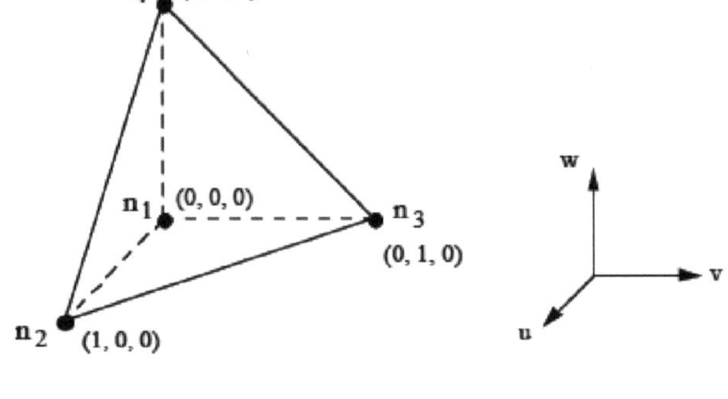

Figure 13.
Reference tetrahedron of type I [6].

Node i∈N	Nodal base function p_i (u, v, w) = s_i (u, v, w)
1	1 − u − v − w
2	u
3	v
4	w

Table 1.
Nodal base functions of the tetrahedron of type I.

Edge	e = {i, j}		$s_e(u)$, u = (u, v, w)		
e∈E	i∈N	j∈N	$s_{e,u}$	$s_{e,v}$	$s_{e,w}$
1	1	2	1 − v −w	u	u
2	1	3	v	1 − u −w	v
3	1	4	w	w	1 − u −v
4	2	3	− v	u	0
5	2	4	− w	0	u
6	3	4	0	− w	v

Table 2.
Notation of the edges of the tetrahedron of type I and associated edge base functions (s_e).

Facet	f = {i, j, k}		
f∈F	i∈N	j∈N	k∈N
1	1	2	4
2	1	3	2
3	1	4	3
4	2	3	4

Table 3.
Notation of the facets of the tetrahedron of type I.

Edge-node incidence matrix

$$G_{AN} = \begin{array}{c|cccc} {}_e\diagdown{}^n & 1 & 2 & 3 & 4 \\ \hline 1 & -1 & 1 & . & . \\ 2 & -1 & . & 1 & . \\ 3 & -1 & . & . & 1 \\ 4 & . & -1 & 1 & . \\ 5 & . & -1 & . & 1 \\ 6 & . & . & -1 & 1 \end{array} \tag{54}$$

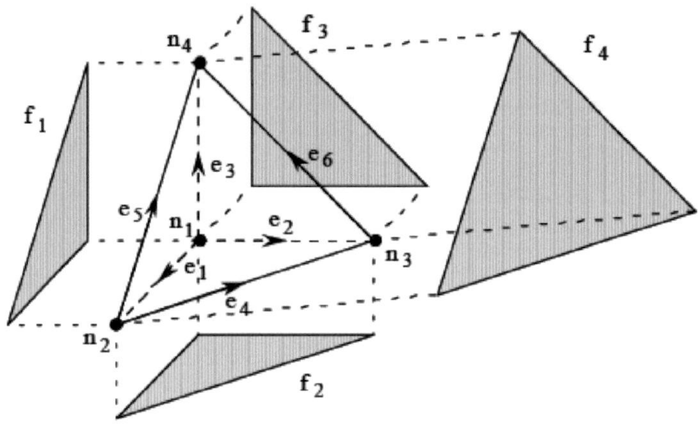

Entity	Node (n_i)	Edge (e_i)	Facet (f_i)	Volume
Number	4	6	4	1

Figure 14.
Geometric entities defined on a tetrahedron of type I [6].

Facet-edge incidence matrix

$$R_{FA} = \begin{array}{c|cccccc} f \diagdown e & 1 & 2 & 3 & 4 & 5 & 6 \\ \hline 1 & 1 & . & -1 & . & 1 & . \\ 2 & -1 & 1 & . & -1 & . & . \\ 3 & . & -1 & 1 & . & . & -1 \\ 4 & . & . & . & 1 & -1 & 1 \end{array}, \tag{55}$$

Volume-facet incidence matrix

$$D_{VF} = \begin{array}{c|cccc} v \diagdown f & 1 & 2 & 3 & 4 \\ \hline 1 & (1 & 1 & 1 & 1) \end{array} \tag{56}$$

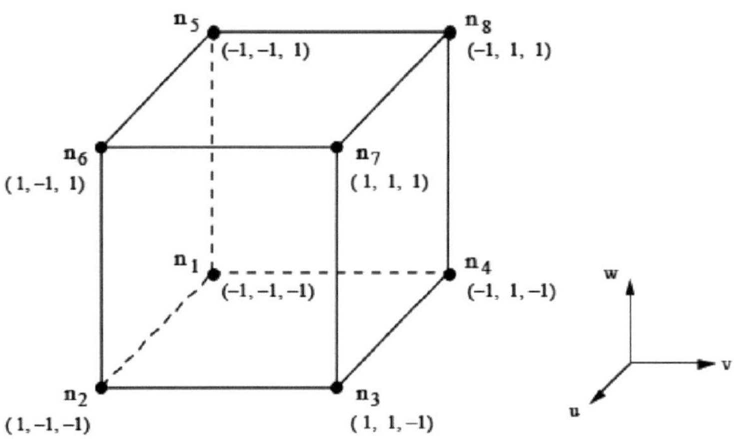

Figure 15.
Reference hexahedron of type I [6].

5.2.2 Reference hexahedron of type I

The reference hexahedron of type I is an element with 8 nodes whose coordinates are given in **Figure 15**. The associated geometric entities, as well as their notation, are shown in **Figure 16**. The nodal and edge base functions of this element are given in **Tables 4** and 5. **Table 6** shows the notation of facets. The incidence matrices are given by (56), (57) and (58).

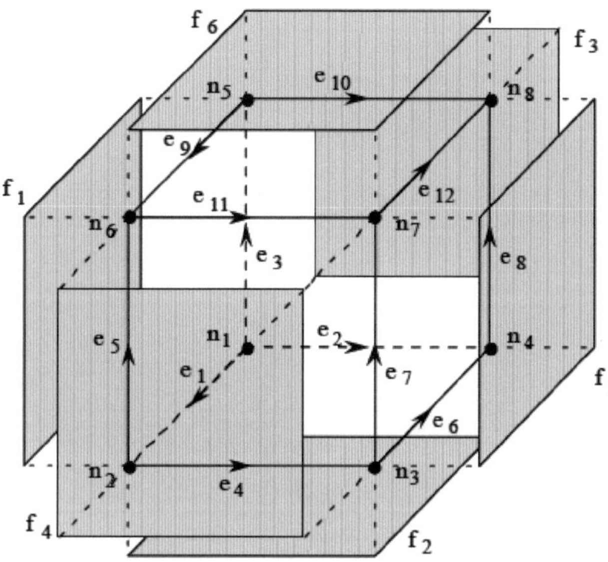

Entity	Node (n_i)	Edge (e_i)	Facet (f_i)	Volume
Number	8	12	6	1

Figure 16.
Geometric entities defined on a hexahedron of type I [4].

Node $i \in N$	**Nodal base function** $p_i(u, v, w) = s_i(u, v, w)$
1	$(1 - u)(1 - v)(1 - w)/8$
2	$(1 + u)(1 - v)(1 - w)/8$
3	$(1 + u)(1 + v)(1 - w)/8$
4	$(1 - u)(1 + v)(1 - w)/8$
5	$(1 - u)(1 - v)(1 + w)/8$
6	$(1 + u)(1 - v)(1 + w)/8$
7	$(1 + u)(1 + v)(1 + w)/8$
8	$(1 - u)(1 + v)(1 + w)/8$

Table 4.
Nodal base functions of the hexahedron of type I.

Edge	**e = {i, j}**		**$s_e(\mathbf{u})$, $\mathbf{u}$ = (u, v, w)**		
e∈E	i∈N	j∈N	$s_{e,u}$	$s_{e,v}$	$s_{e,w}$
1	1	2	(1 – v) (1 – w) / 8	0	0
2	1	4	0	(1 – u) (1 – w) / 8	0
3	1	5	0	0	(1 – u) (1 – v) / 8
4	2	3	0	(1 + u) (1 – w) / 8	0
5	2	6	0	0	(1 + u) (1 – v) / 8
6	3	4	–(1 + v) (1 – w) / 8	0	0
7	3	7	0	0	(1 + u) (1 + v) / 8
8	4	8	0	0	(1 – u) (1 + v) / 8
9	5	6	(1 – v) (1 + w) / 8	0	0
10	5	8	0	(1 – u) (1 + w) / 8	0
11	6	7	0	(1 + u) (1 + w) / 8	0
12	7	8	–(1 + v) (1 + w) / 8	0	0

Table 5.
Notation of the edges of the hexahedron of type I and associated edge base functions (s_e).

Facet	**f = {i, j, k, l}**			
f∈F	i∈N	j∈N	k∈N	l∈N
1	1	2	6	5
2	1	4	3	2
3	1	5	8	4
4	2	3	7	6
5	3	4	8	7
6	5	6	7	8

Table 6.
Notation of the facets of the hexahedron of type I.

Edge-node incidence matrix

$$\mathbf{G}_{EN} = \begin{array}{c|cccccccc} e \backslash n & 1 & 2 & 3 & 4 & 5 & 6 & 7 & 8 \\ \hline 1 & -1 & 1 & . & . & . & . & . & . \\ 2 & -1 & . & . & 1 & . & . & . & . \\ 3 & -1 & . & . & . & 1 & . & . & . \\ 4 & . & -1 & 1 & . & . & . & . & . \\ 5 & . & -1 & . & . & . & 1 & . & . \\ 6 & . & . & -1 & 1 & . & . & . & . \\ 7 & . & . & -1 & . & . & . & 1 & . \\ 8 & . & . & . & -1 & . & . & . & 1 \\ 9 & . & . & . & . & -1 & 1 & . & . \\ 10 & . & . & . & . & -1 & . & . & 1 \\ 11 & . & . & . & . & . & -1 & 1 & . \\ 12 & . & . & . & . & . & . & -1 & 1 \end{array} \quad (57)$$

Facet-edge incidence matrix

$$
R_{FE} = \begin{array}{c|cccccccccccc}
f \backslash e & 1 & 2 & 3 & 4 & 5 & 6 & 7 & 8 & 9 & 10 & 11 & 12 \\
\hline
1 & 1 & . & -1 & . & 1 & . & . & . & -1 & . & . & . \\
2 & -1 & 1 & . & -1 & . & -1 & . & . & . & . & . & . \\
3 & . & -1 & 1 & . & . & . & . & -1 & . & 1 & . & . \\
4 & . & . & . & 1 & -1 & . & 1 & . & . & . & -1 & . \\
5 & . & . & . & . & . & 1 & -1 & 1 & . & . & . & -1 \\
6 & . & . & . & . & . & . & . & . & 1 & -1 & 1 & 1
\end{array} \tag{58}
$$

Volume-facet incidence matrix

$$
D_{VF} = \begin{array}{c|cccccc}
v \backslash f & 1 & 2 & 3 & 4 & 5 & 6 \\
\hline
1 & (1 & 1 & 1 & 1 & 1 & 1)
\end{array} \tag{59}
$$

5.2.3 Reference prism of type I

The reference prism of type I is an element with 6 nodes whose coordinates are given in **Figure 17**. The associated geometric entities, as well as their notation, are shown in **Figure 18**. The nodal and edge base functions of this element are given in **Tables 7** and **8**. **Table 9** shows the notation of facets. The incidence matrices are given by (59), (60) and (61).

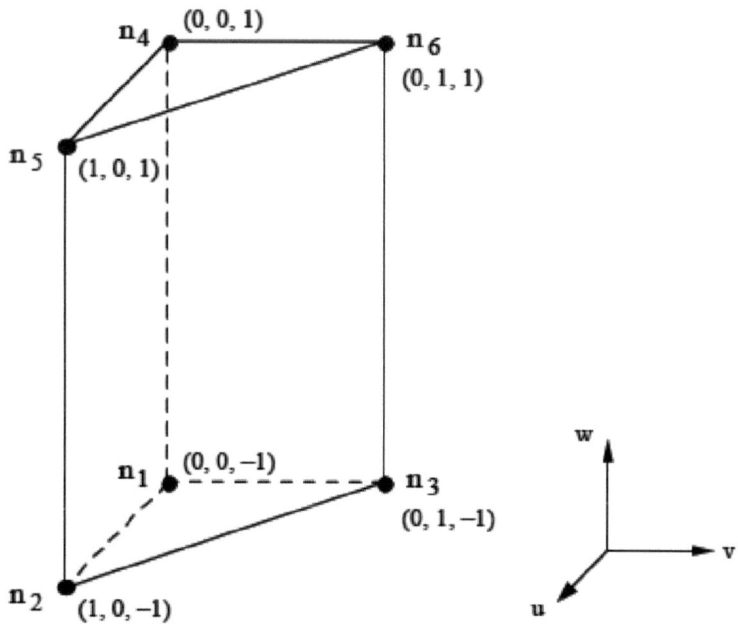

Figure 17.
Reference prism of type I [6].

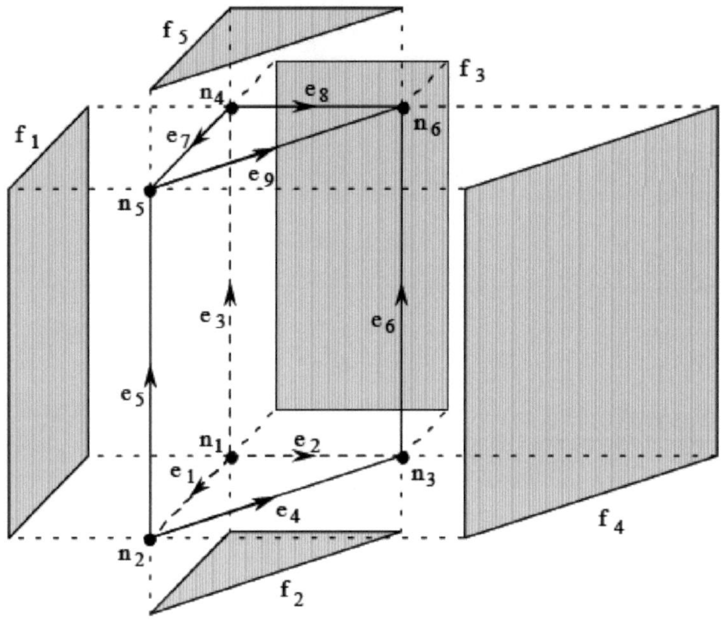

Entity	Node (n_i)	Edge (e_i)	Facet (f_i)	Volume
Number	6	9	5	1

Figure 18.
Geometric entities defined on a prism of type I [6].

Node i∈N	Nodal base function p_i (u, v, w) = s_i (u, v, w)
1	(1 − u − v) (1 − w) / 2
2	u (1 − w) / 2
3	v (1 − w) / 2
4	(1 − u − v) (1 + w) / 2
5	u (1 + w) / 2
6	v (1 + w) / 2

Table 7.
Nodal base functions of the prism of type I.

Edge-node incidence matrix

$$
G_{EN} = \begin{array}{c|cccccc} e \diagdown n & 1 & 2 & 3 & 4 & 5 & 6 \\ \hline 1 & -1 & 1 & . & . & . & . \\ 2 & -1 & . & 1 & . & . & . \\ 3 & -1 & . & . & 1 & . & . \\ 4 & . & -1 & 1 & . & . & . \\ 5 & . & -1 & . & . & 1 & . \\ 6 & . & . & -1 & . & . & 1 \\ 7 & . & . & . & -1 & 1 & . \\ 8 & . & . & . & -1 & . & 1 \\ 9 & . & . & . & . & -1 & 1 \end{array} \tag{60}
$$

Edge	**e = {i, j}**		**$s_e(u)$, u = (u, v, w)**		
e∈E	**i∈N**	**j∈N**	**$s_{e,u}$**	**$s_{e,v}$**	**$s_{e,w}$**
1	1	2	(1 − v) (1 − w) / 2	u (1 − w) / 2	0
2	1	3	v (1 − w) / 2	(1 − u) (1 − w) / 2	0
3	1	4	0	0	(1 − u − v) / 2
4	2	3	− v (1 − w) / 2	u (1 − w) / 2	0
5	2	5	0	0	u / 2
6	3	6	0	0	v / 2
7	4	5	(1 − v) (1 + w) / 2	u (1 + w) / 2	0
8	4	6	v (1 + w) / 2	(1 − u) (1 + w) / 2	0
9	5	6	− v (1 + w) / 2	u (1 + w) / 2	0

Table 8.
Notation of the edges of the prism of type I and associated edge base functions (s_e).

Facette		**f = {i, j, k (, l)}**		
f∈F	**i∈N**	**j∈N**	**k∈N**	**l∈N**
1	1	2	5	4
2	1	3	2	—
3	1	4	6	3
4	2	3	6	5
5	4	5	6	—

Table 9.
Notation of the facets of the prism of type I.

Facet-edge incidence matrix

$$R_{FE} = \begin{array}{c|ccccccccc} f \diagdown e & 1 & 2 & 3 & 4 & 5 & 6 & 7 & 8 & 9 \\ \hline 1 & 1 & . & -1 & . & 1 & . & -1 & . & . \\ 2 & -1 & 1 & . & -1 & . & . & . & . & . \\ 3 & . & -1 & 1 & . & . & -1 & . & 1 & . \\ 4 & . & . & . & 1 & -1 & 1 & . & . & -1 \\ 5 & . & . & . & . & . & . & 1 & -1 & 1 \end{array} \tag{61}$$

Volume-facet incidence matrix

$$D_{VF} = \begin{array}{c|ccccc} v \diagdown f & 1 & 2 & 3 & 4 & 5 \\ \hline 1 & (1 & 1 & 1 & 1 & 1) \end{array} \tag{62}$$

6. Applications

The practical test problem is a 3-D model based on the benchmark problem 19 of the TEAM workshop including a stranded inductor (coil) and an aluminum plate (**Figure 19**) [10].

The coil is excited by a sinusoidal current which generates the distribution of time varying magnetic fields around the coil (**Figure 20**). The relative permeability and electric conductivity of the plate are $\mu_{r,plate} = 1$, $\sigma_{r,plate} = 35.26$ MS/m, respectively. The source of the magnetic field is a sinusoidal current with the maximum ampere turn being 2742AT. The problem is tested with two cases of frequencies of the 50 Hz and 200 Hz.

The 3-D dimensional mesh with edge elements is depicted in **Figure 21** (*left*). The distribution of magnetic flux density generated by the excited electric current in the coil is pointed out in **Figure 21** (*right*). The computed results on the of the z-component of the magnetic flux density along the lines A1-B1 and A2-B2 (**Figure 19**) is checked to be close to the measured results for different frequencies of exciting currents (already proposed by authors in [10]) are shown in **Figure 21**. The mean errors between calculated and measured methods [10] on the magnetic flux density are lower than 10%.

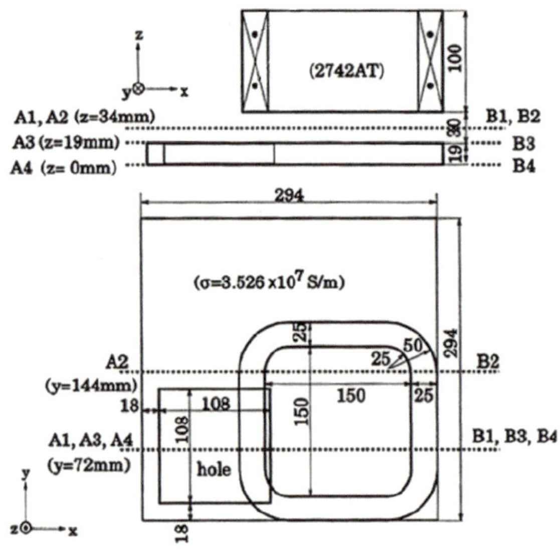

Figure 19.
Modeling of TEAM problem 7: Coil and conducting plate [10].

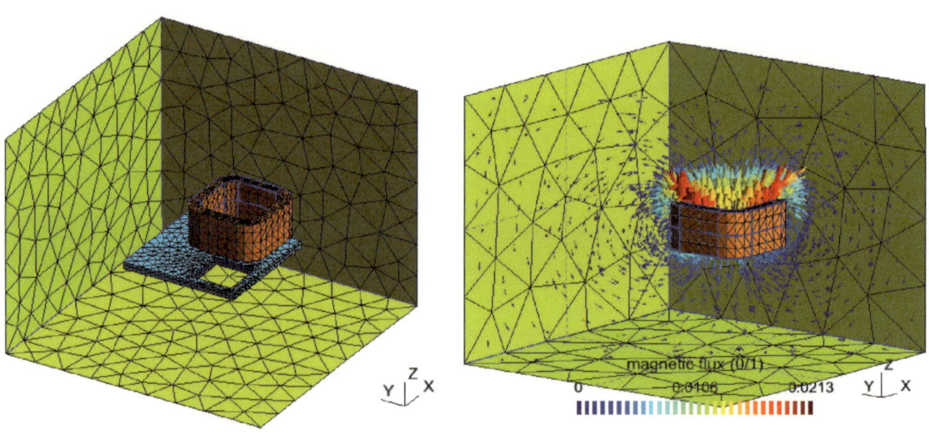

Figure 20.
The 3-D mesh model with edge elements of the coil and conducting plate, and the limited boundary [4] (left), *and distribution of magnetic flux density generated by the excited sinusoidal current in the coil, with* $\mu_{r,plate} = 1$, $\sigma_{r,plate} = 35.26 \frac{MS}{m}$ *and* f = 50 Hz.

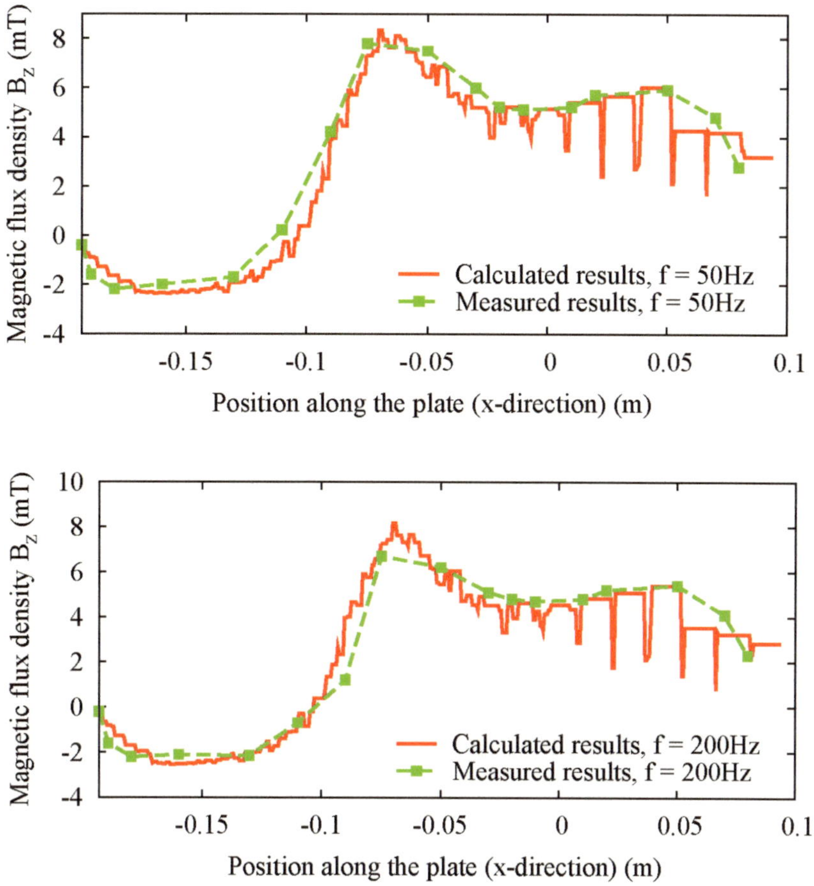

Figure 21.
The comparison of the calculated results with the measured results on magnetic flux densities at y = 72 mm, with $\mu_{r,plate} = 1$, $\sigma_{r,plate} = 35.26\frac{MS}{m}$ *and different frequencies [4].*

The y component of the varying of the eddy current losses with different frequencies (50 Hz and 200 Hz) along the lines A3-B3 and A4-B4 (**Figure 19**) is shown in **Figure 22**. The computed results are also compared with the measured results as well [7]. The obtained results from the theory modeling are quite similar as what measured from the measurements. The maximum error near the end of the conductor plate on the eddy currents between two methods are below 20% for both cases (50 Hz and 200 Hz).

7. Conclusions

In the 3D computation of the magnetic flux density and eddy current, thanks to the set of Maxwell's equations, it has been successfully developed for two weak formulations, where the discretization of the fields is performed by Whitney edge elements [2, 3, 8]: magnetostatic formulation and magnetodynamic formulation. The developments of the method is validated on the actual problem (TEAM problem 7) [10]. The numerical error between simlated and measured results on the magnetic flux densities and eddy current is lower than 10%. This is also proved that there is a very good validation between two methods. The results have been achieved by a detailed study of the magnetodynamic formulation.

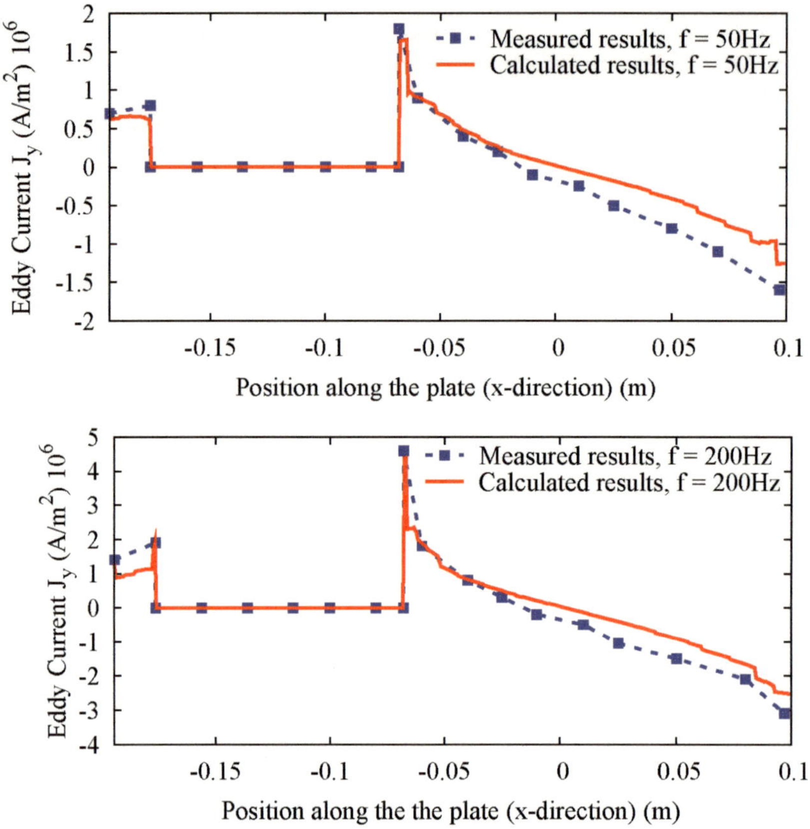

Figure 22.
The comparison of the calculated results with the measured results at z = 19 mm, with $\mu_{r,plate} = 1, \sigma_{r,plate} = 35.26\frac{MS}{m}$ and different frequencies [4].

Author details

Dang Quoc Vuong[1*] and Bui Minh Dinh[2]

1 Department of Electrical Equipment, School of Electrical Engineering, Hanoi University of Science and Technology, Vietnam

2 School of Electrical Engineering, Hanoi University of Science and Technology, Vietnam

*Address all correspondence to: vuong.dangquoc@hust.edu.vn

References

[1] Gerard Meunier, et al,. "The Finite Element Method for Electromagnetic Modeling", John Wiley & Sons Inc, 2008.

[2] A. Bosavit, "Whitney forms: A class of finite elements for three-dimensional computations in electromagnetism", IEEE Proceedings, Vol. 135, Pt. A, no. 8, pp. 493–499, 1998.

[3] A. Bosavit, "Electromagnetisme en vue de la modelisation", Collection Mathematiques et Applications, Springer-Verlag, 1991.

[4] Vuong Dang Quoc and Christophe Geuzaine "Using edge elements for modeling of 3-D magnetodynamic problem via a subproblem method", Science and Technology Development Journal. ; 23(1): 439–445.

[5] S. Koruglu, P. Sergeant, R.V. Sabarieqo, Vuong. Q. Dang, M. De Wulf "Influence of contact resistance on shielding efficiency of shielding gutters for high-voltage cables," IET Electric Power Applications, Vol.5, No.9, (2011), pp. 715–720.

[6] Quoc VD. Modeling of Electromagnetic Systems by Coupling of Subproblems with Application to Thin Shell Models [Ph. D Thesis]. 2013. Belgium: University of Liege

[7] Vuong Q. Dang, P. Dular R.V. Sabariego, L. Krähenbühl, C. Geuzaine, "Subproblem approach for modelding multiply connected thin regions with an h-conformal magnetodynamic finite element formulation", in EPJ AP., vol. 63, no. 1, 2013.

[8] P. Dular, W. Legros, A. Nicolet, "Coupling of local and global quantities in various finite element formulations and its application to electrostatics, magnetostatics and magnetodynamics", IEEE Transactions on Magnetics, vol. 34, no. 5, pp. 3078–3081, 1998.

[9] P. Dular, J.-Y.Hody, A. Nicolet, A. Genon, W. Legros, "Mixed finite elements associated with a collection of tetrahedra, hexahedra and prisms", IEEE Transactions on Magnetics, vol. 30, no. 5, pp. 2980–2983, 1994.

[10] Kovacs G, Kuczmann M. Solution of the TEAM workshop problem No.7 by the finite Element Method. In: Approved by the International Compumag Society Board at Compumag. 2011

Chapter 4

A Combination of Finite Difference and Finite Element Methods for Temperature and Stress Predictions of Early-Age Concrete Members

Tu Anh Do

Abstract

A combination of finite difference and finite element methods was employed to develop a model for predicting the temperature development and thermally induced stresses in early-age concrete members (such as bridge footings, piers, columns, girders, and slabs). A two-dimensional finite difference (FD) scheme was utilized for heat generation and transfer within a hydrating concrete member. A finite element (FE) plane strain model was then established to compute the thermal stresses in the concrete subjected to the temperature changes. The FD-FE model can be easily created using any programing language, and the methodology can be used to predict the temperatures and stresses as well as assess the possibility of early-age cracking in concrete members.

Keywords: finite difference, finite element, early-age concrete, heat of hydration, thermal stress, thermal cracking, insulation layer

1. Introduction

Thermal cracking is one of the biggest concerns regarding early-age concrete. Hydration of a large amount of cement results in higher peak temperatures as well as larger temperature differences between the concrete surface and the core. Such large temperature differentials can cause substantial tensile stresses that might increase the likelihood of early-age cracking in the concrete [1].

In order to control the temperature in early-age concrete structures, thus mitigating the risk of thermal cracking, temperature and stress analyses should be performed beforehand. Different methods have been used for predicting the temperatures and thermal stresses concrete structures at an early age. Among them, the Schmidt's method is a simple approach but has been widely used for computing the temperatures for single nodes in the concrete [1]. The finite difference (FD) method was also employed in spreadsheet programs [2, 3] or in computer programs [4, 5] for calculating temperature–time histories in concrete elements. A two-dimensional model for thermal analysis based on the finite volume method (FVM) was introduced by Yikici and Chen [6]. The finite element (FE) method has been commonly utilized for both thermal and stress analyses of early-age concrete structures [7–13].

This chapter presents a two-dimensional FD scheme for thermal analysis of a concrete element. An FE analysis was then used to calculate the temperature-induced stresses in the concrete. The analysis results were compared with measurements of actual concrete elements. The combined approach can be a simple and useful tool for analyzing temperatures and thermal stresses in early-age concrete elements.

2. FD scheme for solving heat transfer

The heat evolution and temperature in a concrete element can be known by solving the governing differential equation as described in Eq. (1):

$$\rho c_p \frac{\partial T}{\partial t} = k\left(\frac{\partial^2 T}{\partial x^2} + \frac{\partial^2 T}{\partial y^2} + \frac{\partial^2 T}{\partial z^2}\right) + \dot{Q} \tag{1}$$

where ρ is density; c_p is specific heat; T is temperature; t is time; k is thermal conductivity; x, y, and z are coordinates; and Q is heat evolution rate.

The finite difference formulation for any node in the system can be written as [14]:

$$\sum_{All\ sides} \dot{Q}^i + \dot{E}^i = \rho c_p V \frac{T_m^{i+1} - T_m^i}{\Delta t} \tag{2}$$

where Q^i = rate of heat conduction at time step i; $\dot{E}^i$ = rate of heat generation at time step i; $T^i{}_m$ and $T^{i+1}{}_m$ = temperatures of node m at time step i and $i + 1$, respectively; and Δt = time interval.

During the actual construction stage of concrete structures, the concrete is usually covered by formwork and/or insulation materials. Heat generated from cement hydration is conducted through the formwork and/or insulation layer before being dissipated to the surroundings by surface convection (**Figure 1**).

Considering a formwork/insulation layer covering the concrete, and assuming a unit square mesh for the concrete ($\Delta x = \Delta y = l$) and an insulation thickness of d (**Figure 2**), the FD formulation for the interior node can be computed using Eq. (3).

$$T_{m,n}^{i+1} = \tau_F\left(T_{m-1,n}^i + T_{m+1,n}^i + T_{m,n+1}^i + T_{m,n-1}^i\right) + T_{m,n}^i(1 - 4\tau_F) + \tau_F \frac{\dot{e}_{m,n} l^2}{k} \tag{3}$$

where $T^i{}_{m,n}$ = temperatures of node (m,n) at time step i; $T_{m-1,n}^i, T_{m+1,n}^i, T_{m,n+1}^i, T_{m,n-1}^i$ = temperatures at neighboring nodes; and τ_F = dimensionless Fourier number,

$$\tau_F = \frac{k\Delta t}{\rho c_p l^2} \tag{4}$$

Using Eq. (2), the FD equations for each of the four outer corner nodes of the insulation can be derived. For instance, the quarter size volume element of the insulation layer ($d \times d \times 1$) represented by the top left outer corner node (1,N) is

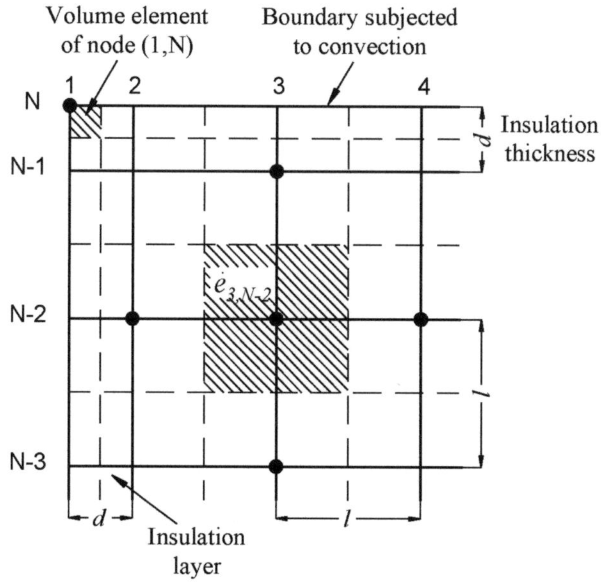

Figure 1.
FD mesh for heat conduction of concrete covered with insulation layer.

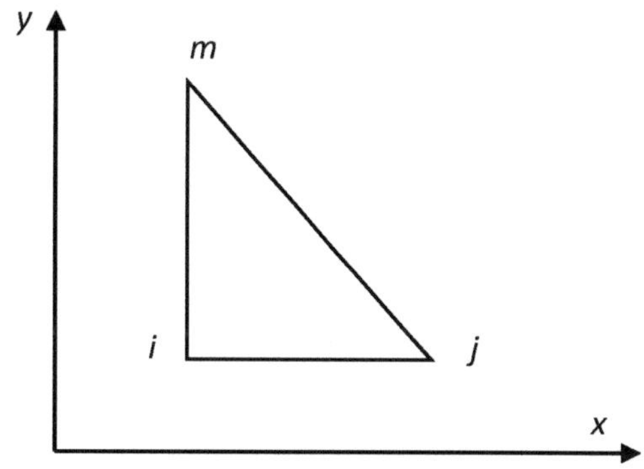

Figure 2.
FE plane triangular element.

subjected to convection on both sides and to conduction from the right and bottom nodes, the energy balance relation above (Eq. (2)) becomes:

$$T_{1,N}^{i+1} = T_{1,N}^{i}\left(1 - 4A_1 - 4A_1\frac{hd}{k_s}\right) + 2A_1\left(T_{1,N-1}^{i} + T_{2,N}^{i} + 2T_a\frac{hd}{k_s}\right) \tag{5}$$

where ρ_s = insulation material density; c_{ps} = specific heat of insulation material; k_s = thermal conductivity of insulation material; T_a is the ambient temperature; h is the convection coefficient; and

$$A_1 = \frac{k_s \Delta t}{\rho_s c_{ps} d^2} \tag{6}$$

The insulation volume element at the surface node (2,N) adjacent to the top left outer corner node is subjected to convection at the top and conduction at the left, right, and bottom surfaces. An energy balance on this element gives:

$$T_{2,N}^{i+1} = T_{2,N}^{i}\left[1 - \frac{2h(d+l)A_2}{k_s} - 2A_2 - 2dlA_2 - \frac{2(d+l)A_2}{d}\right] + 2A_2\left[T_{1,N}^{i} + dlT_{3,N}^{i} + \frac{d+l}{d}T_{2,N-1}^{i} + \frac{h(d+l)}{k_s}T_a\right] \quad (7)$$

where

$$A_2 = \frac{k_s \Delta t}{\rho_s c_{ps}(d+l)d} \quad (8)$$

Similarly, the insulation volume element of half size at a surface node is subjected to convection at the top and conduction at the left, right, and bottom surfaces. An energy balance on this element gives:

$$T_{m,n}^{i+1} = T_{m,n}^{i}\left(1 - \frac{2d}{l}A_3 - \frac{2l}{d}A_3 - \frac{2hl}{k_s}A_3\right) + A_3\left(\frac{d}{l}T_{m-1,n}^{i} + \frac{d}{l}T_{m+1,n}^{i} + \frac{2l}{d}T_{m,n-1}^{i} + \frac{2hl}{k_s}T_a\right) \quad (9)$$

where

$$A_3 = \frac{k_s \Delta t}{\rho_s c_{ps} dl} \quad (10)$$

The "mixed" volume element at the concrete's corner node (2,N-1) is subjected to conduction at the four sides. An energy balance on this element gives:

$$T_{2,N-1}^{i+1} = T_{2,N-1}^{i}\left[1 - \frac{4(l+d)A_4}{d} - \frac{4dA_4}{l} - \frac{4kA_4}{k_s}\right] + A_4\left[\frac{2(l+d)}{d}T_{1,N-1}^{i} + \frac{2(l+d)}{d}T_{2,N}^{i} + 2\left(\frac{d}{l} + \frac{k}{k_s}\right)T_{2,N-2}^{i} + 2\left(\frac{d}{l} + \frac{k}{k_s}\right)T_{3,N-1}^{i} + \dot{e}_{2,N-1}\frac{l^2}{k_s}\right] \quad (11)$$

where

$$A_4 = \frac{k_s \Delta t}{\rho_s d(d+2l)c_{ps} + \rho c_p l^2} \quad (12)$$

The "mixed" volume element at a concrete's top surface node is also subjected to conduction at the four sides. An energy balance on this element gives:

$$T_{\mathrm{m},N-1}^{i+1} = T_{\mathrm{m},N-1}^{i}\left(1 - \frac{2k_s l}{d}A_5 - \frac{2k_s d}{l}A_5 - 4kA_5\right) + A_5\left[\frac{2k_s l}{d}T_{\mathrm{m},N}^{i} + \left(\frac{k_s d}{l} + k\right)T_{m-1,\mathrm{N}-1}^{i} + \left(\frac{k_s d}{l} + k\right)T_{m+1,\mathrm{N}-1}^{i} + 2kT_{\mathrm{m},N-2}^{i} + \dot{e}_{m,N-1}l^2\right] \quad (13)$$

where

$$A_5 = \frac{\Delta t}{(\rho_s dl c_{ps} + \rho l^2 c_p)} \quad (14)$$

It is noted that the stability criterion of the explicit method requires all primary coefficients to be positive or zero for all nodes:

$$\begin{cases} 1-4\tau_F \geq 0 \\ 1-4A_1-4A_1\dfrac{hd}{k_s} \geq 0 \\ 1-\dfrac{2h(d+l)A_2}{k_s}-2A_2-2dlA_2-\dfrac{2(d+l)A_2}{d} \geq 0 \\ 1-\dfrac{2d}{l}A_3-\dfrac{2l}{d}A_3-\dfrac{2hl}{k_s}A_3 \geq 0 \\ 1-\dfrac{4(l+d)A_4}{d}-\dfrac{4dA_4}{l}-\dfrac{4kA_4}{k_s} \geq 0 \\ 1-\dfrac{2k_sl}{d}A_5-\dfrac{2k_sd}{l}A_5-4kA_5 \geq 0 \end{cases} \quad (15)$$

The maximum time step used to solve the problem must satisfy Eq. (15) above.

If an insulation layer is not used, the new corner temperature $T^i_{m,n+1}$ will be simplified to:

$$T^{i+1}_{m,n} = T^i_{m,n}\left(1-4\tau_F-4\tau_F\frac{h_cl}{k}\right)+2\tau_F\left(T^i_{m+1,n}+T^i_{m,n-1}+2T_a\frac{h_cl}{k}+\frac{\dot{e}_{m,n}l^2}{2k}\right) \quad (16)$$

and the next time step temperature of a top surface node will be simplified to:

$$T^{i+1}_{m,n} = T^i_{m,n}\left(1-4\tau_F-2\tau_F\frac{h_cl}{k}\right)+\tau_F\left(T^i_{m-1,n}+T^i_{m+1,n}+2T^i_{m,n-1}+2T_a\frac{h_cl}{k}+\frac{\dot{e}_{m,n}l^2}{k}\right) \quad (17)$$

The maximum time step in this case is as follows:

$$\Delta t \leq \frac{l^2\rho c_p}{4k\left(1+\frac{h_cl}{k}\right)} \quad (18)$$

2.1 Rate of hydration heat

The rate of heat liberated from cement hydration depends on the temperature of the concrete element itself. The heat rate can be experimentally determined using isothermal [10, 15], adiabatic [16, 17], or semi-adiabatic calorimetry [4]. The experimental adiabatic temperature rise (ATR) can be converted into a maturity-based heat rate as presented by Ballim and Graham [18], in which the total heat (Q) liberated at any time (t) is firstly computed from the ATR using the following relationship:

$$Q = c_p(T_t - T_0)\frac{m_s}{m_c} \quad (19)$$

where T_t = sample temperature at time t; T_0 = initial sample temperature; m_s = mass of concrete sample; and m_c = mass of the cementitious materials in the mix. The heat rate in the adiabatic condition is then calculated by differentiating Eq. (19):

$$q_t = \frac{dQ}{dt} \tag{20}$$

The "maturity heat rate" (q_{te}), as shown in Eq. (21), is used in a further thermal analysis of concrete, which considers the maturity of concrete.

$$q_{t_e} = \frac{dQ}{dt_e} \tag{21}$$

where t_e is equivalent age (or maturity) [19]:

$$t_e = \int_0^t \exp\left(\frac{E_a}{R}\left(\frac{1}{T_r} - \frac{1}{T_c(t)}\right)\right) dt \tag{22}$$

where E_a is apparent activation energy (J/mol); R is the universal gas constant (8.314 J/mol-K); $T_c(t)$ is concrete temperature (K); and T_r is reference temperature (K).

The activation energy (E_a) of a cement blend can be estimated from its chemical compositions using the following relationship derived by Poole [20]:

$$\begin{aligned} E_a = {} & 41230 + 1416000\left(p_{C_3A} + p_{C_4AF}\right)p_{cem}p_{SO_3}p_{cem} - 347000p_{Na_2O_{eq}} \\ & - 19.8Blaine + 29600p_{FA}p_{FA\text{-}CaO} + 16200p_{slag} - 51600p_{SF} \end{aligned} \tag{23}$$

where p_{FA} = % fly ash in the cementing blend; $p_{FA\text{-}CaO}$ = % CaO in fly ash; p_{slag} = % slag in the cementing blend; p_{SF} = percentage of silica fume in the cementitious materials; Blaine = cement fineness (m^2/kg); p_i = percentage of i component in the cement (C_3A, C_4AF, SO_3, cem = cement); and p_{Na2Oeq} = % Na_2O_{eq} in cement (= 0.658 × %K_2O + %Na_2O).

The actual heat rate, which will be used in a numerical model, can be reconstructed from the maturity heat rate using the following equation:

$$q_t = q_{t_e}\frac{dt_e}{dt} \tag{24}$$

The maturity-based heat rate curve q_{te} should be built from an isothermal or adiabatic test, before the actual heat rate can be computed at each time step for the analysis [18]. The drawback of this method is that the total time of the constructed maturity-based heat rate is limited by the test duration.

There are several models to mathematically characterize the heat generation from the cement hydration. The 3-parameter exponential degree of hydration model show in Eq. (25) [21] has been widely used for predicting temperature development in concrete since it includes the temperature effect through the equivalent age:

$$\alpha(t_e) = \alpha_u \exp\left(-\left[\frac{\tau}{t_e}\right]^{\beta}\right) \tag{25}$$

where α_u is ultimate degree of hydration; τ and β are hydration parameters.

The total cumulative heat $Q(t_e)$ is proportional to the degree of hydration $\alpha(t_e)$ as expressed in Eq. (26). The rate of heat generation with respect to equivalent age and real age can be determined using Eqs. (27) and (28), respectively.

$$Q(t_e) = Q_c.\alpha(t_e) \tag{26}$$

$$q(t_e) = \frac{dQ}{dt_e} = Q_c.\alpha(t_e).\left(\frac{\tau}{t_e}\right)^{\beta}.\frac{\beta}{t_e} \tag{27}$$

$$q(t) = \frac{dQ}{dt} = \frac{dQ}{dt_e}.\frac{dt_e}{dt} = Q_c.\alpha(t_e).\left(\frac{\tau}{t_e}\right)^{\beta}.\frac{\beta}{t_e}.\exp\left(\frac{E_a}{R}\left(\frac{1}{T_r} - \frac{1}{T_c(t)}\right)\right) \tag{28}$$

where Q_c is the total available heat (J/m^3).

The hydration parameters (α_u, τ and β) can be determined from the fitted curve, Eq. (25), using the experimental ATR data. These parameters can also be calculated from an experimental isothermal cumulative heat curve without converting the real time into the equivalent age because in the isothermal condition (i.e., at a reference temperature of 23°C), the test time is identical to the equivalent age.

3. FE method for solving thermal stresses

Since a common concrete structure has one dimension larger than the other two, the middle cross section should be analyzed; hence, a FE plane strain problem is selected for the stress computation. A triangular element is chosen with nodes *i, j, m* numbered in a counterclockwise order as illustrated in **Figure 2** [22]. The strain at any point within the element is estimated by Eq. (29):

$$\{\varepsilon\} = [B]\{a^e\} \tag{29}$$

where a^e = element displacement vector, and

$$[B] = \frac{1}{2\Delta}\begin{bmatrix} b_i & 0 & b_j & 0 & b_m & 0 \\ 0 & c_i & 0 & c_j & 0 & c_m \\ c_i & b_i & c_j & b_j & c_m & b_m \end{bmatrix} \tag{30}$$

in which $a_i = x_j y_m - x_m y_j$; $b_i = y_j - y_m$; $c_i = x_m - x_j$with the other coefficients obtained by a cycle permutation of the subscripts in the order *i, j, m*; and Δ is area of the triangle.

The stress vector in the element can be calculated as:

$$\{\sigma\} = \begin{bmatrix}\sigma_x & \sigma_y & \tau_{xy}\end{bmatrix}^T = [D](\{\varepsilon\} - \{\varepsilon_0\}) \tag{31}$$

where

$$[D] = \frac{E}{(1+\nu)(1-2\nu)}\begin{bmatrix} 1-\nu & \nu & 0 \\ \nu & 1-\nu & 0 \\ 0 & 0 & (1-2\nu)/2 \end{bmatrix} \tag{32}$$

and the thermal strain is derived as [22]:

$$\{\varepsilon_0\} = (1+\nu)\alpha_c\theta^e[1 \quad 1 \quad 0]^T \tag{33}$$

in which ν = Poisson's ratio, α_c = coefficient of thermal expansion, and θ^e = temperature change (from the previous time step to the current time step) subjected to the element. The element stiffness matrix *ijm* is calculated using the following equation:

$$k_e = \int B^T DBt_t dxdy \text{ or } k_e = B^T DBt_t\Delta \tag{34}$$

where t_t = element thickness.
The nodal forces due to thermal strain is computed as follows:

$$\{f_T\} = \int_v [B]^T[D]\{\varepsilon_0\}dv = \frac{E\alpha_c\theta^e}{2(1-2\nu)}\begin{bmatrix} b_i & c_i & b_j & c_j & b_m & c_m \end{bmatrix}^T \tag{35}$$

in which E = elastic modulus. The nodal displacement vector U is derived by solving the global system of equations:

$$[K]\{U\} = \{f_T\} \tag{36}$$

Computational Procedure

- FD thermal analysis:

1. Define geometry of the structure (including the nodal grid), initial material properties, initial temperature and boundary conditions, and time interval.

2. Compute the nodal degree of hydration and the rate of heat evolution.

3. Compute the new temperature at each node.

4. Iterate (2) & (3) and record the temperatures.

- FE stress analysis:

5. Divide the nodal grid into triangular elements (the vertices coincide with the FD grid nodes).

6. At $t = n$ $(n = 1, 2, \ldots)$, calculate average temperature, equivalent age and degree of hydration of each element.

7. Let $i = 1$, compute each element's effective modulus.

8. Compute element stiffness matrix, global stiffness matrix, and equivalent nodal forces; solve for nodal displacements and element stresses.

9. Let $i = i + 1$ and iterate (7) and (8) till $i = n$. Sum all the stresses at step (8) to get the total stress.

10. Let $n = n + 1$. Iterate (6) through (10) until the final time step is achieved.

4. Temperature analysis of bridge pier footing

A bridge pier footing constructed in Orlando, Florida was monitored for temperature development within 7 days after casting (**Figure 3**). The concrete footing

Figure 3.
Bridge footing for monitoring in Orlando, Florida.

had dimensions of 6.71-m × 3.05-m × 1.75-m and was insulated with 25.4-mm thick polystyrene foam boards at its bottom, top, and sides during 7 days.

The cementitious materials of Mix #1 was experimentally measured for the heat of hydration using an isothermal calorimeter. The hydration parameters and calculated activation energy (E_a) for Mix #1 are presented in **Table 1**.

The concrete had a density of 2238 kg/m^3, specific heat of 1045 J/kg-K, and thermal conductivity of 1.87 W/m-K. The footing was insulated with Styrofoam that has density, thermal conductivity, and specific heat of 16 kg/m^3, 0.04 W/m-K, and 1200 J/kg-K, respectively [14]. The boundary conditions consist of the initial

Mix	τ (h)	β	α_u	Q_c (J/m^3)	E_a (J/mol)
#1	16.73	0.8764	8.314	1.26×10^8	35,451
#2	14.0	0.94	0.703	1.67×10^8	41,800

Table 1.
Hydration parameters and activation energy.

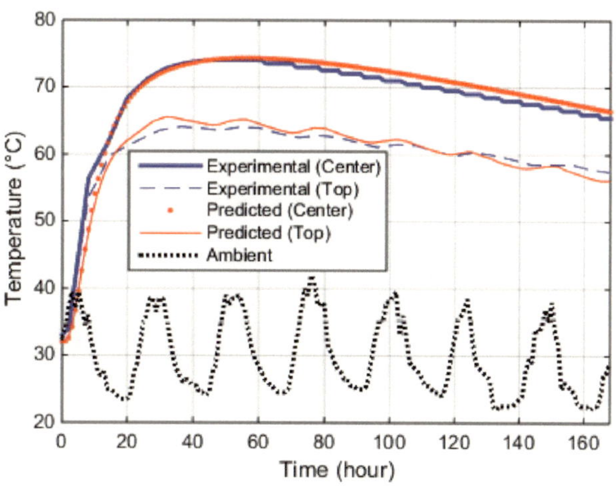

Figure 4.
Predicted and measured temperature profiles in the footing.

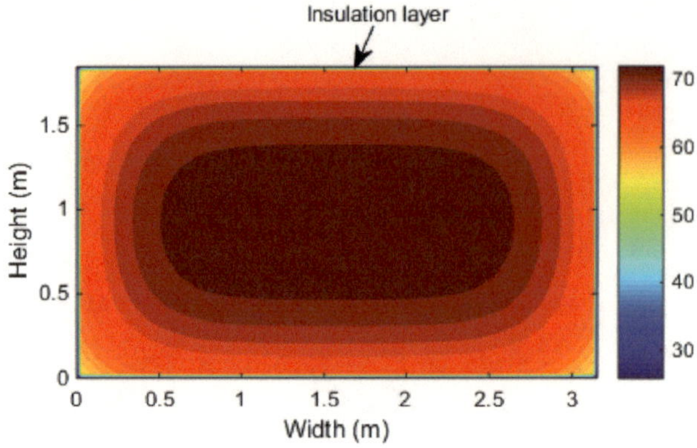

Figure 5.
FD temperature contour in the footing at 40 h (°C).

temperature and the external ambient temperature over time. The air convection coefficient at the insulation surfaces was assumed to be 13.9 W/m^2-K [5, 23].

Two temperature sensors were installed at the center and at the center top surface of the footing to record temperatures within 7 days after placement. The measured temperatures at the center and top, and the ambient temperature are shown in **Figure 4**. A peak measured temperature of 74°C occurred in the middle 42 hours after concrete casting. **Figure 5** shows the temperature distribution in the footing at 40 h calculated by the FD model. The predicted FD temperatures at the center and the top of the footing are also plotted in **Figure 4**. It is clear that the temperature histories computed using the FD model show very close agreement with those collected in the field.

5. Temperature and thermal stress predictions of cap beam

A bridge concrete cap beam (pier cap) was analyzed for temperatures and thermal stresses due to the heat of cement hydration. The cross section of the pier cap was 1.6-m by 2.1-m. The concrete used in the cap beam is Mix #2 with the hydration parameters and activation energy listed in **Table 1**. The concrete coefficient of thermal expansion (CTE) of 8.5×10^{-6}/°C, density of 2287 kg/m^3, the

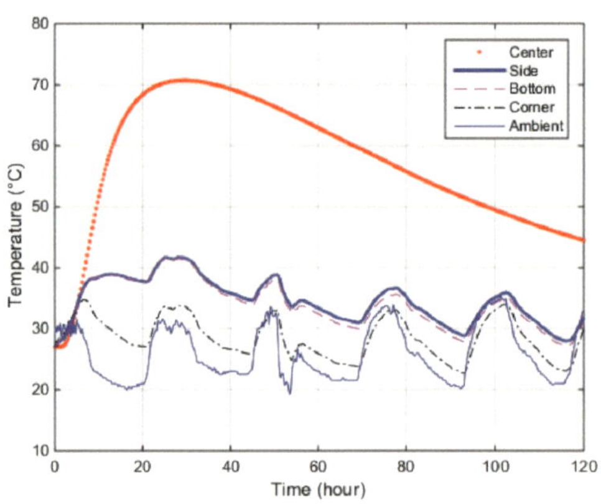

Figure 6.
Temperature profiles at different points in the cap beam.

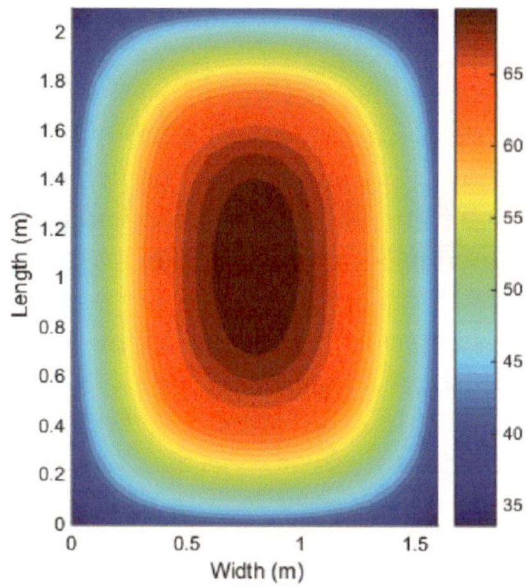

Figure 7.
Temperature distribution of the section at 30 h.

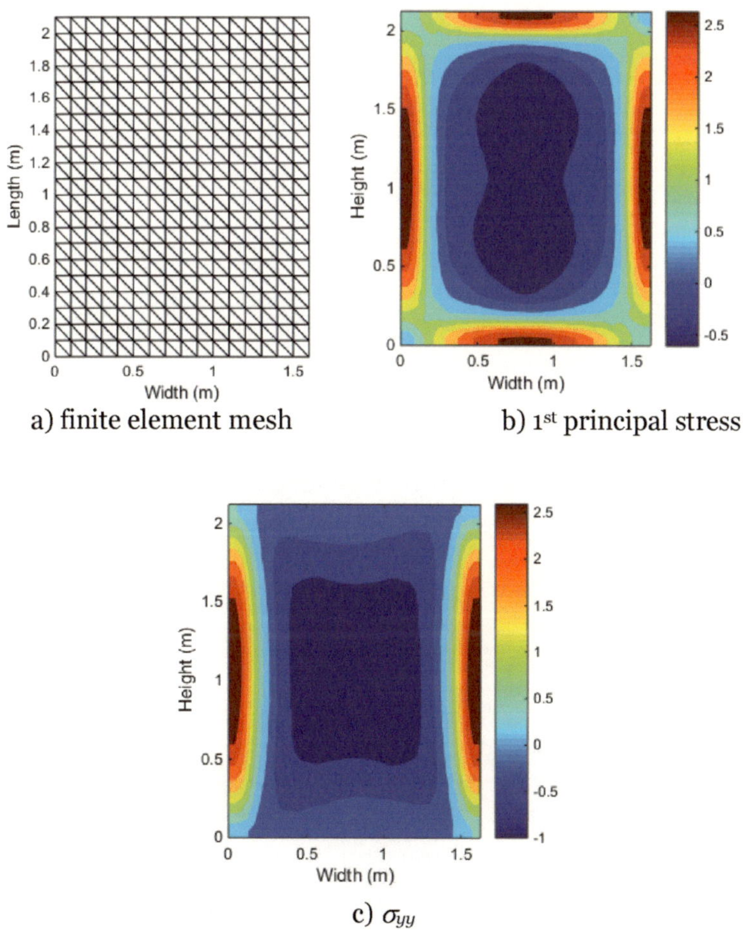

Figure 8.
FE mesh and stress distributions (MPa) in pier cap at 21 h.

specific heat of 1028 J/kg-K, and thermal conductivity of 1.87 W/m-K were assumed in the analysis.

Figure 6 shows the calculated temperature profiles at the core, corner and side of the section. The center temperature peaked at 70.7°C approximately 28 hours after casting. The temperature contours are also depicted in **Figure 7**.

The thermal analysis was followed by a stress calculation using the FE model with the element mesh shown in **Figure 8a**. The 1st principal stress and stress component σ_{yy} contours are shown in **Figure 8b** and **c**, respectively. The figure reveals that the maximum stress is σ_{yy} occurring at the mid-sides and having almost the same magnitude as the 1st principal stress.

The calculated stresses over time at different locations of the pier cap are plotted in **Figure 9**. Clearly, the maximum stress is σ_{yy} at the mid-sides, while σ_{xx} is the maximum stress at the corner (compared to σ_{yy}), thus the middle sides have a higher risk of cracking.

To assess the model's accuracy, the computed stress-time histories are compared with those obtained from the 3-D ABAQUS FE model developed by Lin and Chen [12]. It is worth noting that the ABAQUS model was validated using measurements on 2 concrete blocks. **Figure 9b** shows that the 2-D FE results reasonably match with those of the 3-D ABAQUS model.

The 3-D ABAQUS results reveal that the maximum stress is the component σ_{zz} at the corner. Nevertheless, the 2-D FE model cannot compute this stress component

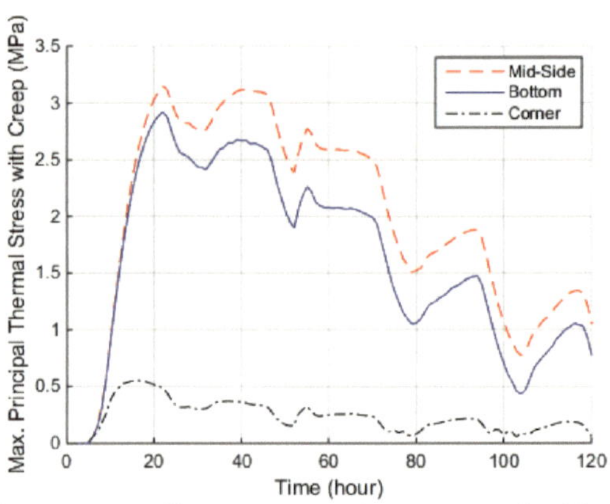

a) Stress profiles at corner, bottom, and mid-side

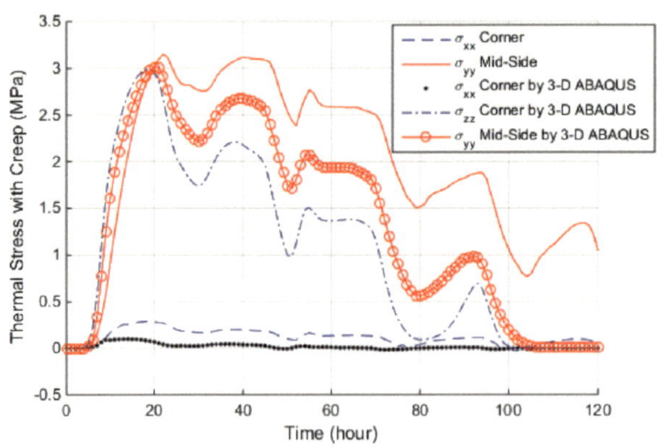

b) 2-D FE plane-strain vs. 3-D ABAQUS stresses

Figure 9.
Calculated stresses in the pier cap.

because it is out-of-plane. The plane element over-predicts σ_{xx} at the corner compared with that of the 3-D ABAQUS. The maximum stress σ_{zz} at the corner predicted using the 3-D ABAQUS model is about the same magnitude as σ_{yy} at the mid-sides, hence the critical stress magnitude as well as cracking risk can still be forecast by the 2-D analysis.

6. Conclusions

In this study, FD and FE formulations were created for solving the transient heat transfer equation and thermal stresses in a concrete element. The results of this study show that the approach that combines the FD and FE methods can be a useful and effective tool for predicting temperature evolution and thermally induced stresses in early-age concrete members with simple geometries. The FD model can analyze thermal behavior of a concrete placement covered with formwork or an insulation layer, thus it can help engineers/contractors control concrete temperature during construction.

Acknowledgements

This work is financially supported by the Ministry of Transport of Vietnam. Special thanks are given to Prof. H. L. Chen and Mr. G. Leon at West Virginia University (WVU) for their contributions to this study.

Author details

Tu Anh Do
University of Transport and Communications, No. 3 Cau Giay Street, Dong Da District, Hanoi, Vietnam

*Address all correspondence to: doanhtu@utc.edu.vn

References

[1] ACI, *207.1 R-05 Guide to Mass Concrete*. 2005.

[2] Ballim, Y., *A numerical model and associated calorimeter for predicting temperature profiles in mass concrete*. Cement and Concrete Composites, 2004. **26**(6): p. 695–703.

[3] Yikici, T.A. and H.-L. Chen, *Numerical prediction model for temperature development in mass concrete structures*. Transportation Research Record: Journal of the Transportation Research Board, 2015(2508): p. 102–110.

[4] Riding, K.A., *Early age concrete thermal stress measurement and modeling*. 2007.

[5] Do, T., et al., *A combined finite difference and finite element model for temperature and stress predictions of cast-in-place cap beam on precast columns*. Construction and Building Materials, 2019. **217**: p. 172–184.

[6] Yikici, T.A. and R.H. Chen, *2D modeling temperature development of mass concrete structures at early age*, in *High Tech Concrete: Where Technology and Engineering Meet*. 2018, Springer. p. 612–620.

[7] Bombich, A.A., S. Garner, and C.D. Norman, *Evaluation of Parameters Affecting Thermal Stresses in Mass Concrete*. 1991, Concrete Technology Information Analysis Center Vicksburg MS.

[8] Waller, V., et al., *Using the maturity method in concrete cracking control at early ages*. Cement and Concrete Composites, 2004. **26**(5): p. 589–599.

[9] Lawrence, A.M., et al., *Effect of early age strength on cracking in mass concrete containing different supplementary cementitious materials: Experimental and finite-element investigation*. Journal of Materials in Civil Engineering, 2011. **24**(4): p. 362–372.

[10] Do, T., et al., *Importance of insulation at the bottom of mass concrete placed on soil with high groundwater*. Transportation Research Record: Journal of the Transportation Research Board, 2013(2342): p. 113–120.

[11] Do, T., et al., *Determination of required insulation for preventing early-age cracking in mass concrete footings*. Transportation Research Record: Journal of the Transportation Research Board, 2014(2441): p. 91–97.

[12] Lin, Y. and H.-L. Chen, *Thermal analysis and adiabatic calorimetry for early-age concrete members*. Journal of Thermal Analysis and Calorimetry, 2016. **124**(1): p. 227–239.

[13] Do, T.A., et al., *Effects of thermal conductivity of soil on temperature development and cracking in mass concrete footings*. Journal of Testing and Evaluation, 2014. **43**(5): p. 1078–1090.

[14] Cengel, Y., *Heat and mass transfer: fundamentals and applications*. 2014: McGraw-Hill Higher Education.

[15] Do, T.A., *Influence of footing dimensions on early-age temperature development and cracking in concrete footings*. Journal of Bridge Engineering, 2014. **20**(3): p. 06014007.

[16] Do, T.A., et al., *Evaluation of heat of hydration, temperature evolution and thermal cracking risk in high-strength concrete at early ages*. Case Studies in Thermal Engineering, 2020: p. 100658.

[17] Lin, Y. and H.-L. Chen, *Thermal analysis and adiabatic calorimetry for early-age concrete members*. Journal of Thermal Analysis and Calorimetry, 2015. **122**(2): p. 937–945.

[18] Ballim, Y. and P. Graham, *A maturity approach to the rate of heat evolution in concrete.* Magazine of Concrete Research, 2003. **55**(3): p. 249–256.

[19] Hansen, P.F. and E. Pedersen, *Curing of concrete structures*. 1984: BKI.

[20] Poole, J.L., *Modeling temperature sensitivity and heat evolution of concrete*. 2007: The University of Texas at Austin.

[21] Hansen, P.F. and E.J. Pedersen, *Maturity computer for controlled curing and hardening of concrete*. 1977.

[22] Zienkiewicz, O. and R. Taylor, *The finite element method 5th edition, volume 1: The Basis*. 2000, Oxford: Butterworth-Heinemann.

[23] Kim, S.G., *Effect of heat generation from cement hydration on mass concrete placement*, in *Civil Engineering*. 2010, Iowa State University.

Chapter 5

Convective Heat and Mass Transfer of Two Fluids in a Vertical Channel

Suresh Babu Baluguri and G. Srinivas

Abstract

A mathematical model for convective heat and mass transfer of two immiscible fluids in a vertical channel of variable width with thermo-diffusion, diffusion-thermal effects is presented. The governing boundary layer equations generated for momentum, angular momentum, energy and species concentration are solved with appropriate boundary conditions using Galeriken finite element method. The effects of the pertinent parameters are studied in detail. Furthermore, the rate of heat transfer, mass transfer and shear stress near both walls is analyzed.

Keywords: vertical channel, immiscible fluids, finite element method, heat and mass transfer

1. Introduction

The developments are carried out in the field of fluid dynamics which was initiated by Euler [1] by proving his famous equations of fluid flow for ideal (inviscid) fluids. Fluid dynamics is a subset of the science that looks at the materials in motion. Hydrodynamics deals with the fluids which are in motion. Fluid dynamics comes under science of fluid mechanics along with the other subcategories as fluid statics, which corresponds to fluids at rest, while fluid dynamics includes fluids in motion. Fluid is defined as a matter in a gas or liquid state. Fluid dynamics is governed by the regulations of preservation of mass, energy and linear momentum. These laws state that the total amount in a closed model remains unchangeable and the energy and mass cannot be formed or demolished. They can deform but will not disappear. Another governing law is the continuum hypothesis which defines that they are uninterrupted and their characteristics fluctuate all over. The history of fluid dynamics can be found in Rouse and Ince [2] and Tokaty [3].

1.1 Micropolar fluid

The subject of micro-polar fluids attained higher degree by many researchers because when the fluid is with the suspended particles we cannot analyze the properties of fluid flow by regular Newtonian fluid characteristics. Generally this fluid is defined as non-Newtonian comprising of small firm cylindrical matters, polymer liquids, liquefied suspensions, animal blood and such related components. The existence of dust or smoke especially, gas are characterized as micro-polar fluids in fluid dynamics. Eringen [4, 5] had taken the initiation in describing the subject of micropolar fluids. In his theory he considered the local impacts emerging from the micro-structure and the inherent movement of the fluid elements. Peddieson and McNitt [6], Ariman et al. [7] addressed many investigations and applications of micropolar fluid mechanics, which are also described in the works of Lukaszewicz [8] and Eringen [9]. A.J. Chamkha et al. [10] analyzed the wholly established free convection of micropolar fluid in a upright passage.

1.2 Magnetohydrodynamics (MHD)

MHD is the science concerned with the motions of electro fluids and their interactions with magnetic fields. It is a vital branch and comparatively new in the field of fluid dynamics. When a conducting fluid is moving along a magnetic field, it results in induction of an electric field current which in turn produces the body forces. According to Faraday's principle, on passage of electric current in a magnetic area, it experiences a force making to direct it at right angles to the electric field. Similarly, if the conductor has electromagnetic forces of the same order as the hydrodynamical and inertial forces, these forces are taken in the equation of motion along with the other forces. The integration of Navier–Stokes relations of fluid dynamics and Maxwell's expressions of electromagnetism describe magneto hydrodynamics which are to be solved simultaneously. There are many scientific & technical applications in the literature: heating and flow control in metal structures, power production from 2-phase models or seeded high temperature gases, magnetic constraints of extreme temperature plasma and dynamo that develop magnetic field in environmental matters.

The concept of MHD flow of the boundary layer in a vertical channel is greatly considered in present metallurgical and metal processing fields. Most of the metallic materials are manufactured from the molten state. It is significant to determine the heat transfer in metals, which are electric conductors. Therefore, a controlled cooling system is required, so that, it can be regulated through an external magnetic field.

1.3 Convective heat and mass transference

Convection is the movement of molecules within the fluids. It belongs to the fundamental means of heat and mass transference that is carried out by ways of diffusion and random Brownian movement of distinct liquid elements. In our context, convection refers to the totality of advective and diffusive transfer. However, it is taken for only advective phenomena. A mechanism of transfer of heat occurring due to bulk motion of fluids is regarded as convective heat transfer. Emphasis is given to heat that is being passed and distributed.

Extensive research has been done over convective heat and mass transference of the fluid flow in vertical channels and other geometries. The existence of temperature and concentration differences or gradients lead to the convective heat and mass transfer and it is regarded as an area of study for broad examination because it is applied in several engineering issues, which are common in atmospheric buoyancy induced actions, liquid and semi-solid bodies and so on. There are quite a large number of application in the heat and mass transfer flows like, rocket nozzles, nuclear power plants, air craft and its re-entry in atmosphere, chemical and process instruments, mist formation and dispersal, temperature and humidity circulation over cultivation farms, plants destruction because of freezing, etc. Packham [11] considered the steady co-current motion of two immiscible viscous fluids in a parallel tube, the fluid interface being ripple-free and plane. Shail [12] considered the Hartmann flow of a conducting fluid in the pass way between two parallel insulating sheets of unbounded length, there exists a sheet of non-conductive fluid between the conductive fluid and the upper passage layer and given the conclusion that considerable increase could be attained in the conductive fluid velocity for appropriate proportions of the depth and viscosity of the two fluids. Beckermann et al. [13] conducted a numerical and experimental study to analyze the fluid motion and heat transfer in a upright rectangular cover which is occupied partly with a vertical layer of a fluid-saturated absorbent structure, where it is determined that the fluid quantity entering the fluid area to the absorbent layer is dependent on the Darcy & Rayleigh numbers. Lohrasbi and Sahai [14] researched in 2-phase MHD flow and heat transfer with the 1-phase conductive fluid.

1.4 Viscous dissipation

Viscous dissipation relates to the conversion of kinetic to internal energy (heating up the fluid) with respect to viscosity. It plays a significant part in normal convection in numerous units which hold huge deviations of gravitational force Gebhart [15]. Gebhart and Mollendorf [16] examined viscous dissipation in peripheral normal convection by taking in account of exponential deviation of wall temperature using resemblance relation. Fand and Brucker [17] stated that the impact of viscous dissipation is important in case of normal/natural convection in absorbent structure with respect to their investigational correlation for the heat transference in peripheral motions. Fand et al. [18] validated the comment for the Darcy method by experimental and analytical means when predicting the heat transfer coefficient from a parallel chamber implanted in an absorbent medium. Viscous dissipation performs as a heat source and heats the medium considerably. Nakayama and Pop [19] evaluated the influence of viscous dissipation on the Darcy's free convection towards an arbitrary shaped non-isothermal matter placed in a permeable medium. Murthy and Singh [20] observed viscous dissipation on non-Darcy normal convection from an erect flat sheet in a permeable medium saturated with Newtonian fluid. It is deduced that heat transfer decreases significantly with the presence of viscous dissipation effect. El-Amin [21] analyzed the impact of viscous dissipation and Joule heating on magneto fluid dynamics forced convection jointly on a non-isothermal straight container fixed in a fluid saturated permeable membrane. Bejan [22] defined that the calculations are limitedin examining the dissipation effect by means of a stable, 1-D energy relation, based on the analogical form with viscous dissipation effect. Pantokratoras [23] evaluated the viscous dissipation effects in a normal convection using a warmed straight plate. Seddeek [24] investigated viscous dissipation effect and thermophoresis on Darcy Forchheimer mixed convection in a fluid saturated permeable medium. Duwairi et al. [25] studied the effects of viscous dissipation and Joule heating employing an isothermal cone in a saturated porous medium. Various non-Newtonian fluids have high viscosity because the irreparable criterion owing to viscous dissipation sometimes becomes vital. Hence it motivates investigators to analyze the effects of viscous dissipation in a non-Newtonian fluid saturated permeable medium. Cortell [26] analyzed viscous dissipation effect and thermal boundary layer radiation on a nonlinear wide plate. Kairi and Murthy [27] analyzed the viscous dissipation impact over normal convection heat and mass transference from an upright cone in a non-Newtonian fluid saturated non-Darcy absorbent structure. Cortell [28] analyzed the influences of suction, viscous dissipation and thermal radiation over heat transfer of a power-law fluid past a boundless permeable sheet.

1.5 Diffusion effects

The diffusion effects namely thermal-diffusion (Soret) and diffusion-thermo (Dufour) are highly important in fluid mechanics. Soret is the transfer of mass formed by temperature gradients, i.e. species diversity evolving in a primary homogeneous blend directed to a thermal gradient. Diffusion-thermo effect is the heat transfer or the heat flux formed by concentration gradient. The problems concerned to heat and mass transference and density variations with temperature and concentration lead to integrated buoyancy convected force. The diffusion impacts influence the flow field in boundary membrane on an upright channel.

Chapman and Cowling [29] developed the diffusion-thermo and thermal-diffused heat and mass effects. Eckert and Drake [30] suggested that Dufour effect has widened magnitude and so this effect should not be ignored. Kafoussias and Williams [31] included the boundary layer flows with Soret and Dufour effects for the combined forced-normal convection problem. Anghel et al. [32] analyzed the Dufour and Soret effects of a free convection boundary layer on a vertical field inserted in a permeable membrane. Postelnicu [33] evaluated the effects of thermal-diffusion and diffusion-thermo on combined heat and mass transference in natural convection boundary layer flow in a Darcian porous media under transverse magnetic effect. Alam and Rahman [34] analyzed the effects of thermal-diffusion and diffusion-thermo on combined and free convection heat and mass transference flow past an erect permeable flat sheet inserted in a porous membrane with or without flexible suction. In many studies, Dufour and Soret effects are ignored based on a minor magnitude order than Fourier's and Fick's laws effect. The effect of thermal-diffusion and diffusion-thermo influences over the motion area in mixed convection boundary-layer on an upright surface kept in a permeable medium and on mixed convection flow past a vertical permeable even sheet with varying suction. Chamkha and Ben-Nakhi [35] studied the combined convection flow in the existence of thermal radiation with an erect porous layer embedded in an absorbent media considering thermal-diffusion and diffusion-thermo effects. El-Aziz [36] examined the Dufour and Soret effects on MHD heat and mass transference on a porous widening layer in the presence of thermal radiation in a combined manner. Maleque [37] considered only the diffusion-thermo effect on convective heat and mass transference past a rotating permeable disk, in where the thermal-diffusion effect is ignored. Anwar Beg et al. [38] described the thermal-diffusion and diffusion-thermo impacts by numerically studying the free convection MHD heat and mass transfer over a stretching layer with saturated permeable structure. Pal and Chatterjee [39] study shows combined convected magneto hydrodynamic heat and mass transference past a stretching plate considering Ohmic dissipated thermal-diffusion and diffusion-thermo impacts with micro-polar fluid. MHD flow of a pair of immiscible and conducting fluids within isothermal and insulated moving sheets under an applied electric and inclined magnetic effect and with an induced magnetic field has been investigated by Stamenkovic et al. [40].

1.6 Two fluid flow

For many years, Scientists and Engineers have been showing interest in two phase flows, which arise in many industrial applications. The two-phase fluid flow phenomena are important in pipe flows, fluidized beds, sedimentation, gas purification, transport processes and shock waves. The study of dynamics of two phase fluid system is concerned with the motion of a liquid or gas containing immiscible, suspended stokesian solid particles. In the equations of motion of two phase fluid flows, which are the modified form of Navier-Stoke's equations, the presence of dust adds an extra force term which represents the interaction between the dust and the fluid particles. The modified form of Navier–Stokes equations coupled with Euler equations of motion for perfect fluids are used as the equations of motion of fluid phase and particle phase, respectively. Practical application of these flows may be found in heat exchanges utilizing liquid metal or liquid sodium coolants in the area of thermal instability in boiling heat transfer studies. Malashetty and Leela [41, 42] studied two-fluid flow and heat transference in a parallel fluid passage in both conductive phases. Such investigations are beneficial to understand the slag layer effects over heat transfer features of a coal-fired magneto hydrodynamic generator. Vajravelu et al. [43] dealt with the hydromagnetic unstable motion of two immiscible conducting fluids between two porous media of different porosity. Malashetty and Umavathi [44] studied 2-phase magnetohydrodynamic flow and heat transference in a sloped passage, where 1-phase is conductive and the transport characteristics of the fluids are assumed to be unvarying. Srinivasan et al. [45] theoretically studied the two immiscible fluid models in a permeable membrane by considering the impacts of non-Darcian boundary and inertia. Malashetty et al. [46] explored the complexities of completely established 2-fluid magneto hydrodynamic flow including and excluding applied electric field in an slant pass way and described the solution of energy and momentum equations, using perturbation method for smaller value product of Prandtl and Eckert number in completely progressed free convection 2-fluid MHD flow of a tilted passage.

2. Mathematical formulation

The two infinite plates are kept at $Y = -h_1$ and $Y = h_2$ initially as shown in **Figure 1** and the two sheets are isothermal with dissimilar temperatures T_1 and T_2 respectively. The distance $-h_1 \le Y \le 0$ represents region-1 and the distance $0 \le Y \le h_2$ represents region-2 where the first one occupies micropolar fluid and the other, viscous fluid. Here the buoyancy force determines the fluid flow.

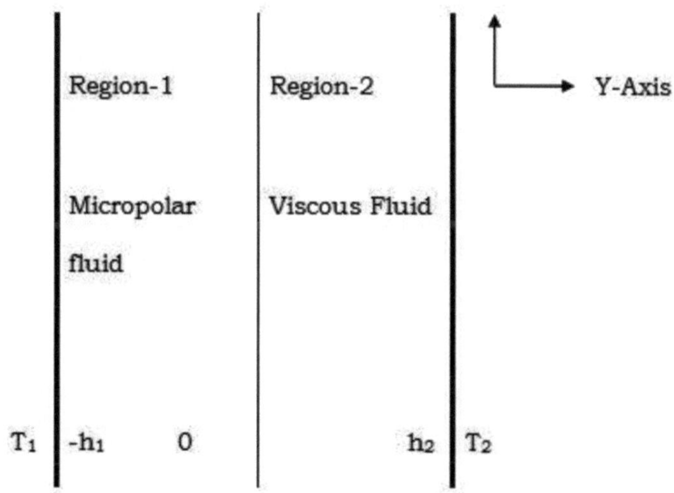

Figure 1.
Schematic diagram of the problem.

The governing conditions for the problem are developed based on the assumptions stated below:

1. The flow is assumed to be 1-D, steady, laminar, immiscible and incompressible.

2. The transport characteristics of the two fluids are presumed to be constant.

3. The fluid flow is fully developed.

4. The temperature and heat flows are continuous at the interface.

5. $T_1 > T_2, C_1 > C_2$.

3. Governing equations

The governing equations which are derived in Chapter-II under the above assumptions yields.

Region-1:

$$\frac{dU_1}{dY} = 0 \text{ [Law of conservation of mass]} \tag{1}$$

$$\rho_1 = \rho_0[1 - \beta_{1T}(T_1 - T_0) - \beta_{1C}(C_1 - C_0)] \text{ [Physicalstate]} \tag{2}$$

$$\frac{\mu_1 + K}{\rho_1}\frac{d^2U_1}{dY^2} + \frac{K}{\rho_1}\frac{dn}{dY} + g\beta_{1T}(T_1 - T_0) + g\beta_{1C}(C_1 - C_0) - \frac{\sigma B_0{}^2 U_1}{\rho_1} = 0 \text{ [Momentum]} \tag{3}$$

$$\gamma\frac{d^2n}{dY^2} - K\left[2n + \frac{dU_1}{dY}\right] = 0 \text{ [Conservation of Angular Momentum]} \tag{4}$$

where $\gamma = \left(\mu_1 + \frac{K}{2}\right)j$

$$\frac{k_1}{\rho_1 C_p}\frac{d^2T_1}{dY^2} + \frac{1}{\rho_1 C_p}\left[\mu_1\left(\frac{dU_1}{dY}\right)^2 + \frac{\rho_1 D_1 K_{T1}}{C_{S1}}\frac{d^2C_1}{dY^2}\right] = 0 \text{ [Energy]} \tag{5}$$

$$D_1\frac{d^2C_1}{dY^2} + \frac{D_1 K_{T1}}{T_M}\frac{d^2T_1}{dY^2} = 0 \text{ [Diffusion]} \tag{6}$$

Region-2:

$$\frac{dU_2}{dY} = 0 \text{ [Continuity]} \tag{7}$$

$$\rho_2 = \rho_0[1 - \beta_{2T}(T_2 - T_0) - \beta_{2C}(C_2 - C_0)] \text{ [State]} \tag{8}$$

$$\frac{\mu_2}{\rho_2}\frac{d^2U_2}{dY^2} + g\beta_{2T}(T_2 - T_0) + g\beta_{2C}(C_2 - C_0) - \frac{\sigma B_0{}^2 U_2}{\rho_2} = 0 \text{ [Momentum]} \tag{9}$$

$$\frac{k_2}{\rho_2 C_p}\frac{d^2T_2}{dY^2} + \frac{1}{\rho_2 C_p}\left[\mu_2\left(\frac{dU_2}{dY}\right)^2 + \frac{\rho_2 D_2 K_{T2}}{C_{S2}}\frac{d^2C_2}{dY^2}\right] = 0 \text{ [Energy]} \tag{10}$$

$$D_2\frac{d^2C_2}{dY^2} + \frac{D_2 K_{T2}}{T_M}\frac{d^2T_2}{dY^2} = 0 \text{ [Diffusion]} \tag{11}$$

The above equation models (1) to (11) are solved by the following boundary and interface parameters.

$U_1 = 0$ at $Y = -h_1$, $U_2 = 0$ at $Y = h_2$,$U_1(0) = U_2(0)$,
$T = T_1$ at $Y = -h_1$, $T = T_2$ at $Y = h_2$,$T_1(0) = T_2(0)$,
$C = C_1$ at $Y = -h_1$, $C = C_2$ at $Y = h_2$,$C_1(0) = C_2(0)$,
$n = 0$ at $Y = -h_1$,$(\mu_1 + K)\frac{dU_1}{dY} + Kn = \mu_2\frac{dU_2}{dY}$ at $Y = 0$,
$\frac{dn}{dY} = 0$ at $Y = 0$, $k_1\frac{dT_1}{dY} = k_2\frac{dT_2}{dY}$ at $Y = 0$,$D_1\frac{dC_1}{dY} = D_2\frac{dC_2}{dY}$ at $Y = 0$.

The following non dimensional variables form the equation systems (1) to (11) in to dimensionless form:

$y = \frac{Y}{h_1}$(region-1), $y = \frac{Y}{h_2}$ (region-2), $u_1 = \frac{U_1}{U_0}$, $u_2 = \frac{U_2}{U_0}$, $\theta_1 = \frac{T_1 - T_0}{\Delta T}$,
$\theta_2 = \frac{T_2 - T_0}{\Delta T}$, $N = \frac{h_1}{U_0}n$, $j = h^2$ (Characteristic length), $K' = \frac{K}{\mu_1}$, $c_1 = \frac{C_1 - C_0}{\Delta C}$, $c_2 = \frac{C_2 - C_0}{\Delta T}$, $Gr = \frac{g\beta_{1T}\Delta T h_1^3}{\nu_1^2}$, $Gc = \frac{g\beta_{1C}\Delta C h_1^3}{\nu_1^2}$, $R = \frac{U_0 h_1}{\upsilon_1}$, $Sr = \frac{D_1 K_{T1}\Delta T}{T_M \Delta C U_0 h_1}$, $Sc = \frac{\upsilon_1}{D_1}$, $Du = \frac{D_1 K_{T1}\Delta C}{C_P C_{S1}\nu_1\Delta T}$,
$M = \frac{\sigma B_0^2 h_1^2}{\mu_1}$, $Pr = \frac{\mu_1 Cp}{k_1}$, $Ec = \frac{U_0^2}{Cp\ \Delta T}$,
$C_S = \frac{C_{S1}}{C_{S2}}$, $K_T = \frac{K_{T1}}{K_{T2}}$, $D = \frac{D_1}{D_2}$, $h = \frac{h_1}{h_2}$, $m = \frac{\mu_1}{\mu_2}$, $\alpha = \frac{k_1}{k_2}$, $\rho = \frac{\rho_1}{\rho_2}$, $b_1 = \frac{\beta_{1T}}{\beta_{2T}}$, $b_2 = \frac{\beta_{1C}}{\beta_{2C}}$,$\nu = \frac{\nu_1}{\nu_2}$.

The dimensionless forms of governing equations thus obtained are:

Region-1:

$$\frac{d^2N}{dy^2} - \frac{2K'}{2+K'}\left(2N + \frac{du_1}{dy}\right) = 0 \tag{12}$$

$$(1+K')\frac{d^2u_1}{dy^2} + K'\frac{dN}{dy} + \frac{Gr}{R}\theta_1 + \frac{Gc}{R}c_1 - Mu_1 = 0 \tag{13}$$

$$\frac{1}{PrR}\frac{d^2\theta_1}{dy^2} + \frac{Ec}{R}\left(\frac{du_1}{dy}\right)^2 + \frac{Du}{R}\frac{d^2c_1}{dy^2} = 0 \tag{14}$$

$$\frac{1}{ScR}\frac{d^2c_1}{dy^2} + Sr\frac{d^2\theta_1}{dy^2} = 0 \tag{15}$$

Region −2

$$\frac{d^2u_2}{dy^2} + \frac{m}{b_1\rho h^2}\frac{Gr}{R}\theta_2 + \frac{m}{b_2\rho h^2}\frac{Gc}{R}c_2 - \frac{mM}{h^2}u_2 = 0 \tag{16}$$

$$\frac{\rho h}{\alpha}\frac{1}{PrR}\frac{d^2\theta_2}{dy^2} + \frac{\rho h}{m}\frac{Ec}{R}\left(\frac{du_2}{dy}\right)^2 + \frac{c_s h}{DK_T}\frac{D_u}{R}\frac{d^2c_2}{dy^2} = 0 \tag{17}$$

$$\frac{h}{D}\left(\frac{1}{ScR}\right)\frac{d^2c_2}{dy^2} + \frac{h}{K_T D}Sr\frac{d^2\theta_2}{dy^2} = 0 \tag{18}$$

The dimensionless boundary and interface conditions thus formed are:

$$\begin{aligned}
&u_1 = 0 \text{ at } y = -1, u_2 = 0 \text{ at } y = 1, u_1(0) = u_2(0),\\
&\theta_1 = 1 \text{ at } y = -1, \theta_2 = 0 \text{ at } y = 1, \theta_1(0) = \theta_2(0),\\
&c_1 = 1 \text{ at } y = -1, c_2 = 0 \text{ at } y = 1, c_1(0) = c_2(0),\\
&N = 0 \text{ at } y = -1, \frac{du_1}{dy} + \frac{K'}{1+K'}N = \frac{1}{mh(1+K')}\frac{du_2}{dy} \text{ at } y = 0,\\
&\frac{dN}{dy} = 0 \text{ at } y = 0, \frac{d\theta_1}{dy} = \frac{1}{h\alpha}\frac{d\theta_2}{dy} \text{ at } y = 0, \frac{dc_1}{dy} = \frac{1}{hD}\frac{dc_2}{dy} \text{ at } y = 0
\end{aligned} \tag{19}$$

4. Solution of the problem

The finite element method as described in Chapter-II is applied in solving the dimensionless coupled differential equations generated by the fluid flows. For the problem discussed here, it is considered that each region is classified into 100 linear elements and each element is 3 nodded.

The element equations associated with Eqs. (12) to (18) is as follows:

$$\int_{y_i}^{y_{i+1}} \left(\frac{d^2N}{dy^2} - \frac{2K'}{2+K'} \left[2N + \frac{du_1}{dy} \right] \right) \eta_k dy = 0 \tag{20}$$

$$\int_{y_i}^{y_{i+1}} \left((1+K') \frac{d^2u_1}{dy^2} + K' \frac{dN}{dy} + \frac{Gr}{R} \theta_1 + \frac{Gc}{R} c_1 - Mu_1 \right) \eta_k dy = 0 \tag{21}$$

$$\int_{y_i}^{y_{i+1}} \left(\frac{1}{\Pr R} \frac{d^2\theta_1}{dy^2} + \frac{Ec}{R} \left(\frac{du_1}{dy} \right)^2 + \frac{Du}{R} \frac{d^2c_1}{dy^2} \right) \eta_k dy = 0 \tag{22}$$

$$\int_{y_i}^{y_{i+1}} \left(\frac{1}{ScR} \frac{d^2c_1}{dy^2} + Sr \frac{d^2\theta_1}{dy^2} \right) \eta_k dy = 0 \tag{23}$$

$$\int_{y_i}^{y_{i+1}} \left(\frac{d^2u_2}{dy^2} + \frac{m}{b_1\rho h^2} \frac{Gr}{R} \theta_2 + \frac{m}{b_2\rho h^2} \frac{Gc}{R} c_2 - \frac{mM}{h^2} u_2 \right) \chi_k dy = 0 \tag{24}$$

$$\int_{y_i}^{y_{i+1}} \left(\frac{\rho h}{\alpha} \frac{1}{\Pr R} \frac{d^2\theta_2}{dy^2} + \frac{\rho h}{m} \frac{Ec}{R} \left(\frac{du_2}{dy} \right)^2 + \frac{c_s h}{DK_T} \frac{D_u}{R} \frac{d^2c_2}{dy^2} \right) \chi_k dy = 0 \tag{25}$$

$$\int_{y_i}^{y_{i+1}} \left(\frac{h}{D} \left(\frac{1}{ScR} \right) \frac{d^2c_2}{dy^2} + \frac{h}{K_T D} Sr \frac{d^2\theta_2}{dy^2} \right) \chi_k dy = 0 \tag{26}$$

Where η_k and χ_k denotes the shape functions of a typical element (y_i, y_{i+1}) in the region 1 and 2 correspondingly.

On integrating the above equations and by replacing the finite element Galerkin calculations,

$$u_1^i = \sum_{j=1}^{3} u_j^i \eta_j^i,\ c_1^i = \sum_{j=1}^{3} c_j^i \eta_j^i,\ N^i = \sum_{j=1}^{3} N_j^i \eta_j^i,\ \theta_1^i = \sum_{j=1}^{3} \theta_j^i \eta_j^i,$$

$$u_2^i = \sum_{j=1}^{3} u_j^i \chi_j^i,\ c_2^i = \sum_{j=1}^{3} c_j^i \chi_j^i,\ \theta_2^i = \sum_{j=1}^{3} \theta_j^i \chi_j^i.$$

From Eq. (20) we get

$$\int_{y_i}^{y_{i+1}} \frac{dN^i}{dy} \frac{d\eta_k}{dy} dy + \frac{2K'}{2+K'} \int_{y_i}^{y_{i+1}} \left[2N^i \eta_k dy - \frac{d\eta_k}{dy} u_1^i \right] dy = \left[\eta_k \frac{dN^i}{dy} + \eta_k u_1^i \right]_{y_i}^{y_{i+1}}$$

The stiffness matrix equation corresponding to the above is

$$\left[a_{\mathrm{kj}}^{i}\right]\left[N_{k}^{i}\right]+\left[b_{\mathrm{kj}}^{i}\right]\left[u_{k}^{i}\right]=\left[Q_{1j}^{i}\right] \tag{27}$$

where $a_{\mathrm{kj}}^{i}=\int_{y_i}^{y_{i+1}}\frac{d\eta_k}{\mathrm{dy}}\frac{d\eta_j^i}{\mathrm{dy}}\mathrm{dy}+\frac{2K'}{2+K'}\int_{y_i}^{y_{i+1}}\left[2\eta_k\eta_j^i\right]\mathrm{dy}$

$$b_{\mathrm{kj}}^{i}=-\frac{2K'}{2+K'}\int_{y_i}^{y_{i+1}}\left[\frac{d\eta_j^i}{\mathrm{dy}}\eta_k\right]\mathrm{dy}$$

$$Q_{1j}^{i}=\left[\eta_k\frac{\mathrm{dN}^i}{\mathrm{dy}}+\eta_k u_1^i\right]_{y_i}^{y_{i+1}}$$

From Eq. (21) we get

$$\int_{y_i}^{y_{i+1}}(1+K')\frac{d\eta_k}{\mathrm{dy}}\frac{\mathrm{du}_1^i}{\mathrm{dy}}\mathrm{dy}+\int_{y_i}^{y_{i+1}}K'\frac{d\eta_k}{\mathrm{dy}}N^i\mathrm{dy}\text{-}\frac{\mathrm{Gr}}{R}\int_{y_i}^{y_{i+1}}\eta_k\theta_1^i\mathrm{dy}\text{-}$$
$$\frac{\mathrm{Gc}}{R}\int_{y_i}^{y_{i+1}}\eta_k c_1^i\,\mathrm{dy}+M\int_{y_i}^{y_{i+1}}\eta_k u_1^i\,\mathrm{dy}=\left[\text{-}(1+K')\eta_k\frac{\mathrm{du}_1^i}{\mathrm{dy}}-K'\eta_k N^i\right]_{y_i}^{y_{i+1}}$$

The stiffness matrix equation corresponding to the above is

$$\left[c_{\mathrm{kj}}^{i}\right]\left[u_{k}^{i}\right]+\left[d_{\mathrm{kj}}^{i}\right]\left[N_{k}^{i}\right]+\left[e_{\mathrm{kj}}^{i}\right]\left[\theta_{k}^{i}\right]+\left[f_{\mathrm{kj}}^{i}\right]\left[c_{k}^{i}\right]=\left[Q_{2j}^{i}\right] \tag{28}$$

where $c_{\mathrm{kj}}^{i}=\int_{y_i}^{y_{i+1}}(1+K')\frac{d\eta_j^i}{\mathrm{dy}}\frac{d\eta_k}{\mathrm{dy}}\mathrm{dy}+M\int_{y_i}^{y_{i+1}}\eta_j^i\eta_k\mathrm{dy}$, $d_{\mathrm{kj}}^{i}=K'\int_{y_i}^{y_{i+1}}\left[\frac{d\eta_j^i}{\mathrm{dy}}\eta_k\right]\mathrm{dy}$

$$e_{\mathrm{kj}}^{i}=-\frac{\mathrm{Gr}}{R}\int_{y_i}^{y_{i+1}}\left[\eta_j^i\eta_k\right]\mathrm{dy},\ f_{\mathrm{kj}}^{i}=-\frac{\mathrm{Gc}}{R}\int_{y_i}^{y_{i+1}}\left[\eta_j^i\eta_k\right]\mathrm{dy}$$

$$Q_{2j}^{i}=\left[\text{-}(1+K')\eta_k\frac{\mathrm{du}_1^i}{\mathrm{dy}}-K'\eta_k N^i\right]_{y_i}^{y_{i+1}}$$

From Eq. (22) we get

$$\frac{1}{\mathrm{Pr}R}\int_{y_i}^{y_{i+1}}\frac{d\eta_k}{\mathrm{dy}}\frac{d\theta_1^i}{\mathrm{dy}}\mathrm{dy}-\frac{\mathrm{Ec}}{R}\int_{y_i}^{y_{i+1}}\eta_k\left(\frac{\mathrm{du}_1^i}{\mathrm{dy}}\right)^2\mathrm{dy}\text{-}\frac{\mathrm{Du}}{R}\int_{y_i}^{y_{i+1}}\frac{d\eta_k}{\mathrm{dy}}\frac{\mathrm{dc}_1^i}{\mathrm{dy}}\mathrm{dy}$$
$$=\left[\frac{1}{\mathrm{Pr}R}\eta_k\frac{d\theta_1^i}{\mathrm{dy}}-\frac{\mathrm{Du}}{R}\eta_k\frac{\mathrm{dc}_1^i}{\mathrm{dy}}\right]_{y_i}^{y_{i+1}}$$

The stiffness matrix equation corresponding to the above is

$$\left[g_{\mathrm{kj}}^{i}\right]\left[\theta_{k}^{i}\right]+\left[u_{k}^{i}\right]^{T}\left[h_{\mathrm{kj}}^{i}\right]\left[u_{k}^{i}\right]+\left[m_{\mathrm{kj}}^{i}\right]\left[c_{k}^{i}\right]=\left[Q_{3j}^{i}\right] \tag{29}$$

where

$$g_{\mathrm{kj}}^{i}=\frac{1}{\mathrm{Pr}R}\int_{y_i}^{y_{i+1}}\frac{d\eta_k}{\mathrm{dy}}\frac{d\eta_j^i}{\mathrm{dy}}\mathrm{dy},h_{\mathrm{kj}}^{i}$$

$$=\frac{-\mathrm{Ec}}{R}\int_{y_i}^{y_{i+1}}\begin{bmatrix}\left(\frac{d\eta_1^i}{\mathrm{dy}}\right)^2 & \left(\frac{d\eta_1^i}{\mathrm{dy}}\right)\left(\frac{d\eta_2^i}{\mathrm{dy}}\right) & \left(\frac{d\eta_1^i}{\mathrm{dy}}\right)\left(\frac{d\eta_3^i}{\mathrm{dy}}\right)\\ \left(\frac{d\eta_2^i}{\mathrm{dy}}\right)\left(\frac{d\eta_1^i}{\mathrm{dy}}\right) & \left(\frac{d\eta_2^i}{\mathrm{dy}}\right)^2 & \left(\frac{d\eta_2^i}{\mathrm{dy}}\right)\left(\frac{d\eta_3^i}{\mathrm{dy}}\right)\\ \left(\frac{d\eta_3^i}{\mathrm{dy}}\right)\left(\frac{d\eta_1^i}{\mathrm{dy}}\right) & \left(\frac{d\eta_3^i}{\mathrm{dy}}\right)\left(\frac{d\eta_2^i}{\mathrm{dy}}\right) & \left(\frac{d\eta_3^i}{\mathrm{dy}}\right)^2\end{bmatrix}[\eta_k]\mathrm{dy}.$$

$$m_{\mathrm{kj}}^{i}=\frac{-\mathrm{Du}}{R}\int_{y_i}^{y_{i+1}}\frac{d\eta_k}{\mathrm{dy}}\frac{d\eta_j^i}{\mathrm{dy}}\mathrm{dy},Q_{3j}^{\ i}=\left[\frac{1}{\mathrm{Pr}R}\eta_k\frac{d\theta_1^i}{\mathrm{dy}}-\frac{\mathrm{Du}}{R}\eta_k\frac{\mathrm{dc}_1^i}{\mathrm{dy}}\right]_{y_i}^{y_{i+1}}.$$

From Eq. (23) we get

$$\frac{1}{\mathrm{Sc}R}\int_{y_i}^{y_{i+1}}\frac{d\eta_k}{\mathrm{dy}}\frac{\mathrm{dc}_1^i}{\mathrm{dy}}\mathrm{dy}+\mathrm{Sr}\int_{y_i}^{y_{i+1}}\frac{d\eta_k}{\mathrm{dy}}\frac{d\theta_1^i}{\mathrm{dy}}\mathrm{dy}=\left[\frac{1}{\mathrm{Sc}R}\eta_k\frac{\mathrm{dc}_1^i}{\mathrm{dy}}+\mathrm{Sr}\eta_k\frac{d\theta_1^i}{\mathrm{dy}}\right]_{y_i}^{y_{i+1}}$$

The stiffness matrix equation corresponding to the above is

$$\left[n_{\mathrm{kj}}^{i}\right]\left[c_k^i\right]+\left[p_{\mathrm{kj}}^{i}\right]\left[\theta_k^i\right]=\left[Q_{4j}^{\ i}\right]\tag{30}$$

where $n_{\mathrm{kj}}^{i}=\frac{1}{\mathrm{Sc}R}\int_{y_i}^{y_{i+1}}\frac{d\eta_k}{\mathrm{dy}}\frac{d\eta_j^i}{\mathrm{dy}}\mathrm{dy},\ p_{\mathrm{kj}}^{i}=\mathrm{Sr}\int_{y_i}^{y_{i+1}}\frac{d\eta_k}{\mathrm{dy}}\frac{d\eta_j^i}{\mathrm{dy}}\mathrm{dy}$

$$Q_{4j}^{\ i}=\left[\frac{1}{\mathrm{Sc}R}\eta_k\frac{\mathrm{dc}_1^i}{\mathrm{dy}}+\mathrm{Sr}\eta_k\frac{d\theta_1^i}{\mathrm{dy}}\right]_{y_i}^{y_{i+1}}$$

From Eq. (24) we get

$$\int_{y_i}^{y_{i+1}}\frac{d\chi_k}{\mathrm{dy}}\frac{\mathrm{du}_2^i}{\mathrm{dy}}\mathrm{dy}\text{-}\frac{m}{b_1\rho h^2}\frac{\mathrm{Gr}}{R}\int_{y_i}^{y_{i+1}}\chi_k\theta_2^i\mathrm{dy}\text{-}\frac{m}{b_2\rho h^2}\frac{\mathrm{Gc}}{R}\int_{y_i}^{y_{i+1}}\chi_k c_2^i\,\mathrm{dy}$$

$$+\frac{\mathrm{mM}}{h^2}\int_{y_i}^{y_{i+1}}\chi_k u_2^i\,\mathrm{dy}=\left[\text{-}\chi_k\frac{\mathrm{du}_2^i}{\mathrm{dy}}\right]_{y_i}^{y_{i+1}}$$

The stiffness matrix equation corresponding to the above is

$$\left[C_{\mathrm{kj}}^{i}\right]\left[u_k^i\right]+\left[D_{\mathrm{kj}}^{i}\right]\left[\theta_k^i\right]+\left[E_{\mathrm{kj}}^{i}\right]\left[c_k^i\right]=\left[Q_{5j}^{\ i}\right]\tag{31}$$

where $C^i_{kj} = \int_{y_i}^{y_{i+1}} \frac{d\chi^i_j}{dy}\frac{d\chi_k}{dy}dy + \frac{mM}{h^2}\int_{y_i}^{y_{i+1}} \chi^i_j\chi_k dy$, $D^i_{kj} = -\frac{m}{b_1\rho h^2}\frac{Gr}{R}\int_{y_i}^{y_{i+1}}\left[\chi^i_j\chi_k\right]dy$,

$E^i_{kj} = -\frac{m}{b_2\rho h^2}\frac{Gc}{R}\int_{y_i}^{y_{i+1}}\left[\chi^i_j\chi_k\right]dy$, $Q_{5j}^i = \left[-\chi_k\frac{du^i_2}{dy}\right]_{y_i}^{y_{i+1}}$.

From Eq. (25) we get

$$\frac{\rho h}{\alpha}\frac{1}{PrR}\int_{y_i}^{y_{i+1}}\frac{d\chi_k}{dy}\frac{d\theta^i_2}{dy}dy - \frac{\rho h}{m}\frac{Ec}{R}\int_{y_i}^{y_{i+1}}\chi_k\left(\frac{du^i_2}{dy}\right)^2 dy\text{-}\frac{c_s h}{DK_T}\frac{Du}{R}\int_{y_i}^{y_{i+1}}\frac{d\chi_k}{dy}\frac{dc^i_2}{dy}dy =$$

$$\left[\frac{\rho h}{\alpha}\frac{1}{PrR}\chi_k\frac{d\theta^i_2}{dy} - \frac{c_s h}{DK_T}\frac{Du}{R}\chi_k\frac{dc^i_2}{dy}\right]_{y_i}^{y_{i+1}}$$

The stiffness matrix equation corresponding to the above is

$$\left[F^i_{kj}\right]\left[\theta^i_k\right] + \left[u^i_k\right]^T\left[G^i_{kj}\right]\left[u^i_k\right] + \left[H^i_{kj}\right]\left[c^i_k\right] = \left[Q_{6j}^i\right] \tag{32}$$

where $F^i_{kj} = \frac{\rho h}{\alpha}\frac{1}{PrR}\int_{y_i}^{y_{i+1}}\frac{d\chi_k}{dy}\frac{d\chi^i_j}{dy}dy$,

$$G^i_{kj} = -\frac{\rho h}{m}\frac{Ec}{R}\int_{y_i}^{y_{i+1}}\begin{bmatrix} \left(\frac{d\chi^i_1}{dy}\right)^2 & \left(\frac{d\chi^i_1}{dy}\right)\left(\frac{d\chi^i_2}{dy}\right) & \left(\frac{d\chi^i_1}{dy}\right)\left(\frac{d\chi^i_3}{dy}\right) \\ \left(\frac{d\chi^i_2}{dy}\right)\left(\frac{d\chi^i_1}{dy}\right) & \left(\frac{d\chi^i_2}{dy}\right)^2 & \left(\frac{d\chi^i_2}{dy}\right)\left(\frac{d\chi^i_3}{dy}\right) \\ \left(\frac{d\chi^i_3}{dy}\right)\left(\frac{d\chi^i_1}{dy}\right) & \left(\frac{d\chi^i_3}{dy}\right)\left(\frac{d\chi^i_2}{dy}\right) & \left(\frac{d\chi^i_3}{dy}\right)^2 \end{bmatrix}[\chi_k]dy$$

$H^i_{kj} = \text{-}\frac{c_s h}{DK_T}\frac{Du}{R}\int_{y_i}^{y_{i+1}}\frac{d\chi_k}{dy}\frac{d\chi^i_j}{dy}dy$, $Q_{6j}^i = \left[\frac{\rho h}{\alpha}\frac{1}{PrR}\chi_k\frac{d\theta^i_2}{dy} - \frac{c_s h}{DK_T}\frac{Du}{R}\chi_k\frac{dc^i_2}{dy}\right]_{y_i}^{y_{i+1}}$.

From Eq. (26) we get

$$\frac{h}{D}\frac{1}{ScR}\int_{y_i}^{y_{i+1}}\frac{d\chi_k}{dy}\frac{dc^i_2}{dy}dy + \frac{h}{K_T D}Sr\int_{y_i}^{y_{i+1}}\frac{d\chi_k}{dy}\frac{d\theta^i_2}{dy}dy = \left[\frac{h}{D}\frac{1}{ScR}\chi_k\frac{dc^i_2}{dy} + \frac{h}{K_T D}Sr\chi_k\frac{d\theta^i_2}{dy}\right]_{y_i}^{y_{i+1}}$$

The stiffness matrix equation corresponding to the above is

$$\left[L^i_{kj}\right]\left[c^i_k\right] + \left[M^i_{kj}\right]\left[\theta^i_k\right] = \left[Q_{7j}^i\right] \tag{33}$$

where $L^i_{kj} = \frac{h}{D}\frac{1}{ScR}\int_{y_i}^{y_{i+1}}\frac{d\chi_k}{dy}\frac{d\chi^i_j}{dy}dy$, $M^i_{kj} = \frac{h}{K_T D}Sr\int_{y_i}^{y_{i+1}}\frac{d\chi_k}{dy}\frac{d\chi^i_j}{dy}dy$

$$Q_{7j}^i = \left[\frac{h}{D}\frac{1}{ScR}\chi_k\frac{dc^i_2}{dy} + \frac{h}{K_T D}Sr\chi_k\frac{d\theta^i_2}{dy}\right]_{y_i}^{y_{i+1}}$$

The Langrange's interpolation polynomials are used as the shape functions at each of the nodes are considered as follows:

$$\eta_i^1 = \frac{\left(y-\left(\frac{2i-101}{100}\right)\right)\left(y-\left(\frac{2i-100}{100}\right)\right)}{\left(\left(\frac{2i-102}{100}\right)-\left(\frac{2i-101}{100}\right)\right)\left(\left(\frac{2i-102}{100}\right)-\left(\frac{2i-100}{100}\right)\right)},$$

$$\eta_i^2 = \frac{\left(y-\left(\frac{2i-102}{100}\right)\right)\left(y-\left(\frac{2i-100}{100}\right)\right)}{\left(\left(\frac{2i-101}{100}\right)-\left(\frac{2i-102}{100}\right)\right)\left(\left(\frac{2i-101}{100}\right)-\left(\frac{2i-100}{100}\right)\right)},$$

$$\eta_i^3 = \frac{\left(y-\left(\frac{2i-101}{100}\right)\right)\left(y-\left(\frac{2i-102}{100}\right)\right)}{\left(\left(\frac{2i-100}{100}\right)-\left(\frac{2i-102}{100}\right)\right)\left(\left(\frac{2i-100}{100}\right)-\left(\frac{2i-101}{100}\right)\right)}.$$

and similarly for $\chi_i^1, \chi_i^2, \chi_i^3$.

The shear stress values, heat (Nusselt number) and mass transfer rate (Sherwood number) are calculated at both walls as per the following relations:

$$\mathrm{St}_1 = \left[\frac{\partial u_1}{\partial y}\right]_{y=-1}, \mathrm{St}_2 = \left[\frac{\partial u_2}{\partial y}\right]_{y=1}, \mathrm{Nu}_1 = \left[\frac{\partial \theta_1}{\partial y}\right]_{y=-1}, \mathrm{Nu}_2 = \left[\frac{\partial \theta_2}{\partial y}\right]_{y=1}, \mathrm{Sh}_1 = \left[\frac{\partial c_1}{\partial y}\right]_{y=-1}, \mathrm{Sh}_2 = \left[\frac{\partial c_2}{\partial y}\right]_{y=1}.$$

5. Results and discussion

The numerical solution of the system of equations is analyzed for several values of the governing factors and its corresponding graphical representations are resulted. Thermal Grashof (Gr), Molecular Grashof (Gc) and Reynolds numbers (R), Magnetic field (M) and Material parameters (K′) and Dufour (Du), Schmidt (Sc), Soret (Sr) and Eckert numbers (Ec) are fixed as Gr = 5, Gc = 5, R = 3, M = 3,

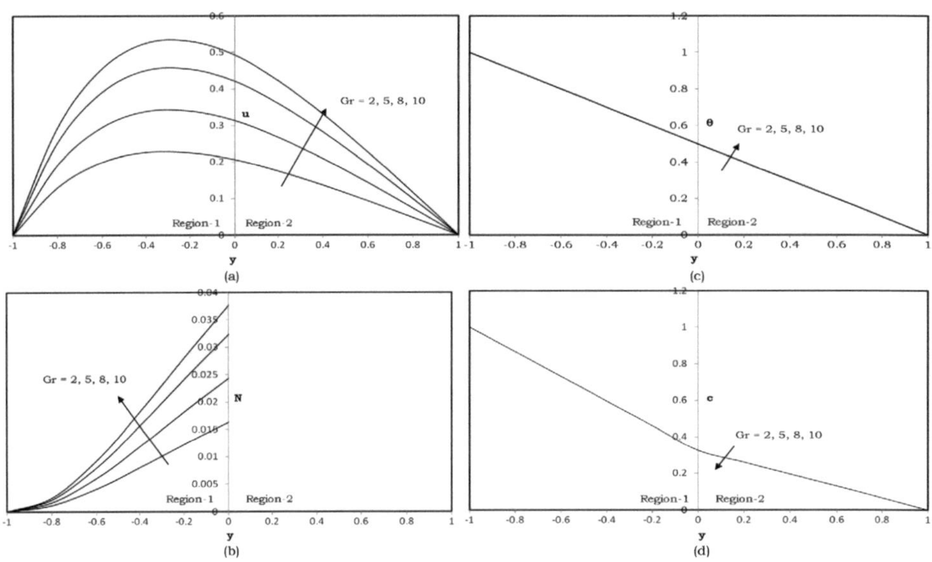

Figure 2.
(a) Represents behavior of u. (b) Represents behavior of N for Gr. (c) Represents behavior of θ. (d) Represents behavior of c for Gr.

K′ = 0.1, Du = 0.08, Sr. = 0.1, Sc = 0.66, Sr. = 0.001 for all the profiles excepting the varying parameter.

The profiles of all the governing parameters are depicted from **Figures 2–10**. The flow in micropolar region is found to be more than the flow in viscous region. The variations of linear momentum and angular momentum are clear for each and every governing parameter. The variation of temperature and diffusion are very narrow except for the parameters R, Du, Sr., and Sc. The temperature and diffusion are uniform across the channel and are found to be significant at the mid region of

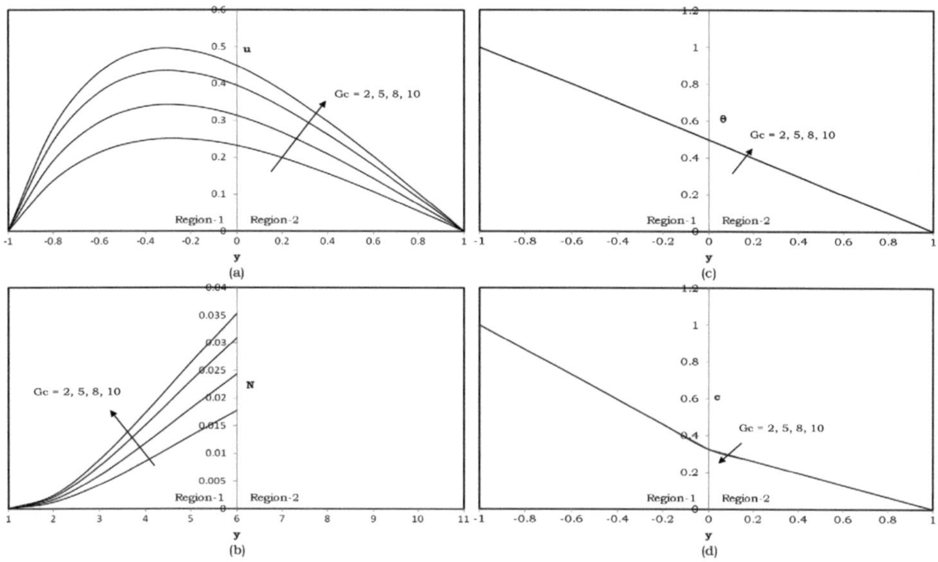

Figure 3.
(a) Represents behavior of u. (b) Represents behavior of N for Gc. (c) Represents behavior of θ. (d) Represents behavior of c for Gc.

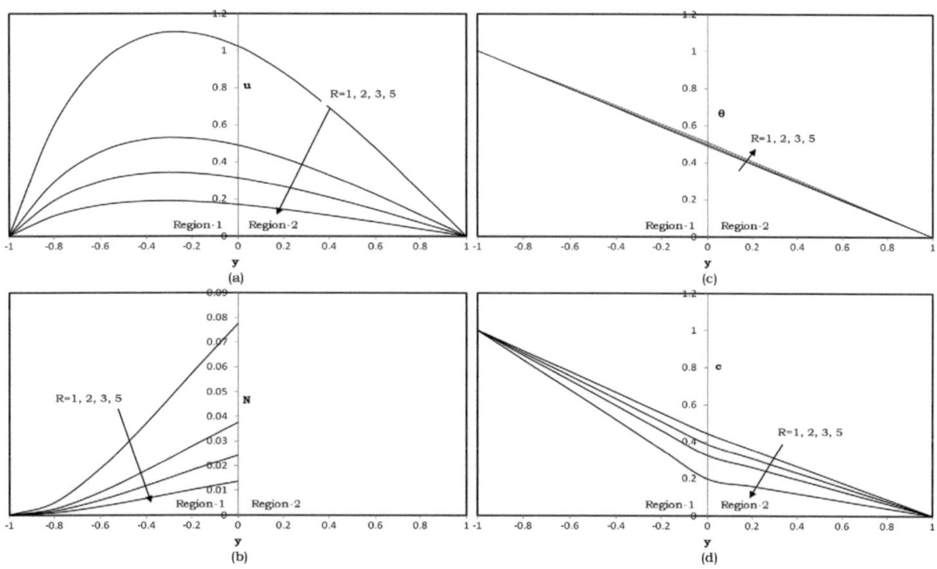

Figure 4.
(a) Represents behavior of u. (b) Represents behavior of N for R. (c) Represents behavior of θ. (d) Represents behavior of c for R.

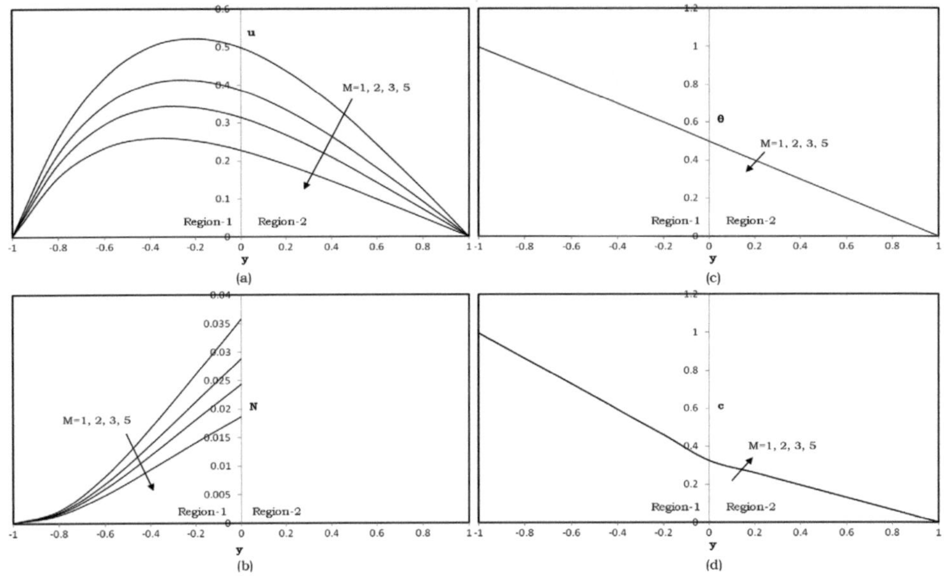

Figure 5.
(a) Represents behavior of u. (b) Represents behavior of N for M. (c) Represents behavior of θ. (d) Represents behavior of c for M.

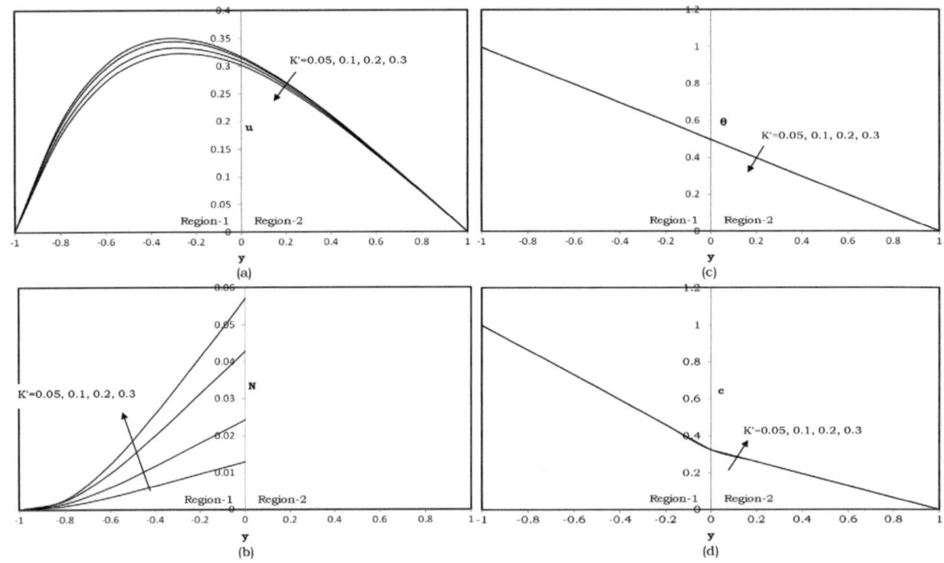

Figure 6.
(a) Represents behavior of u. (b) Represents behavior of N for K′. (c) Represents behavior of θ. (d) Concentration profiles for K′.

the channel. The diffusion is slightly effected at the interface due to two fluids. Hence the two fluid flow model has much importance in the real time systems. All our results are compared with earlier studies and they are validated.

Figure 2(a)–(d) illustrate the effect of Grashof numbers on velocity, angular velocity, temperature and diffusion. As Gr increases the velocity and angular velocity increases substantially. The buoyancy enhances the flow in both regions i.e. thermal buoyancy force dominates the viscous force in both regions of the channel and it is found to be more in micropolar region. The lowest velocity corresponds to

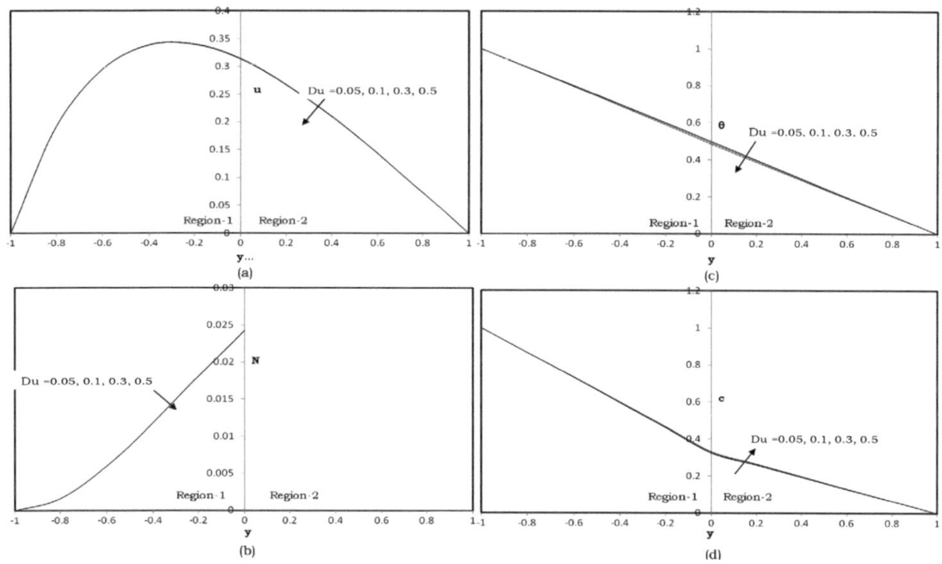

Figure 7.
(a) Represents behavior of u. (b) Represents behavior of N for Du. (c) Represents behavior of θ. (d) Represents behavior of c for Du.

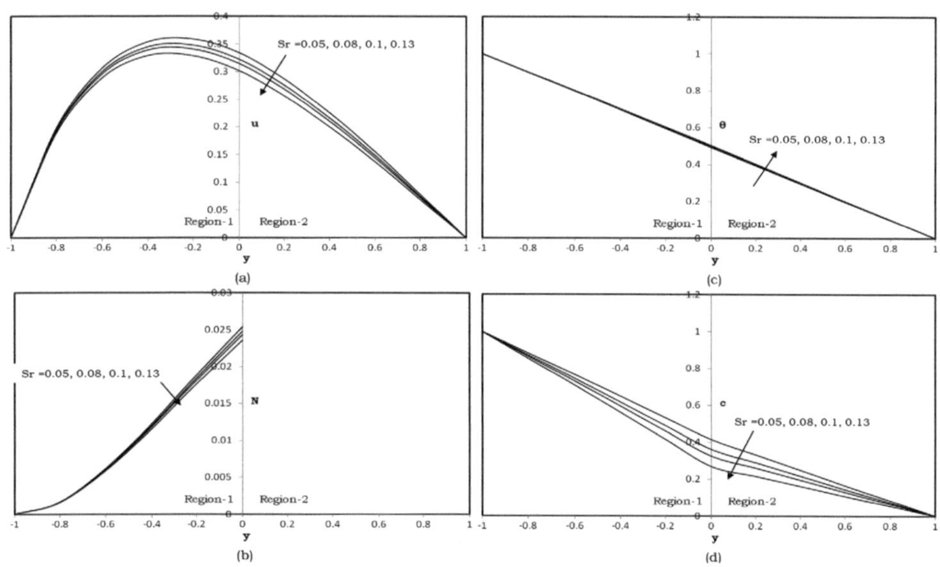

Figure 8.
(a) Represents behavior of u. (b) Represents behavior of N for Sr. (c) Represents behavior of θ. (d) Represents behavior of c for Sr.

Gr = 2. Higher Gr values boost up the flow in both regions. As Gr increases the minute enhancement of temperature and diffusion are observed. Similar observations are noticed with all the variations of Gc which are plotted in **Figure 3(a)–(d)**.

Figure 4(a)–(d) describe the Reynolds number (R) impact on velocity, angular velocity, temperature and diffusion. The reduction of velocity is found with increase of Reynolds number due to domination of inertial force on viscous force in both regions of the channel and found more drastic in viscous region. Also reduces the micro rotation with increase of Reynolds number. The effect of inertial forces enhances the temperature and reduction of the diffusion is shown with increase of R.

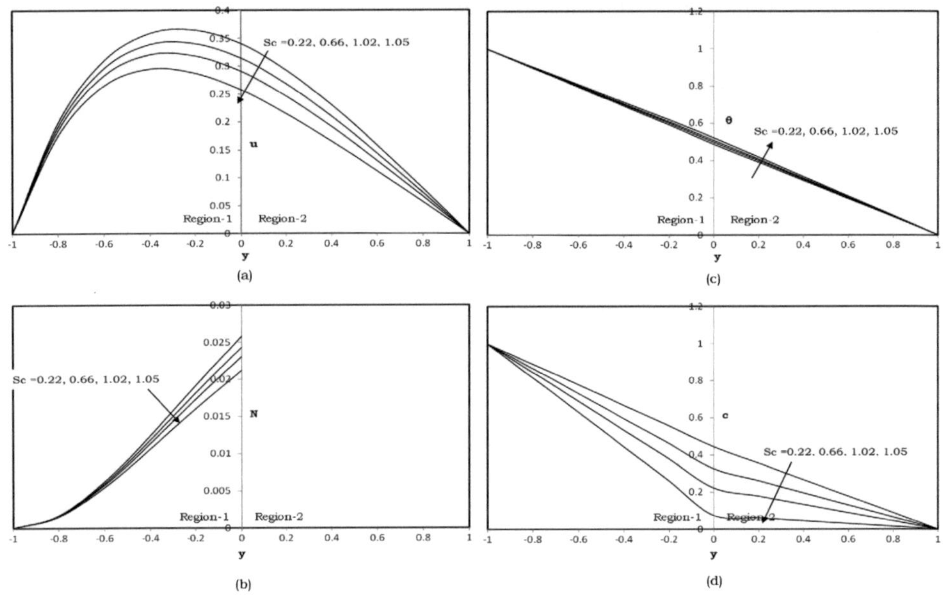

Figure 9.
(a) Represents behavior of u. (b) Represents behavior of N for Sc. (c) Represents behavior of θ. (d) Represents behavior of c for Sc.

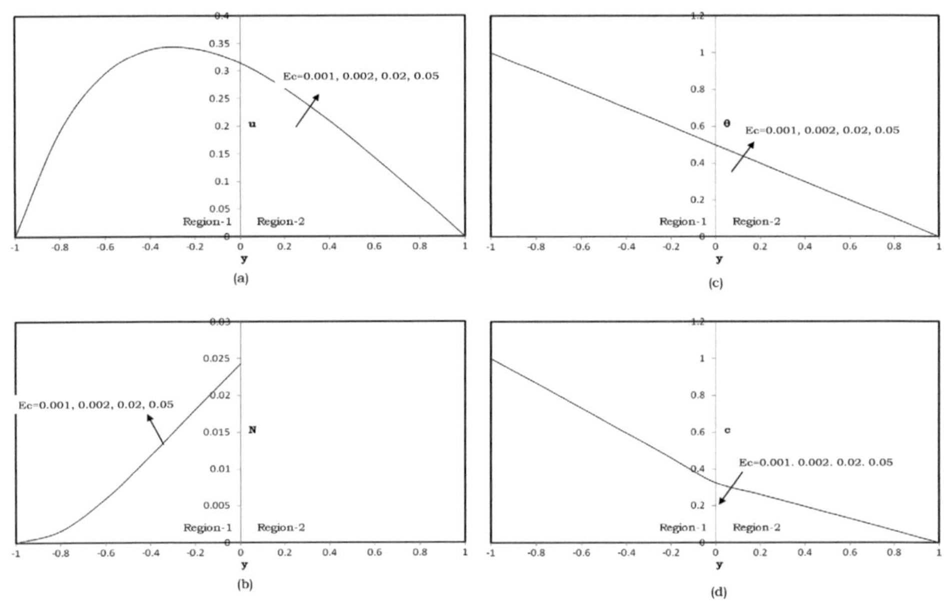

Figure 10.
(a) Represents behavior of u. (b) Represents behavior of N for Ec. (c) Represents behavior of θ. (d) Represents behavior of c for Ec.

Figure 5(a)–(d) describe the magnetic field (M) effect on velocity, angular velocity, temperature and diffusion. They portray that there could be seen reduction in velocity and angular velocity as M increases. It shows that magnetic field has a tendency to retard fluid velocity and angular velocity due to the formation of resistive Lorentz force, where when magnetic effect is applied to the fluid, it tends to retard the fluid motion. The magnetic field parametric impact on temperature and diffusion is minute.

Figure 6(a)–(d) explain the effect of material parameter (K′) on velocity, angular velocity, temperature and diffusion. The effect of this parameter is very significant in both velocity and angular velocity, as K′ increases the velocity decreases significantly and this is reversed with respect to angular velocity. Minute effect is observed for both temperature and diffusion.

Figure 7(a)–(d) explain the effect of Dufour number (Du) on velocity, angular velocity, temperature and diffusion. **Figure 7(a)** depict the fact that as Du increases, i.e. when molecular diffusivity increases, the velocity is reduced and leads to reduction of micro rotation also from **Figure 7(b)** it clearly indicates the influence of the concentration gradients to the thermal energy flux in the flow. **Figure 7(c)** specifies that the temperature reduces with increase of molecular diffusivity over the thermal diffusivity. It is clear that the diffusion profiles increase with increase of dufour number as observed in **Figure 7(d)**.

Figure 8(a)–(d) relates the Soret number (Sr) impact on velocity, angular velocity, temperature and diffusion. **Figure 8(a)** shows that as Sr. increases i.e. thermal diffusivity increases the decrease in velocity is found and micro rotation also decreases with increase of Sr. from **Figure 8(b)**. Soret number states the impact of temperature gradients stimulating considerable mass diffusion effects. Here, as Soret number increases it leads to rise in temperature and shows the decay in the fluid concentration from **Figure 8(c)** and **(d)**.

Figure (a)–(d) specify the Schmidt number (Sc) effect on velocity, angular velocity, temperature and diffusion. **Figure 9(a)** and **(b)** show that as Sc increases, velocity and angular velocity decreases significantly. From **Figure 9(c)** and **(d)** it is found that the temperature increases with increase of Sc and fluid concentration reduces with increase in Schmidt number.

Figure 10(a)–(d) define the effect of Eckert number (Ec) on velocity, angular velocity, temperature and diffusion. From all the Figures it is concluded that the enthalpy is not having much influence over the flow for small variation of enthalpy. **Figure 10(a)** and **(b)** show that the velocity and angular velocity increase when Eckert number increases. So it is observed that momentum and angular momentum are inversely proportional to enthalpy. From **Figure 10(c)** the temperature increases with increase of kinetic energy. The kinetic energy reduces the concentration of the fluid as shown from **Figure 10(d)**.

Table 1 shows the Shear stress and Nusselt and Sherwood numbers values with all the effects of all governing functions. From this table, it is observed that the absolute Shear stress enhances with increase in Gr on both the boundaries $y = -1$ and $y = 1$ because of buoyancy forces and similar nature is observed for Gc also. For increase of Reynolds number, magnetic field and Material parameter and Dufour, Soret and Schmidt numbers, the stress reduces on both the boundaries. This case is reversed for dissipation effect. The Nusselt number i.e. rate of heat transfer decreases on the boundary at $y = -1$ and increases on the other boundary $y = 1$ for the parameters Gr, Gc, R, Sr., Sc and Ec. The rise in convection is leading to reduction of heat transfer rate on the plate bounding the region -1, the reverse effect is observed for boundary of the region-2. Drastic heat transfer rate is observed for the variations of the Reynolds Number. The increase in the Reynolds number decreases the heat transfer rate on the left plate and enhances on the right plate. For the other parameters M, K′, Du the effect is reversal. The Sherwood number i.e. rate of mass transfer increase on the boundary at $y = -1$ and decrease at the boundary $y = 1$ for the parameters Gr, Gc, R, Sr., Sc, Ec. This is because the rise in convection and inertial forces leading to enhance the concentration. For the other parameters M, K′, Du the effect is reversal i.e. mass transfer increases at the left boundary and decreases at the right.

Gr	St-I	St-II	Nu-I	Nu-II	Sh-I	Sh-II
2	−0.830728	0.238773	0.501802	0.498151	0.674487	0.325508
5	−1.21308	0.367972	0.501705	0.498196	0.674512	0.325473
8	−1.59544	0.497188	0.501571	0.49826	0.674547	0.325425
10	−1.85037	0.583344	0.50146	0.498312	0.674576	0.325385
Gc	***St-I***	***St-II***	***Nu-I***	***Nu-II***	***Sh-I***	***Sh-II***
2	−0.867571	0.276385	0.50179	0.498161	0.674491	0.3255
5	−1.21308	0.367972	0.501705	0.498196	0.674512	0.325473
8	−1.55858	0.459553	0.501593	0.498243	0.674541	0.325438
10	−1.78891	0.520603	0.501504	0.498279	0.674564	0.32541
R	***St-I***	***St-II***	***Nu-I***	***Nu-II***	***Sh-I***	***Sh-II***
1	−3.75411	1.2211	0.51129	0.487749	0.556265	0.443686
2	−1.84876	0.581703	0.507141	0.492627	0.614305	0.385672
3	−1.21308	0.367972	0.501705	0.498196	0.674512	0.325473
5	−0.703233	0.195659	0.489827	0.51014	0.80178	0.198213
M	***St-I***	***St-II***	***Nu-I***	***Nu-II***	***Sh-I***	***Sh-II***
1	−1.56381	0.632044	0.50152	0.498322	0.674563	0.325375
2	−1.35498	0.468803	0.50164	0.498238	0.67453	0.325441
3	−1.21308	0.367972	0.501705	0.498196	0.674512	0.325473
5	−1.02772	0.252261	0.501772	0.498159	0.674494	0.325503
K′	***St-I***	***St-II***	***Nu-I***	***Nu-II***	***Sh-I***	***Sh-II***
0.05	−1.25054	0.369786	0.501697	0.498198	0.674514	0.325472
0.1	−1.21308	0.367972	0.501705	0.498196	0.674512	0.325473
0.2	−1.1461	0.364131	0.501719	0.498193	0.674509	0.325476
0.3	−1.08773	0.360146	0.501732	0.49819	0.674506	0.325478
Du	***St-I***	***St-II***	***Nu-I***	***Nu-II***	***Sh-I***	***Sh-II***
0.08	−1.21308	0.367972	0.501705	0.498196	0.674512	0.325473
0.1	−1.213	0.36789	0.502207	0.497695	0.674236	0.325748
0.3	−1.21205	0.366921	0.50812	0.491804	0.670989	0.328973
0.5	−1.21073	0.365583	0.516295	0.483705	0.666499	0.333392
Sr	***St-I***	***St-II***	***Nu-I***	***Nu-II***	***Sh-I***	***Sh-II***
0.05	−1.24193	0.397433	0.510133	0.489762	0.584941	0.415051
0.08	−1.22481	0.379948	0.505131	0.494768	0.638104	0.361884
0.1	−1.21308	0.367972	0.501705	0.498196	0.674512	0.325473
0.13	−1.19499	0.349512	0.496423	0.503481	0.730638	0.269345
Sc	***St-I***	***St-II***	***Nu-I***	***Nu-II***	***Sh-I***	***Sh-II***
0.22	−1.25121	0.406909	0.512844	0.487049	0.556131	0.443864
0.66	−1.21308	0.367972	0.501705	0.498196	0.674512	0.325473
1.02	−1.17975	0.333945	0.491969	0.507938	0.777967	0.222014
1.5	−1.13199	0.285191	0.478018	0.521897	0.926194	0.073786
Ec	***St-I***	***St-II***	***Nu-I***	***Nu-II***	***Sh-I***	***Sh-II***

Gr	St-I	St-II	Nu-I	Nu-II	Sh-I	Sh-II
0.001	−1.21308	0.367972	0.501705	0.498196	0.674512	0.325473
0.002	−1.21309	0.367979	0.501527	0.498276	0.674559	0.325413
0.02	−1.21331	0.368093	0.49832	0.499703	0.675391	0.324325
0.05	−1.21367	0.368285	0.492968	0.502086	0.676779	0.322509

Table 1.
Shear stress, Nusselt number, Sherwood numbers.

6. Conclusions

- Significant effect of Soret number on momentum, diffusion are observed
- Diffusion effects are reducing the momentum and are more pronounced in concentration.
- For enhancement of inertial force the momentum and diffusion reduces to a higher extent in case of micropolar region than viscous region.
- The diffusion parameters are reducing the magnitude of the shear stress on both the boundaries.
- For dufour heat transfer rate effect is enhancing on the hot plate and reducing on the cold plate but reverse effect is observed for rate on mass transfer. Exactly opposite is observed for Soret number.
- Reduction of heat transference rate on hot plate and enhancement on the cold plate is observed due to viscous dissipation.

Author details

Suresh Babu Baluguri[1]* and G. Srinivas[2]

1 Department of Mathematics, Sreyas Institute of Engineering and Technology, Hyderabad, 500 068, Telangana, India

2 Department of Mathematics, Guru Nanak Institute of Engineering and Technology, Hyderabad, 500 009, Telangana, India

*Address all correspondence to: bsureshmaths@gmail.com

References

[1] L. Euler, General principles of the motion of fluids, Physica. D, Vol. 237 (14), 2008, pp. 1825–1839.

[2] H. Rouse, S. Ince, History of Hydraulics, Iowa Inst, Hydraulic Research, Vol. 269, 1957.

[3] G.A.Tokaty, A History and Philosophy of Fluid Mechanics, GT Foulis & Co. Ltd., Oxfordshire, 1971.

[4] A.C.Eringen, Simple Microfluids, International Journal of Engineering Science,Vol 2(2), 1964, pp. 205–217.

[5] A.C.Eringen, Theory Of Thermomicrofluids, Journal of Mathematical Analysis and Applications, Vol. 38(2), 1972, pp. 480–496.

[6] J. Peddieson and R.P. McNitt, Boundary layer theory for a micropolar fluid, Adv. Eng. Sci.Vol.5,1970, pp. 405.

[7] T.Ariman, M.A.Turk, N.D.Sylvester, Application of microcontinum fluid mechanics, International Journal of Engineering Science, Vol.12, 1974, pp. 273–293.

[8] G. Lukaszewicz, Micropolar fluids: theory and application. Birkhäuser, Basel. 1999.

[9] A.C. Eringen, Microcontinuum field theories II: fluent media. Springer, NewYork, 2001.

[10] A.J. Chamkha, T.Groşan and I. Pop, Fully developed free convection of a micropolar fluid in a vertical channel, International Communications in Heat and Mass Transfer, Vol 29(8), 2002, pp. 1119–1127.

[11] Packham, B. A. and R. Shall, Stratified laminar flow of two immiscible fluids, Mathematical Proceedings of the Cambridge Philosophical Society. Vol. 69(3), Cambridge University Press, 1971.

[12] R. Shail, On laminar two-phase flows in magnetohydrodynamics, International Journal of Engineering Science, Vol.11 (10), 1973, pp. 1103–1108.

[13] C. Beckermann, S.Ramadhyani, R. Viskanta, Natural convection flow and heat transfer between a fluid layer and a porous layer inside a rectangular enclosure, ASME Journal of Heat Transfer Vol.109(2), 1987, pp. 363–370.

[14] J.Lohrasbi and V. Sahai, Magnetohydrodynamic heat transfer in two-phase flow between parallel plates, Applied Scientific Research, Vol. 45(1), 1988, pp. 53–66.

[15] B.Gebhart, Effects of viscous dissipation in natural convection, Journal of Fluid Mechanics, Vol. 14(02), 1962, pp. 225–232.

[16] B. Gebhart and J. Mollendorf, Viscous dissipation in external natural convection flows, Journal of fluid Mechanics, Vol.38(01), 1969, pp. 97–107.

[17] M.Fand and James Brucker, A correlation for heat transfer by natural convection from horizontal cylinders that account for viscous dissipation, International Journal of Heat and Mass Transfer, Vol. 26(5), 1983, pp. 709–716.

[18] M.Fand, T.E.Steinberger and P. Cheng, Natural convection heat transfer from a horizontal cylinder embedded in a porous medium, International Journal of Heat and Mass Transfer, Vol. 29(1), 1986, pp.119–133.

[19] A. Nakayama and I. Pop, Free convection over a non isothermal body in a porous medium with viscous dissipation, International

communications in heat and mass transfer, Vol.16(2),1989, pp.173–180.

[20] P.V.S.N.Murthy and P. Singh, Effect of viscous dissipation on a non-Darcy natural convection regime, International journal of heat and mass transfer, Vol 40(6), 1997, pp. 1251–1260.

[21] El-Amin. M. F., Combined effect of viscous dissipation and Joule heating on MHD forced convection over a non-isothermal horizontal cylinder embedded in a fluid saturated porous medium, Journal of Magnetism and Magnetic materials, Vol. 263(3), 2003, pp.337–343.

[22] Bejan.A and S.Lorente, The constructal law and the thermodynamics of flow systems with configuration, International journal of heat and mass transfer, Vol. 47(14), 2004, pp. 3203–3214.

[23] A.Pantokratoras, Effect of viscous dissipation in natural convection along a heated vertical plate, Applied Mathematical Modelling, Vol. 29(6), 2005, pp. 553–564.

[24] Seddeek. M. A. and M. S. Abdelmeguid, Effects of radiation and thermal diffusivity on heat transfer over a stretching surface with variable heat flux, Physics Letters A, Vol.348 (3), 2006, pp. 172–179.

[25] Duwairi.H.M, Osama Abu-Zeid and A. Damesh Rebhi, Viscous and Joule heating effects over an isothermal cone in saturated porous media, Jordan Journal of Mechanical and Industrial Engineering, Vol.1(2), 2007, pp.113–118.

[26] Cortell Rafael, Effects of viscous dissipation and radiation on the thermal boundary layer over a nonlinearly stretching sheet, Physics Letters A, Vol. 372(5), 2008, pp. 631–636.

[27] Kairi.R. R. and P. V. S. N. Murthy, Effect of viscous dissipation on natural convection heat and mass transfer from vertical cone in a non-Newtonian fluid saturated non-Darcy porous medium, Applied mathematics and computation, Vol. 217(20), 2011, pp. 8100–8114.

[28] Cortell Rafael, Heat and fluid flow due to non-linearly stretching surfaces, Applied Mathematics and Computation, Vol. 217(19), 2011, pp. 7564–7572.

[29] Chapman Sydney and Thomas George Cowling, The Mathematical Theory of Non-uniform Gases: An Account of the Kinetic Theory of Viscosity, Thermal Conduction and Diffusion of Gases, Notes Added in 1951. Cambridge university press, 1952.

[30] Eckert RG and Robert M Drake, Analysis of Heat Transfer, McGraw-Hill, 1972.

[31] Kafoussias N.G and E.W.Williams, Thermal-diffusion and diffusion-thermo effects on mixed free-forced convective and mass transfer boundary layer flow with temperature dependent viscosity, International Journal of Engineering Science, Vol. 33(9), 1995, pp. 1369–1384.

[32] Anghel M, H. S. Takhar and I. Pop. Dufour and Soret effects on free convection boundary layer over a vertical surface embedded in a porous medium. Studia Universitatis Babes-Bolyai, Mathematica, Vol. 45(4), 2000, pp.11–21.

[33] Postelnicu Adrian, Influence of a magnetic field on heat and mass transfer by natural convection from vertical surfaces in porous media considering Soret and Dufour effects, International Journal of Heat and Mass Transfer, Vol. 47(6), 2004, pp. 1467–1472.

[34] Alam M. S and M. M. Rahman, Dufour and Soret effects on mixed convection flow past a vertical porous flat plate with variable suction, Nonlinear Analysis: Modelling and Control, Vol.11(1),2006, pp.3–12.

[35] A.J. Chamkha and Abdullatif Ben-Nakhi, MHD mixed convection–radiation interaction along a permeable surface immersed in a porous medium in the presence of Soret and Dufour's effects, Heat and Mass transfer, Vol. 44 (7), 2008, pp. 845.

[36] El-Aziz, Thermal-diffusion and diffusion-thermo effects on combined heat and mass transfer by hydromagnetic three-dimensional free convection over a permeable stretching surface with radiation, Physics Letters A, Vol . 372(3), 2008, pp.263–272.

[37] Maleque Abdul, Effects of combined temperature-and depth-dependent viscosity and Hall current on an unsteady MHD laminar convective flow due to a rotating disk, Chemical Engineering Communications, Vol. 197 (4), 2009, pp. 506–521.

[38] O.Anwar Beg, A .Y. Bakier and V. R. Prasad, Numerical study of free convection magnetohydrodynamic heat and mass transfer from a stretching surface to a saturated porous medium with Soret and Dufour effects, Computational Materials Science, Vol. 46(1), 2009, pp. 57–65.

[39] Pal Dulal and Sewli Chatterjee, Mixed convection magnetohydrodynamic heat and mass transfer past a stretching surface in a micropolar fluid-saturated porous medium under the influence of Ohmic heating, Soret and Dufour effects, Communications in Nonlinear Science and Numerical Simulation, Vol. 16 (3), 2011, pp. 1329–1346.

[40] Z.M.Stamenkovic, D.D.Nikodijevic, M.M. Kocic and J.D. Nikodijević, Mhd flow and heat transfer of two immiscible fluids with induced magnetic field effects, Thermal Science, 16(suppl. 2),2012, pp.323–336.

[41] M.S. Malashetty and V. Leela, Magnetohydrodynamic heat transfer in two fluid flow, Proc. of National Heat Transfer conferences sponsored by AIChE and ASME– HTD, Phase Change Heat Transfer, Vol. 159, 1991, pp.171–175.

[42] M.S. Malashetty and V. Leela, Magnetohydrodynamic heat transfer in two phase flow, International Journal of Engineering Science, Vol. 30, 1992, pp.371–377.

[43] K .Vajravelu, P.V.Arunachalam and S.Sreenadh, Unsteady flow of two immiscible conducting fluids between two permeable beds, Journal of mathematical analysis and applications, Vol.196(3), 1995, pp. 1105–1116.

[44] M. S. Malashetty, J. C. Umavathi, Two-phase magneto hydrodynamicflow and heat transfer in an inclined channel. Int J Multiphase Flow, Vol. 22, 1997, pp. 545–560.

[45] V. Srinivasan and K.Vafai, Analysis of Linear Encroachment in Two-Immiscible Fluid Systems, ASME Journal of Fluid Engineering, Vol. 116,1994, pp. 135–139.

[46] M. S. Malashetty, J. C. Umavathi and J. Prathap Kumar, Convective magnetohydrodynamic two fluid flow and heat transfer in an inclined channel, Heat and Mass Transfer Vol. 37(2), 2001, pp. 259–264.

Chapter 6

A Dynamic Finite Element Cellular Model and Its Application on Cell Migration

Jieling Zhao

Abstract

While the tissue is formed or regenerated, cells migrate collectively and remained adherent. However, it is still unclear what are the roles of cell-substrate and intercellular interactions in regulating collective cell migration. In this chapter, we introduce our newly developed finite element cellular model to simulate the collective cell migration and explore the effects of mechanical feedback between cells and between cell and substrate. Our viscoelastic model represents one cell with many triangular elements. Intercellular adhesions between cells are represented as linear springs. Furthermore, we include a mechano-chemical feedback loop between cell-substrate mechanics and cell migration. Our results reproduce a set of experimental observation of patterns of collective cell migration during epithelial wound healing. In addition, we demonstrate that cell-substrate determined mechanics play an important role in regulating persistent and oriented collective cell migration. This chapter illustrates that our finite element cellular model can be applied to study a number of tissue related problems regarding cellular dynamic changes at subcellular level.

Keywords: finite element model, collective cell migration, cell-substrate mechanics, intercellular adhesion, model developing

1. Introduction

Thanks to the accurate description of changes in material mechanics, finite element method has been widely used in the field of bioengineering to study cellular tissue related problems such as neurulation and epithelial mechanics [1, 2]. However, majority of current finite element models are only restricted on tissues undergoing changes of shapes and displacements at small scale. In addition, during the simulation, the cellular tissue is required to be remained as one integrity. These limitations restrict the traditional finite element method to be applied to study the essential physiological processes such as morphogenesis, tissue regeneration, tumor metastasis, and cancer invasion, where cells often migrate collectively as large coherent strands or tubes. Such large scale of collective cell movement is recognized as the hallmark of tissue-remodeling events. During the past decade, to overcome the limitation of traditional finite element method, dynamic finite element method such as PFEM has been developed to extend the traditional FEM to study mechanics of materials with more flexibility or undergoing larger scale of motility. The object domain (either fluid or solid) is represented as nodes tessellated by triangular mesh.

The mathematical equations governing the physical rules of the mechanical property of the discretized domain defined by the mesh connecting nodes are subsequentially solved in the standard FEM. Under the analysis using dynamic finite element method, the motion of sub-domain of the object can freely move and even separate from the main domain [3]. The advancement of dynamic finite element in achieving both accurate description of material mechanics and large scale of geometric and topological changes makes it suitable to simulate the physiological processes such as wound healing and cancer invasion. During these physiological processes, cells move in collective fashion and respond with chemical and mechanical signals through cell–cell junctions and interactions between cells and their micro-environment.

In this chapter, we introduced our newly developed dynamic finite element cellular model and its application to study the influence of cell-substrate mechanics and intercellular adhesions on collective cell migration. Our model represents each cell as a mesh of triangular elements at sub-cellular level [4]. Each triangular element exhibits viscoelastic characteristic using a Maxwellian model [5]. The effects of line tension forces along the cell boundary according to the local curvature is incorporated [6]. The intercellular adhesions are modeled as elastic springs at sub-cellular scale [7]. In addition, a mechano-chemical feedback pathway including focal adhesion, proteins of Paxillin, Rac, PAK, and Merlin, which are all responsible for cell protrusion [8] is embedded in individual cell. This pathway is collaborated with another mechano-chemical pathway, which is responsible for transmitting mechanical cue through intercellular adhesions [9]. Our model is used to study collective cell migration using a simplified wound tissue. We then compare our simulation results to an *in vitro* study [10]. Finally, we discussed and made the conclusion that the mechanics between cell-substrate play a crucial role in guiding highly efficient collective cell migration. This guidance cue is well maintained and transmitted between cells through the intercellular adhesions.

2. Methods

2.1 Cell geometry

In our model, a cell in 2D $\Omega \subset \mathbb{R}^2$ is represented as an oriented polygon including a number of boundary vertices $V_{\partial\Omega} \equiv \{v_i \in \partial\Omega \subset \mathbb{R}^2\}$, where the location of the vertex $\boldsymbol{v}_i$ is denoted as $\boldsymbol{x}_i$. The set of boundary vertices $V_{\partial\Omega}$, together with a set of internal vertices V_{Int} and a set of triangular elements $T_\Omega \equiv \{\tau_{i,j,k} : \boldsymbol{v}_i, \boldsymbol{v}_j, \boldsymbol{v}_k \in V_{\partial\Omega} \cup V_{\text{Int}}\}$ define the geometry of cell Ω (**Figure 1a**). If two cells are closely in contact, a set of adhesive springs are generated between them (**Figure 1a**, red bars in the dashed blue box). There are several interior vertice on each cell boundary edge. They are evenly distributed along that edge. These interior vertice are the potential locations for newly generated adhesive spring to attach on. Any force applied on that interior vertex through the attached adhesive spring will be mapped onto its nearest end-node vertex of the corresponding boundary edge.

2.2 Viscoelasticity of the cell

Previous researches have demonstrated that the cell cytoskeleton exhibits viscoelastic characteristic [14, 15]. Following the studies of [16, 17], we assume that, during cell deformation and cell migration, linear viscoelasticity is adequate to describe the mechanical properties of the cell.

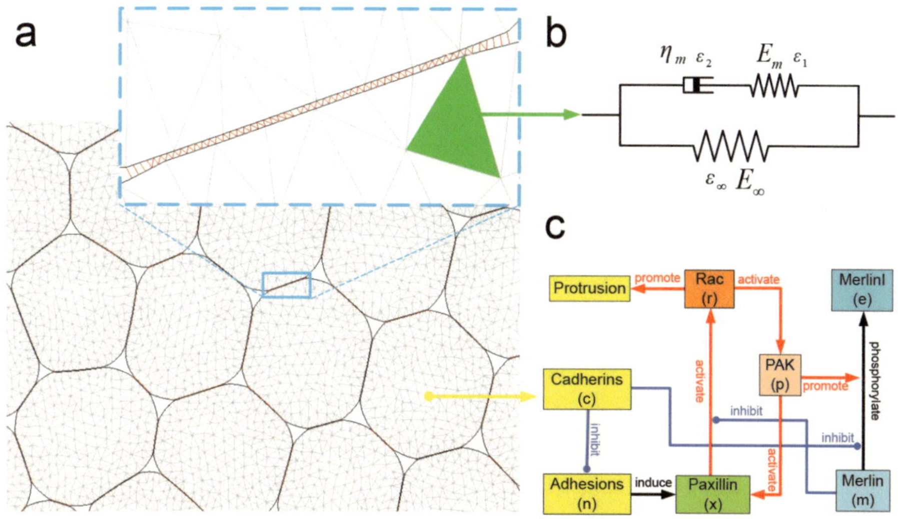

Figure 1.
The cell geometry and the chemical pathway between cell-substrate and intercellular adhesion. (a) the cell in our model is represented as following: The cell boundary is defined by an oriented polygon including a number of boundary vertices. A triangular mesh tiling up a cell is generated based on the method of farthest sampling [11]. The E-cadhesion type of intercellular adhesions between two neighboring cells are represented as elastic springs (red bars in the blue box, the dashed blue box is for a closer view). (b) each triangular element exhibits viscoelastic characteristic using a generalized Maxwell model following [5, 12, 13]. (c) the positive feedback loop between focal adhesion and cell protrusion is built up in each vertex of the triangular mesh following [8]. Such network includes the proteins of integrin, Paxillin, Rac, and PAK. The protein Merlin on the cadherin is also included to count the effects of intercellular adhesion on cell migration [9].

2.2.1 Strain and stress tensors

We use the strain tensor $\boldsymbol{\varepsilon}(\boldsymbol{x},t)$ to describe the local cell deformation at $\boldsymbol{x}$ at time t. $\boldsymbol{\varepsilon}(\boldsymbol{x},t)$ takes the form of $\varepsilon_{1,1} = \partial u_1/\partial x_1$, $\varepsilon_{2,2} = \partial u_2/\partial x_2$, and $\varepsilon_{1,2} = \varepsilon_{2,1} = \frac{1}{2}(\partial u_1/\partial x_2 + \partial u_2/\partial x_1)$, where $\boldsymbol{u}(\boldsymbol{x},t)$ defined as $(u_1(\boldsymbol{x},t), u_2(\boldsymbol{x},t))^T \subset \mathbb{R}^2$ is the displacement of $\boldsymbol{x}$ at time t. We use the stress tensor $\boldsymbol{\sigma}(\boldsymbol{x},t)$ to describe the local forces at $\boldsymbol{x}$ at time t. Here $\boldsymbol{\sigma}$ is correlated with $\boldsymbol{\varepsilon}$ by a generalized Maxwell model: $\boldsymbol{\sigma}(\boldsymbol{x},t) = \boldsymbol{\sigma}_\infty(\boldsymbol{x},t) + \boldsymbol{\sigma}_m(\boldsymbol{x},t)$ [5, 12, 13], where $\boldsymbol{\sigma}_\infty(\boldsymbol{x},t)$ is the stress of the long-term elastic element and $\boldsymbol{\sigma}_m(\boldsymbol{x},t)$ is the stress of the Maxwell elastic element. E_∞, E_m, and η_m denote the long-term elastic modulus, elastic modulus of the Maxwell elastic element, and viscous coefficient of the Maxwell viscous element, respectively (**Figure 1b**). The strain of the Maxwell elastic element $\boldsymbol{\varepsilon}_1(\boldsymbol{x},t)$ and the strain of the viscous element $\boldsymbol{\varepsilon}_2(\boldsymbol{x},t)$ sum up to the strain tensor $\boldsymbol{\varepsilon}(\boldsymbol{x},t)$: $\boldsymbol{\varepsilon}_1(\boldsymbol{x},t) + \boldsymbol{\varepsilon}_2(\boldsymbol{x},t) = \boldsymbol{\varepsilon}(\boldsymbol{x},t)$.

We assume that the total free energy of a cell is the summation of its elastic energy, its adhesion energy due to the contact with the substrate, its elastic energy due to the intercellular adhesions with neighboring cells, and its energy due to the forces exerting on the boundary.

2.2.2 Cell elastic energy

The elastic energy due to the deformation of the cell Ω is given by

$$E_\Omega(t) = \frac{1}{2}\int_\Omega \left(\boldsymbol{\sigma}_\infty(\boldsymbol{x},t) + \sigma_a \boldsymbol{\delta}_{ij}(\boldsymbol{x})\right)^T \boldsymbol{\varepsilon}(\boldsymbol{x},t) d\boldsymbol{x} + \frac{1}{2}\int_\Omega \boldsymbol{\sigma}_m(\boldsymbol{x},t)^T \boldsymbol{\varepsilon}_1(\boldsymbol{x},t) d\boldsymbol{x}, \tag{1}$$

where σ_a is a homogeneous contractile pressure following [6].

2.2.3 Cell adhesion energy due to the contact with the substrate

The energy due to the adhesion between the cell and the substrate is given by [6].

$$\frac{Y(\boldsymbol{x},t)}{2}\int_{\Omega}\boldsymbol{u}(\boldsymbol{x},t)^2 d\boldsymbol{x}, \tag{2}$$

where $Y(\boldsymbol{x},t)$ is the adhesion coefficient at time t and is proportional to the strength of local focal adhesions [18]: $Y(\boldsymbol{x},t) = \frac{n_{x,t}}{n_0}E_{st}Y_a$, where $n_{x,t}$ is the number of bound integrins at location $\boldsymbol{x}$ at t (more details of calculating $n_{x,t}$ in **Model of focal adhesion**), n_0 is a normalized constant number, E_{st} is the stiffness of the substrate and Y_a is the basic adhesion constant following [18].

2.2.4 Cell adhesion energy due to intercellular adhesion

The energy due to the intercellular adhesions, which are modeled as elastic springs, is given by $\frac{1}{2}\sum_l k_l \boldsymbol{u}_l(t)^2$, where k_l is the spring constant of the spring l. Its orientation angle at time t is denoted as $\theta_l(t)$. Its transformation vector $T(\theta)$ is denoted as $(\cos(\theta_l), \sin(\theta_l), -\cos(\theta_l), -\sin(\theta_l))$. So $\boldsymbol{u}_l(t)$ can be written as $\boldsymbol{u}_l(t) = T(\theta_l)(\boldsymbol{u}_{l1}(t), \boldsymbol{u}_{l2}(t))$, where $\boldsymbol{u}_{l1}(t)$ and $\boldsymbol{u}_{l2}(t)$ are the displacements of the two end-node vertice $\boldsymbol{x}_1$ and $\boldsymbol{x}_2$ of l at time t. The elastic force of l due to displacement of Δl is applied on $\boldsymbol{x}_i$ and $\boldsymbol{x}_j$ as $\boldsymbol{f}_l = f(\Delta l)\boldsymbol{e}_l$ and $-f(\Delta l)\boldsymbol{e}_l$, respectively, where $f(\Delta l)$ is the magnitude of $\boldsymbol{f}_l$, $\boldsymbol{e}_l$ is the unit vector of the orientation of l.

2.2.5 Boundary and protrusion forces

Furthermore, the local forces applied on the cell boundary also contribute to the energy. Following [6], the tension force along the cell boundary is considered. In addition, we also incorporate the protrusion force on the leading edge of migrating cell. The contribution of these two forces can be written as

$$\int_{\partial\Omega}(\boldsymbol{\lambda}(\boldsymbol{x},t)+\boldsymbol{f}(\boldsymbol{x},t))\boldsymbol{u}(\boldsymbol{x},t)d\boldsymbol{x}, \tag{3}$$

where $\boldsymbol{\lambda}(\boldsymbol{x},t)$ is the line tension force and $\boldsymbol{f}(\boldsymbol{x},t)$ is the protrusion force. Line tension force is written as $\boldsymbol{\lambda}(\boldsymbol{x},t) = -f_m\kappa(\boldsymbol{x},t)\boldsymbol{n}(\boldsymbol{x},t)$, where f_m is a contractile force per unit length, $\kappa(\boldsymbol{x},t)$ is the curvature, and $\boldsymbol{n}(\boldsymbol{x},t)$ is the outward unit normal at $\boldsymbol{x}$ at time t [6]. Protrusion force is denoted as $f(\boldsymbol{x},t) = -f_a\boldsymbol{n}(\boldsymbol{x},t)$, where f_a is the protrusion force per unit length.

2.2.6 The total free energy and its dissipation

In summary, the total free energy of cell Ω at time t is given by

$$E_\Omega(t) = \frac{1}{2}\int_\Omega\left(\left(\boldsymbol{\sigma}_\infty(\boldsymbol{x},t)+\boldsymbol{\sigma}_a\delta_{ij}\right)^T\boldsymbol{\varepsilon}(\boldsymbol{x},t)+\boldsymbol{\sigma}_m^T(\boldsymbol{x},t)\boldsymbol{\varepsilon}_1(\boldsymbol{x},t)\right)d\boldsymbol{x}+\frac{1}{2}\int_\Omega(\sigma_a,\sigma_a,0)\boldsymbol{\varepsilon}(\boldsymbol{x},t)d\boldsymbol{x}$$
$$+\frac{Y(\boldsymbol{x},t)}{2}\int_\Omega\boldsymbol{u}(\boldsymbol{x},t)^2d\boldsymbol{x}+\frac{1}{2}\sum_l k_l\boldsymbol{u}_l(t)^2+\int_{\partial\Omega}(\boldsymbol{\lambda}(\boldsymbol{x},t)+\boldsymbol{f}(\boldsymbol{x},t)+\boldsymbol{f}_l(\boldsymbol{x},t))\boldsymbol{u}(\boldsymbol{x},t)d\boldsymbol{x}). \tag{4}$$

The energy dissipation of $E_\Omega(t)$ due to cell viscosity is determined by the viscous coefficient η_m and the strain of the viscous element $\boldsymbol{\varepsilon}_2(\boldsymbol{x},t)$ [19]: $-\int_\Omega \eta_m \left(\frac{\partial \varepsilon_2}{\partial t}\right)^2 d\boldsymbol{x}$. The dissipation of the total free energy of the cell can be written as

$$\begin{aligned}\frac{\partial}{\partial t}E_\Omega(t) &= \frac{\partial}{\partial t}\Big(\frac{1}{2}\int_\Omega\Big(\left(\boldsymbol{\sigma}_\infty(\boldsymbol{x},t)+\boldsymbol{\sigma}_a\delta_{ij}\right)^T\boldsymbol{\varepsilon}(\boldsymbol{x},t)+\boldsymbol{\sigma}_m^T(\boldsymbol{x},t)\boldsymbol{\varepsilon}_1(\boldsymbol{x},t)\Big)d\boldsymbol{x}+\frac{1}{2}\int_\Omega(\sigma_a,\sigma_a,0)\boldsymbol{\varepsilon}(\boldsymbol{x},t)d\boldsymbol{x}\\ &+\frac{Y(\boldsymbol{x},t)}{2}\int_\Omega \boldsymbol{u}(\boldsymbol{x},t)^2 d\boldsymbol{x}+\frac{1}{2}\sum_l k_l\boldsymbol{u}_l(t)^2+\int_{\partial\Omega}\left(\boldsymbol{\lambda}(\boldsymbol{x},t)+\boldsymbol{f}(\boldsymbol{x},t)+\boldsymbol{f}_l(\boldsymbol{x},t)\right)\boldsymbol{u}(\boldsymbol{x},t)d\boldsymbol{x}\Big)\\ &= -\int_\Omega \eta_m\left(\frac{\partial\boldsymbol{\varepsilon}_2(\boldsymbol{x},t)}{\partial t}\right)^2 d\boldsymbol{x}.\end{aligned} \tag{5}$$

Since $\eta_m\frac{\partial \varepsilon_2}{\partial t} = E_m\boldsymbol{\varepsilon}_1$ and $\boldsymbol{\sigma}_\infty = E_\infty\boldsymbol{\varepsilon}$, and $\boldsymbol{u}_l(t) = T(\theta_l)(\boldsymbol{u}_{l1},\boldsymbol{u}_{l2})$. Eq. (5) can be rewritten as

$$\begin{aligned}&\int_\Omega E_\infty\boldsymbol{\varepsilon}(\boldsymbol{x},t)\frac{\partial\boldsymbol{\varepsilon}(\boldsymbol{x},t)}{\partial t}d\boldsymbol{x}+\int_\Omega \boldsymbol{\sigma}_m^T(\boldsymbol{x},t)\frac{\partial\boldsymbol{\varepsilon}(\boldsymbol{x},t)}{\partial t}d\boldsymbol{x}+\int_\Omega(\sigma_a,\sigma_a,0)\frac{\partial\boldsymbol{\varepsilon}(\boldsymbol{x},t)}{\partial t}d\boldsymbol{x}\\ &+Y\int_\Omega \boldsymbol{u}(\boldsymbol{x},t)\frac{\partial\boldsymbol{u}(\boldsymbol{x},t)}{\partial t}d\boldsymbol{x}+\sum_l T(\theta_l)^T T(\theta_l)k_l(\boldsymbol{u}_{l1}(t),\boldsymbol{u}_{l2}(t))\frac{\partial\boldsymbol{u}(\boldsymbol{x},t)}{\partial t}\\ &+\int_{\partial\Omega}\left(\boldsymbol{\lambda}(\boldsymbol{x},t)+\boldsymbol{f}(\boldsymbol{x},t)+\boldsymbol{f}_l(\boldsymbol{x},t)\right)\frac{\partial\boldsymbol{u}(\boldsymbol{x},t)}{\partial t}dx = 0.\end{aligned} \tag{6}$$

Denoting $\boldsymbol{B} = \begin{pmatrix}\partial/\partial x_1 & 0\\ 0 & \partial/\partial x_2\\ \partial/\partial x_2 & \partial/\partial x_1\end{pmatrix}$, then $\boldsymbol{\varepsilon}(\boldsymbol{x},t) = \boldsymbol{B}\boldsymbol{u}(\boldsymbol{x},t)$. According to Gauss' divergence theorem, we rewrote $\int_\Omega(\sigma_a,\sigma_a,0)\boldsymbol{B}\frac{\partial \boldsymbol{u}(\boldsymbol{x},t)}{\partial t}$ as $\int_\Omega \sigma_a\nabla\cdot\frac{\partial \boldsymbol{u}(\boldsymbol{x},t)}{\partial t}d\boldsymbol{x}$, which leads to $\int_{\partial\Omega}\sigma_a\boldsymbol{n}(\boldsymbol{x},t)\frac{\partial \boldsymbol{u}(\boldsymbol{x},t)}{\partial t}d\boldsymbol{x}$.

Denoting $\boldsymbol{A}_l = \begin{pmatrix}c_l^2 & c_ls_l & -c_l^2 & -c_ls_l\\ c_ls_l & s_l^2 & -c_ls_l & -s_l^2\\ -c_l^2 & -c_ls_l & c_l^2 & c_ls_l\\ -c_ls_l & -s_l^2 & c_ls_l & s_l^2\end{pmatrix}$, where $c_l = cos(\theta_l)$ and $s_l = sin(\theta_l)$. Eq. (6) can be rewritten as

$$\begin{aligned}&\int_\Omega \boldsymbol{B}^T\boldsymbol{\sigma}(\boldsymbol{x},t)^T d\boldsymbol{x}+Y\int_\Omega \boldsymbol{u}(\boldsymbol{x},t)d\boldsymbol{x}+\sum_l \boldsymbol{A}_l k_l(\boldsymbol{u}_{l1}(t),\boldsymbol{u}_{l2}(t)) =\\ &-\int_{\partial\Omega}\sigma_a\boldsymbol{n}(\boldsymbol{x},t)+\boldsymbol{\lambda}(\boldsymbol{x},t)+\boldsymbol{f}(\boldsymbol{x},t)+\boldsymbol{f}_l(\boldsymbol{x},t)d\boldsymbol{x}\end{aligned} \tag{7}$$

2.2.7 Stress of viscoelastic cell and its update

By using general Maxwell model, the stress $\boldsymbol{\sigma}(\boldsymbol{x},t)$ can be written as [12, 13, 19]:

$$E_\infty\boldsymbol{\varepsilon}(\boldsymbol{x},t)+\int_0^t E_m e^{\frac{t-s}{\eta_m/E_\infty}}\frac{\partial\boldsymbol{\varepsilon}(\boldsymbol{x},s)}{\partial s}ds = \boldsymbol{\sigma}_\infty(\boldsymbol{x},t)+\boldsymbol{\sigma}_m(\boldsymbol{x},t). \tag{8}$$

During the time interval $\Delta t = t_{n+1} - t_n$, where t_n is the n-th time step, $\boldsymbol{\sigma}_m^{n+1}(\boldsymbol{x})$ can be written as [19]:

$$e^{-\frac{\Delta t}{\eta_m/E_\infty}}\boldsymbol{\sigma}_m^n(\boldsymbol{x}) + \frac{E_m}{E_\infty}\int_{t_n}^{t_{n+1}} e^{-\frac{t_{n+1}-s}{\eta_m/E_\infty}} ds \frac{\boldsymbol{\sigma}_\infty^{n+1}(\boldsymbol{x}) - \boldsymbol{\sigma}_\infty^n(\boldsymbol{x})}{\Delta t}. \tag{9}$$

Therefore, the stress $\boldsymbol{\sigma}^n(\boldsymbol{x})$ at t_n can be written as

$$\boldsymbol{\sigma}^n(\boldsymbol{x}) = \boldsymbol{\sigma}_\infty^n(\boldsymbol{x}) + \boldsymbol{\sigma}_m^n(\boldsymbol{x}) \tag{10}$$

2.2.8 Force balance equation for discretized time step

For each triangular element $\boldsymbol{\tau}_{i,j,k}$, Eq. (7) at time step t_{n+1} can be rewritten using Eq. (10) as

$$\begin{aligned}&\int_{\boldsymbol{\tau}_{i,j,k}} \boldsymbol{B}^T\left(E_\infty \boldsymbol{B}\boldsymbol{u}^{n+1}(\boldsymbol{x}) + e^{-\frac{\Delta t}{\eta_m/E_\infty}}\boldsymbol{\sigma}_m^n + \gamma_m A_m\left(E_\infty \boldsymbol{B}\boldsymbol{u}^{n+1}(\boldsymbol{x}) - E_\infty \boldsymbol{B}\boldsymbol{u}^n(\boldsymbol{x})\right)\right)d\boldsymbol{x}\\ &+Y\int_{\boldsymbol{\tau}_{i,j,k}} \boldsymbol{u}^{n+1}d\boldsymbol{x} + \sum_{l\in\boldsymbol{\tau}_{i,j,k}} \boldsymbol{A}_l k_l\left(\boldsymbol{u}_{l1}^{n+1}, \boldsymbol{u}_{l2}^{n+1}\right) = \boldsymbol{F}^{n+1}(\boldsymbol{x}),\end{aligned} \tag{11}$$

where $\gamma_m = E_m/E_\infty$, $A_m = \frac{1-e^{-\frac{\Delta t}{\eta_m/E_\infty}}}{\frac{\Delta t}{\eta_m/E_\infty}}$, and $\boldsymbol{F}^{n+1} = -\int_{\partial\Omega}\sigma_a \boldsymbol{n}(\boldsymbol{x},t) + \boldsymbol{\lambda}(\boldsymbol{x},t) + \boldsymbol{f}(\boldsymbol{x},t) + \boldsymbol{f}_l(\boldsymbol{x},t)d\boldsymbol{x}$. Eq. (11) leads to the following linear force-balance equation

$$\boldsymbol{K}_{\boldsymbol{\tau}_{i,j,k}}^{n+1}\boldsymbol{u}_{\boldsymbol{\tau}_{i,j,k}}^{n+1} = \boldsymbol{f}_{\boldsymbol{\tau}_{i,j,k}}^{n+1}, \tag{12}$$

where $\boldsymbol{K}_{\boldsymbol{\tau}_{i,j,k}}^{n+1}$, $\boldsymbol{u}_{\boldsymbol{\tau}_{i,j,k}}^{n+1}$, and $\boldsymbol{f}_{\boldsymbol{\tau}_{i,j,k}}^{n+1}$ are the stiffness matrix, displacement vector, and integrated force vector of $\boldsymbol{\tau}_{i,j,k}$ at time step t_{n+1} (see more details of derivation of (Eq. 12) in [11]).

We can then assemble the element stiffness matrices of all triangular elements into one big global stiffness matrix $\boldsymbol{K}^{n+1}$. Therefore, the linear relationship between the concatenated displacement vector $\boldsymbol{u}^{n+1}$ of all cell vertice and the force vector $\boldsymbol{f}^{n+1}$ on them is given by

$$\boldsymbol{K}^{n+1}\boldsymbol{u}^{n+1} = \boldsymbol{f}^{n+1}. \tag{13}$$

Changes in the cell shape at time step t_{n+1} can be obtained by solving Eq. (13). For vertex $\boldsymbol{v}_i$ at $\boldsymbol{x}_i$, its new location at next time step is then updated as $\boldsymbol{x}_i^{n+1} = \boldsymbol{x}_i^n + \boldsymbol{u}^{n+1}(\boldsymbol{v}_i)$.

2.3 Mechano-chemical pathway in the cell

Upon contact with the environment, cells can transfer the mechanical cues into biochemical signals, which can trigger the initiation of further cellular behaviors [20]. In our model, we considered a mechano-chemical pathway consisting of two parts, where one is to regulate the feedback loop between focal adhesion and cell protrusion and the other is to regulate the transmission of mechanical signal between adjacent cells through intercellular adhesions.

2.3.1 Model of focal adhesion

For each vertex $\boldsymbol{v}_i$ in cell Ω, we assign a constant number of integrin ligand on it. These integrin molecules can bind or unbind with fibronectin molecules on the substrate underneath. Following [21], the numbers of bound and unbound integrin ligand molecules are determined by

$$\frac{dR_b}{dt} = k_f n_s R_u - k_r R_b, \tag{14}$$

where R_u and R_b are the numbers of unbound and bound integrin ligand, respectively; k_f is the binding rate coefficient; n_s is the concentration of fibronectin per cell vertex; k_r is the unbinding rate coefficient. k_r depends on the magnitude of the traction force $\boldsymbol{f}_r$ applied on $\boldsymbol{v}_i$. Traction force $\boldsymbol{f}_r(\boldsymbol{x}, t)$ on $\boldsymbol{x}$ at time t is given by $Y(\boldsymbol{x}, t)\boldsymbol{u}(\boldsymbol{x}, t)$ following [6]. k_r is determined by $k_r = k_{r_0}\left(e^{-0.04f_r} + 4e - 7e^{0.2f_r}\right)$ following [22], where k_{r_0} is a constant. k_f is related with the substrate stiffness by $k_f = k_{f0}E_{st}^2/\left(E_{st}^2 + E_{st0}^2\right)$ [16, 23], where k_{f0} and E_{st0} are constants (see Appendix for choosing E_{st0}).

2.3.2 Model of feedback loop between focal adhesion and cell protrusion

We introduced a simplified model of a positive feedback loop to control the spatial distribution of the focal adhesions, which governs the direction of cell protrusion [8, 24]. In our model, this feedback loop involves proteins of Paxillin, Rac, and PAK (**Figure 1c**). Upon formation of focal adhesion, Paxillin is activated by active PAK. The active Paxillin then activates Rac, which in turn triggers the activation of PAK. The activated Rac is responsible for protruding cells [25]. Since the protein Merlin on the intercellular cadherin complex also plays a role in activating Rac [9], we include Merlin in our feedback loop. The protein concentration over time is updated through a set of differential equations following [8]:

$$\frac{dr}{dt} = k_{x,r}\left(k_m \frac{C_r^2}{C_r^2 + m^2}x - r\right) \tag{15}$$

$$\frac{dp}{dt} = k_{r,p}(r - p) \tag{16}$$

$$\frac{dx}{dt} = k_{p,x}\left(k_x \frac{p^2}{p^2 + C_x^2}n - x\right) \tag{17}$$

where x, r, p, m and n are the concentrations of activated Paxillin, activated Rac, activated PAK, Merlin and bound integrins, respectively. $k_{x,r}$, $k_{r,p}$, $k_{p,x}$, k_m, k_x, C_r, C_x are the parameters of corresponding rates. The level of activated Rac was used to determine the protrusion force on the leading edge of the migrating cell (see details of cell protrusion model in Appendix).

2.3.3 Model of mechanosensing through intercellular adhesion

We added Merlin in the feedback loop (**Figure 1c**) following a previous study reporting that Merlin on the intercellular cadherin complex regulates the Rac activity [26]. As illustrated in **Figure 2a**, for two adjacent cells C_1 and C_2 where C_1 is the leader cell and C_2 is the follower cell, if both cells are at static state, Merlin molecules only locate on the cadherin spring (**Figure 2b** top). As reported by [9], Merlin suppressed the binding of integrin. Due to such suppression, Rac turns to inactivated on the Merlin-expressed site. Once cell C_1 starts to migrate, tension force is generated on the cadherin spring between C_1 and C_2. Merlin is therefore

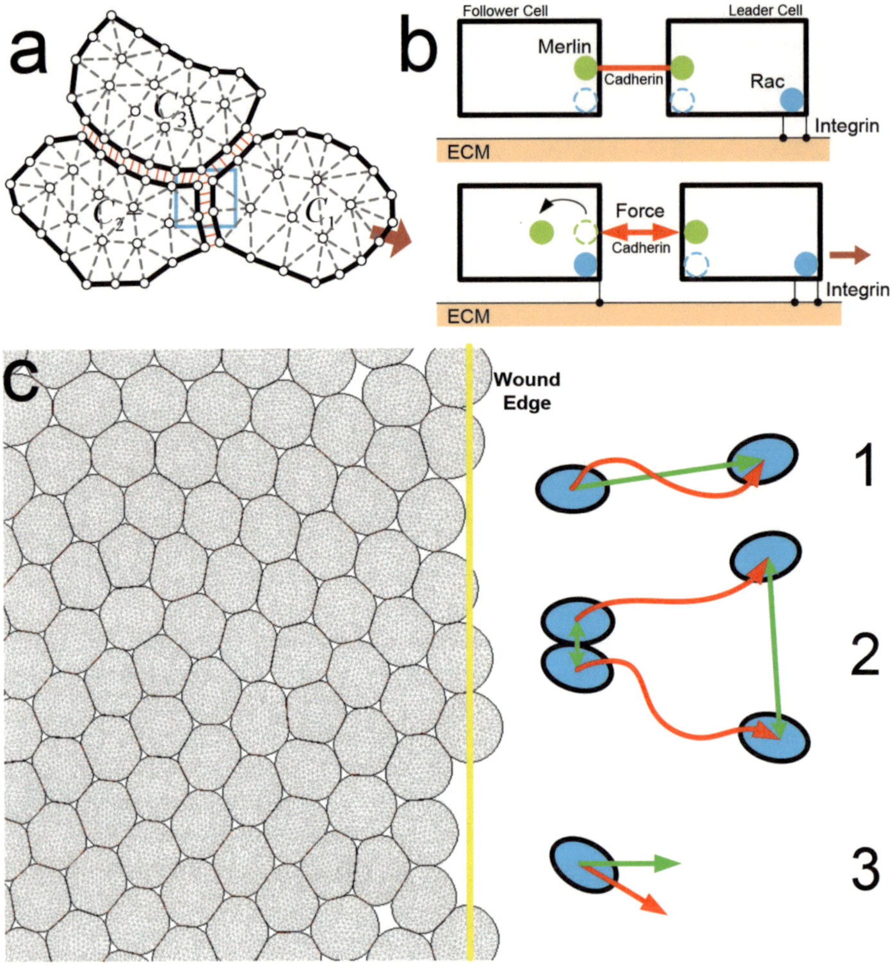

Figure 2.
Mechanosensing through intercellular adhesion and tissue model for collective cell migration. (a) the intercellular adhesion of the cadherin spring (red springs in the blue box) are responsible for transmitting mechanical stimulus from leader cell (C_1) to its follower cells (C_2 and C_3). (b) When cells are static, Merlin which inhibits the Rac activation is bound on the cadherin spring. Once leader cell migrates, stretch is generated on the cadherin spring, Merlin on the follower cell is delocated. Therefore, Rac is activated on the follower cell. (c) the size of the wound epithelial tissue is 720 μm × 240 μm. The right boundary is set as the wound edge (yellow line). Cells can migrate towards the open space on the right. Three measurements are introduced to measure the collective cell migration: (1) migration persistence $p(t_n)$, the ratio of the distance from the current position at time t_n to its initial position (green line), over the length of the traversed path (red curve); (2) normalized pair separation distance $d_{i,j}(t_n)$ is the separation distance between a pair of cells at time t_n (green lines) divided by the average length of the two cells' traversed path (red curves); (3) migration direction angle $\alpha(t_n)$ is the angle between the migration direction (red arrow) and the direction towards the wound (green arrow).

delocalized from cadherin-attached site in response to the generated tension force. As a consequence, Rac is activated there to generate protrusion force to follow the leader cell (**Figure 2b** bottom [26]). For simplicity, we introduced the inactive Merlin phenotype along with the active Merlin phenotype on the two end vertice of one intercellular cadherin adhesion. The negative feedback loop of Merlin-Rac is modeled through a set of differential equations following [9, 26], where active Merlin and inactive Merlin can switch their phenotype, but only active Merlin can suppress the Rac activity (**Figure 1c**). The delocalization of Merlin was simply modeled as Merlin switching to inactive phenotype:

$$\frac{dm}{dt} = e - \delta(f_t > f_{t-thr})k_{m,e}\left(k_p\left(\frac{p^2}{p^2 + C_e^2} + k_e\right)m\right) \quad (18)$$

$$\frac{de}{dt} = \delta(f_t > f_{t-thr})k_{m,e}\left(k_p\left(\frac{p^2}{p^2 + C_e^2} + k_e\right)m\right) - e \quad (19)$$

where m, e and p are the concentrations of Merlin, inactive Merlin, and PAK, respectively. $k_{m,e}$, k_p, k_e, C_e are the corresponding rate parameters. $\delta(x)$ is a kronecker function that $\delta(TRUE) = 1$ and $\delta(FALSE) = 0$. f_t is the tension force through the cadherin spring and f_{t-thr} is a force threshold.

2.4 Cellular tissue for collective cell migration

In our model, the collective cell migration was modeled using a wound tissue of epithelial cells. The tissue size is $720\ \mu m \times 240\ \mu m$. The epithelial cell type is set to MCF-10A, which is used in the *in vitro* study [10]. The corresponding epithelial-specific parameters can be found in **Table 1**. We arbitrarily set the right boundary of the tissue as the wound edge, and cells can migrate towards the open space to the right of the wound edge (**Figure 2c**). The mechano-chemical pathway was initiated first in the cells on the wound edge after they migrate. We followed a previous study [10] to divide the location of cells into four sub-regions according to their distance to the wound edge: Regions I, II, III, and IV whose distance to the wound edge is 0–160 μm, 160–320 μm, 320–480 μm, and 480–640 μm, respectively (**Figure 2c**). We ran the simulation for 12 biological hours, the same experimental duration time in the *in vitro* study [10].

2.5 Measurements of the collective cell migration

In our model, the collective cell migration are measured using four measurements following [10]:

The migration persistence. At time step t_n, the length of the straight line between cell positions at t_n and initial time step t_0 over the length of the migrating trajectory:

$$p(t_n) = \frac{|\boldsymbol{x}(t_n) - \boldsymbol{x}(t_0)|}{\sum_{k=0}^{n-1}|\boldsymbol{x}(t_{k+1}) - \boldsymbol{x}(t_k)|}, \quad (20)$$

where t_0 is the initial time, $\boldsymbol{x}(t_i)$ is the position at time step t_i (**Figure 2c.1**).

The normalized separation distance. At time step t_n, the separation distance of a pair of two adjacent cells 1-2, divided by the average length of their migrating trajectories:

$$d_{1,2}(t_i) = \frac{\left||\boldsymbol{x}_1(t_i) - \boldsymbol{x}_2(t_i)| - |\boldsymbol{x}_1(t_0) - \boldsymbol{x}_2(t_0)|\right|}{\frac{1}{2}\left(\sum_{k=0}^{i-1}|\boldsymbol{x}_1(t_{k+1}) - \boldsymbol{x}_1(t_k)| + \sum_{k=0}^{i-1}|\boldsymbol{x}_2(t_{k+1}) - \boldsymbol{x}_2(t_k)|\right)}, \quad (21)$$

where the numerator is the separation distance between cells i and j at time t_n, and the denominator is the average path length of cells i and j at time t_n (**Figure 2c.2**).

The direction angle. The angle between the cell migration direction and the direction towards the wound.

Definition	Value	Reference
Time step lapse	0.1 sec	NA[a]
Cell radius	10 μm	[27]
Young's modulus of cell	5 kPa	[28]
Poisson ratio of cell	0.40	[29]
Contractile pressure σ_a	2 kPa	[6]
Adhesion energy constant Y_a	0.9 $/\mu m$	[6]
Spring constant of cadherin spring	3.0 $nN/\mu m$	NA
Default length of cadherin spring	100 nm	[30]
Maximum length of cadherin spring	400 nm	[31]
Protrusion force constant f_a	2.0 $nN/\mu m$	[11]
Integrin bound rate k_{f0}	0.5	[22]
Integrin unbound rate k_{r0}	0.4	[22]
Reference substrate stiffness E_{st0}	40 kPa	NA
Rac deactivation rate $k_{x,r}$	4/min	[8]
PAK deactivation rate $k_{r,p}$	10/min	[8]
Paxillin dephosphorylation rate $k_{p,x}$	10/min	[8]
Saturation of phosphorylated Paxillin k_x	1	[8]
Saturation of PAK activation k_p	1	[8]
Saturation of Merlin k_m	1	NA
Merlin phosphorylation rate $k_{m,e}$	10/min	NA
Saturation of phosphorylated Merlin k_e	$1e^{-3}$	NA
Force threshold of delocating Merlin f_{t-thr}	0.15 nN	NA

[a]*Estimated value marked as NA.*

Table 1.
Parameters used in the model.

$$\alpha(t_n) = \arccos(\boldsymbol{u}_c \cdot \boldsymbol{u}_w) \cdot \operatorname{sgn}(\|\boldsymbol{u}_c \times \boldsymbol{u}_w\|), \tag{22}$$

where $\boldsymbol{u}_c$ is the unit vector of the cell migration direction, $\boldsymbol{u}_w$ is the unit vector of direction from the cell mass center towards the wound, $\operatorname{sgn}(x)$ is the sign of x (**Figure 2c.3**).

3. Results

3.1 Mechanics of cell-substrate is crucial to regulate collective cell migration

3.1.1 Morphology and migration pattern under different substrate stiffness

We fist studied collective cell migration under the mechano-chemical mechanism. The trajectories of our simulation showed that cells migrate faster and more persistently on stiffer substrate (**Figure 3a** and **b**). This is compatible with the observed pattern from the *in vitro* study of collective cell migration (**Figure 3c** and **d**). Furthermore, the shape of cell also changes with different substrate stiffness. Cells

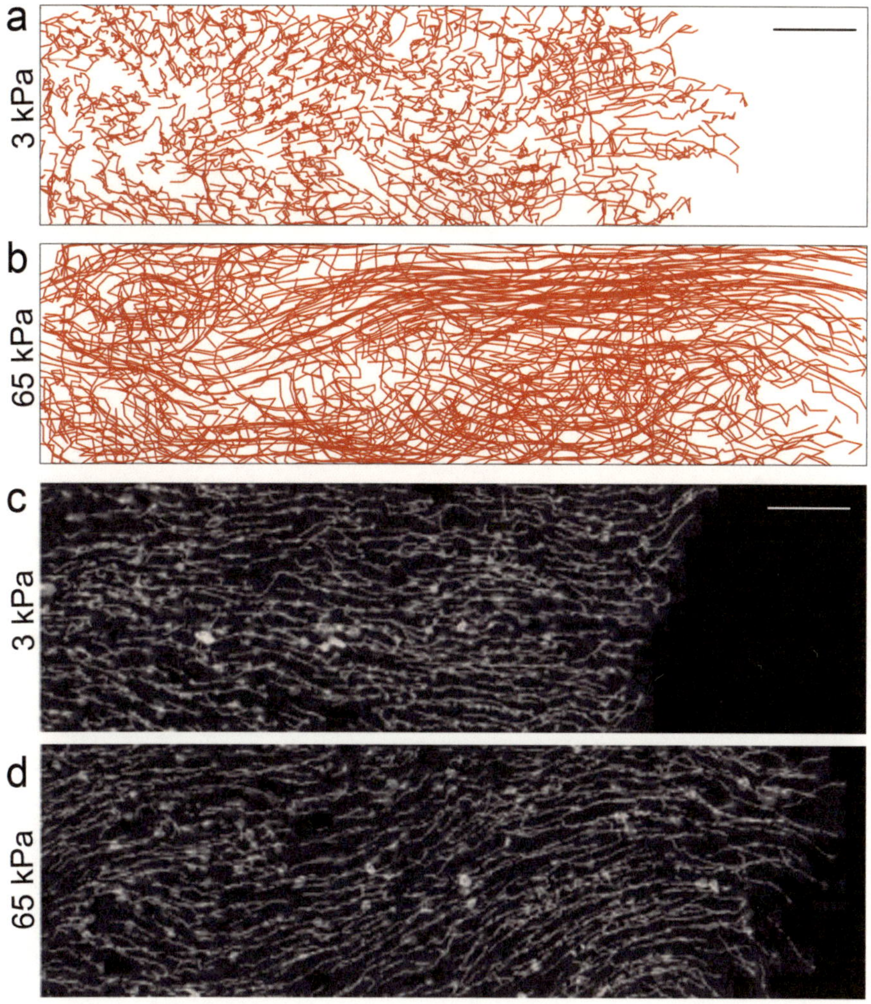

Figure 3.
Cell migrating trajectories. (a–b) the migrating trajectory in our simulation using two substrate stiffness: 3 and 65 kPa. (c–d) the migrating trajectory from the in vitro study using the same substrate stiffness: 3 and 65 kPa [10]. The scale bar is 100 μm.

adopted a more spherical shape on softer substrate (**Figure 4a**) while cells were more elongated on stiffer substrate (**Figure 4c**). The same pattern of cell morphology was observed in [10], where cell extended its protrusions in all directions on softer substrate (**Figure 4b**) while cell protruded only on the leading edge with a long tail on stiffer substrate (**Figure 4d**).

Overall, the patterns of cell trajectory and cell morphology of our simulation are consistent with that from *in vitro* study. This indicated that our mechano-chemical model is valid.

3.1.2 The mechanical signal has long-distance impacts on collective cell migration

We then quantified the cell migration to explore the role of mechanics of cell-substrate and cell–cell mechanics on collective cell migration using the three measurements: persistent ratio $p(t_n)$, normalized separation distance $d_{i,j}(t_n)$ and direction angle $\alpha(t_n)$.

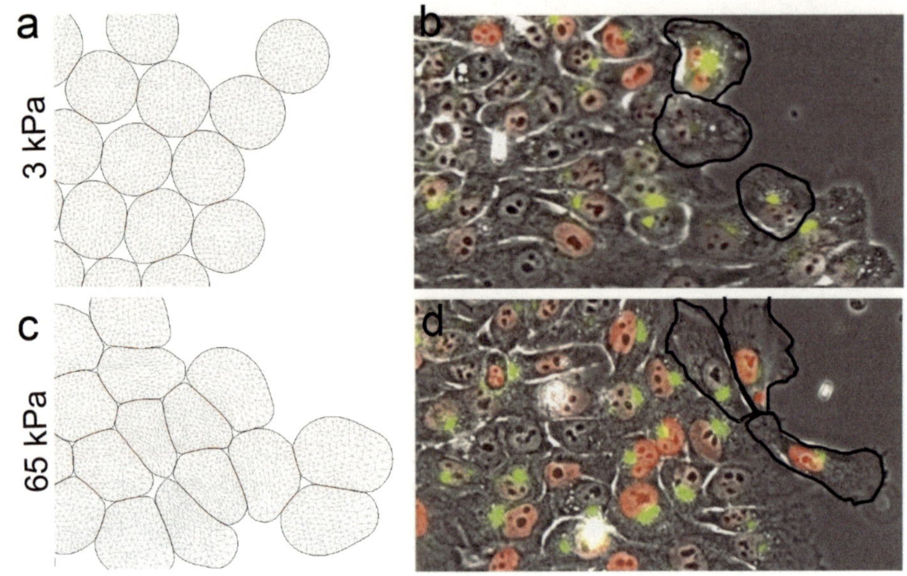

Figure 4.
Cell morphology. (a, c) the cell morphology in our simulation using two substrate stiffness: 3 and 65 kPa. (b, d) the cell morphology from the in vitro study [10] using the same substrate stiffness: 3 and 65 kPa. The cell boundary is highlighted in black.

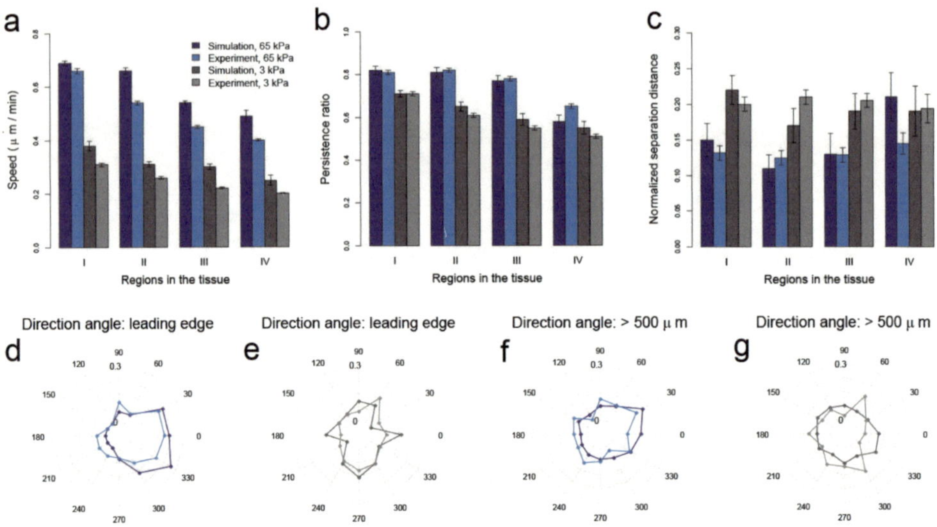

Figure 5.
Measurements of the collective cell migration. (a–c) the cell migration speed, persistence ratio and normalized separation distance of our simulation and in the in vitro study [10]. (d–g) the migration direction angle of the cells on the leading edge and more than 500 μm from the wound edge in our simulation and that in the in vitro study [10] on the substrate with stiffness 65 kPa (d–e) and 3 kPa (f–g). The colors indicating simulation and experiment are shown in (a). The error bars of our simulation depict the standard deviations of four runs of simulation.

We first examined the migrating speed of the cell. In general, cells migrate with higher speed on stiffer substrate (**Figure 5a**, more details of cell migration speed can be found in Appendix). In addition, cells close to the wound edge migrated with higher speed on both stiffer and softer substrate. This speed decreased gradually as the distance to the wound edge increased. On substrate with stiffness of 65 kPa, the migration speed decreased from $0.69 \pm 0.01 \mu m/min$ in Region I to $0.49 \pm 0.02 \mu m/min$ in Region IV, while on substrate with stiffness of 3 kPa, the migration speed decreased from $0.38 \pm 0.02\mu m/min$ in Region I to $0.25 \pm 0.02\mu m/min$ in Region IV (**Figure 5a**). The cell migration speed of our simulation was consistent with that from the *in vitro* study [10]. It is easy to interpret such pattern of cell migration speed. For cells in Region I, especially on the wound edge, there are fewer or even no cells ahead. As the distance to the wound edge increased, it was more crowded and more difficult for cells to migrate forward.

We next examined the migration persistence of the cells. As shown in **Figure 5b**, cells migrate more persistently on stiffer substrate. In addition, cells close to the wound edge migrated with higher migration persistence. For cells on substrate with stiffness of 65 kPa, the persistence ratio decreased from $82 \pm 2\%$ in Region I to $58 \pm 3\%$ in Region IV, while for cells on substrate with stiffness of 3 kPa, the persistence ratio decreased from $71 \pm 1\%$ in Region I to $55 \pm 3\%$ in Region IV (**Figure 5b**). As shown in **Figure 5c**, collective cell migration was coordinated better on stiffer substrate.

In addition, we examined the normalized separation distance of the pairs of migrating cells. As shown in **Figure 5c**, the normalized separation distance increased as the distance to the wound edge increased. In our simulation, for cells on substrate at stiffness of 65 kPa, the separation distance decreased from 0.15 ± 0.02 in Region I to 0.11 ± 0.02 in Region II and then increased to 0.21 ± 0.03 in Region IV, while for cells on substrate at stiffness of 3 kPa, the separation distance decreased from 0.22 ± 0.02 in Region I to 0.17 ± 0.02 in Region II and then increased to 0.19 ± 0.04 in Region IV (**Figure 5c**). This pattern of separation distance in our simulation was also observed in the *in vitro* study [10].

Furthermore, we examined the migration direction angle. We compared this angle for cells on the leading edge of the tissue and cells 500 μm away. Since the cell migration direction is usually along the cell polarity direction [32], we also compared this direction angle to the cell polarization direction reported in [10]. As shown in **Figure 5d–g**, cells exhibit more accurate migration direction towards the wound on stiffer substrate (65 kPa). Only about 10 % of the cells on the leading edge had migration direction opposite to the wound (**Figure 5d**, 90°–270°). For cells $> 500\ \mu m$ away from the wound edge, 30 % of them had migration direction opposite to the wound (**Figure 5f**, 90°–270°). However, for cells on softer substrate (3 kPa), cell migration deviated more from the direction towards the wound where 35 % of the cells on leading edge had migration direction opposite to the wound direction (**Figure 5e**, 90°–270°), while for cells $> 500\ \mu m$ away from the wound edge, this fraction increased to 45 % (**Figure 5g**, 90°–270°).

These measurements implied that substrate stiffness is important to guide collective cell migration. Cells on stiffer substrate can migrate with high persistence, good coordination between cell pairs, and accurate migration direction. Our simulation suggests that the mechano-chemical feedback loop in each cell ensured it to dictate its migration direction. Furthermore, the individual cell movements were organized into a global migrative wave through intercellular adhesions.

4. Conclusions

In this chapter, we introduced our novel finite element cellular model to explore the mechanism behind collective cell migration using a simplified tissue model. This model includes a detailed mechano-chemical feedback loop, which takes into account of formation of focal adhesion and cell protrusion initiated by Rac signaling. In addition, our model incorporates the mechanical cue transmitting between the follower cell and the leader cell. We further examined the effects of cell-substrate contact and intercellular adhesions on collective cell migration.

An important result of this study is that we find the cell-substrate mechanics plays crucial role in guiding collective cell migration with higher persistence, more accurate direction, and better coordination between cell pairs (**Figure 5**). Previous *in vitro* study has shown that cells tend to have elongated shape on stiffer substrate while cells tend to have spherical shape on softer substrate [33]. This is compatible with our simulation (**Figure 4a** and **c**). We anticipate that our finite element cellular model can be applied to a broad of studies of cellular tissue problems.

Appendix A: cell migration model

In our model, cell migration is initiated and maintained by the protrusion force on the leading edge and the cell migration speed varies with the cell-substrate friction following [16].

A.1 Cell-substrate depending on substrate stiffness

The adhesion coefficient $Y(\boldsymbol{x},t)$ of a cell vertex $\boldsymbol{x}$ at time t and set to be proportional to the strength of focal adhesions [18]: $Y(\boldsymbol{x},t) = \frac{n_{x,t}}{n_0} E_{st} Y_a$, where $n_{x,t}$ is the number of binding integrins at location $\boldsymbol{x}$ at t, n_0 is a normalizing constant number, E_st is the stiffness of the substrate, Y_a is the basic adhesion constant taken from [18]. In this way, the cell-substrate friction is related with the stiffness of the substrate.

A.2 Cell protrusion depends on substrate stiffness

In our model, there is a mechano-chemical pathway dictating the cell protrusion. The bound integrin initiates the activation of Rac which regulates the cell protrusion. At time t, the migration direction of the cell C is sampled from all the boundary vertice according to their Rac concentration. One vertex $\boldsymbol{v}_i$ is stochastically selected with the probability $\frac{Rac(\boldsymbol{v}_i)}{\sum_i Rac(\boldsymbol{v}_i)}$. The outward unit normal vector $\boldsymbol{n}(\boldsymbol{v}_i)$ of $\boldsymbol{v}_i$ is chosen as the cell migration direction. Any vertex $\boldsymbol{v}_j$ whose outward unit normal vector $\boldsymbol{n}(\boldsymbol{v}_j)$ is positively aligned with $\boldsymbol{n}(\boldsymbol{v}_i)$, is treated as leading edge vertex. The protrusion force $\boldsymbol{f}_a$ is then applied on each leading edge vertex $\boldsymbol{v}_i$ as $\boldsymbol{f}_a(\boldsymbol{v}_i) = f_a R(\boldsymbol{v}_i) \boldsymbol{n}(\boldsymbol{v}_i)$, where f_a is a constant, $R(\boldsymbol{v}_i)$ is the normalized Rac concentration at $\boldsymbol{v}_i$.

A.3 Calibrating the cell protrusion parameter

As shown in **Figure 6a–c**, the cell leading edge has higher level of bound integrin, along with higher level of Rac due to the effect of positive feedback loop. The Merlin expression is also mechano-dependent. As shown of the pair of cells in the green box of **Figure 6b**, after the right cell migrates, the stretch force on the cadherin spring between them make the Merlin delocate from the left cell. As a result, the left cell can express Rac to protrude following the right cell. If the pair of static cell are simply in contact (**Figure 6b**), the Merlin is expressed on both of them. Therefore, the Rac expression is inhibited. Both of the two cells do not

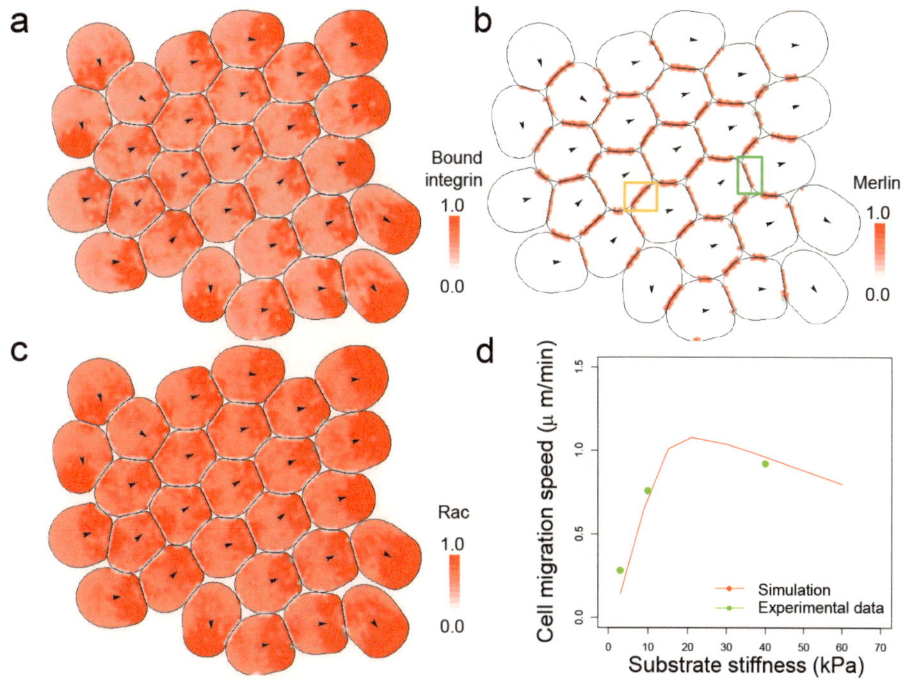

Figure 6.
The cell protrusion depends on mechano-chemical process. (a–c) the spatial distribution of the normalized concentration of bound integrin, Merlin, and Rac. The black arrows indicate the migration direction. The pattern of Merlin expression depends on cell status. Green box in (b): The left cell follows the right one. Merlin is expressed only on the right cell; Orange box in (b): The two static cells are in contact. Merlin is expressed on both of them. (d) Cell migration speed of our simulation is consistent with the experimental observation [33, 34].

protrude against each other. To fit our cell protrusion model to the in vitro data, we calibrate the parameter of E_{st0}: when $E_{st0} = 40kPa$, the cell migration speed of our simulation has the best match with the in vitro studies [33, 34] (**Figure 6d**).

Author details

Jieling Zhao
Inria de Paris, Paris, France

*Address all correspondence to: jieling.zhao@inria.fr

References

[1] Chen X., Brodland G.W. Multi-scale finite element modeling allows the mechanics of amphibian neurulation to be elucidated. Physical Biology. 2008;5: 015003.

[2] Hutson M.S., Veldhuis J., Ma X., Lynch H.E., Cranston P.G., Brodland G. W. Combining laser microsurgery and finite element modeling to assess cell-level epithelial mechanics. Biophysical Journal. 2009;97:3075–3085.

[3] Oñate E., Idelsohn S., Del Pin F., Aubry R. The particle finite element methodan overview. International Journal of Computational Methods. 2004;1:267–307.

[4] Zhao J., Manuchehrfar F., Liang J. Cell-substrate mechanics guide collective cell migration through intercellular adhesion: a dynamic finite element cellular model. Biomechanics and Modeling in Mechanobiology. 2020; 19(5):1781–1796.

[5] Karcher H., Lammerding J., Huang H., Lee R.T., Kamm R.D., Kaazempur-Mofrad M.R. A three-dimensional viscoelastic model for cell deformation with experimental verification. Biophysical Journal. 2003;85(5): 3336–3349.

[6] Oakes P., Banerje S., Marchetti M. Gardel M. Geometry regulates traction stresses in adherent cells. Biophysical Journal. 2014;107:825–833.

[7] Jamali Y., Azimi M., Mofrad M.R. A sub-cellular viscoelastic model for cell population mechanics. PLoS One. 2010; 5(8):e12097.

[8] Cirit M., Krajcovic M., Choi C.K., Welf E.S., Horwitz A.F., Haugh J.M. Stochastic model of integrin-mediated signaling and adhesion dynamics at the leading edges of migrating cells. PLoS Comput. Biol. 2010;6(2):e1000688.

[9] Okada T., Lopez-Lago M., Giancotti F.G. Merlin/nf-2 mediates contact inhibition of growth by suppressing recruitment of Rac to the plasma membrane. J. Cell Biol. 2005;171(2): 361–371.

[10] Ng M.R., Besser A., Danuser G., Brugge J.S. Substrate stiffness regulates cadherin-dependent collective migration through myosin-II contractility. The Journal of Cell Biology. 2012;199(3):545–563.

[11] Zhao J., Cao Y., DiPietro L.A., Liang J. Dynamic cellular finite-element method for modelling large-scale cell migration and proliferation under the control of mechanical and biochemical cues: a study of re-epithelialization. Journal of The Royal Society Interface. 2017;14(129):20160959.

[12] Schoner J.L., Lang J., Seidel H.P. Measurement-based interactive simulation of viscoelastic solids. Computer Graphics Forum. 2004;23: 547–556.

[13] Schwartz J.M., Denninger M., Rancourt D., Moisan C. Laurendeau D. Modelling liver tissue properties using a non-linear visco-elastic model for surgery simulation. Medical Image Analysis. 2005;9(2):103–112.

[14] Rubinstein B., Fournier M.F., Jacobson K., Verkhovsky A.B., Mogilner A. Actin-myosin viscoelastic flow in the keratocyte lamellipod. Biophysical Journal. 2009;97(7):1853–1863.

[15] Ladoux B., Mège R.M., Trepat X. Front–rear polarization by mechanical cues: From single cells to tissues. Trends in Cell Biology. 2016;26(6):420–433.

[16] Dokukina I.V. Gracheva M.E. A model of fibroblast motility on substrates with different rigidities. Biophysical Journal. 2010;98(12): 2794–2803.

[17] Barnhart E., Lee K.C., Keren K., Mogilner A., Theriot J. An adhesion-dependent switch between mechanisms that determine motile cell shape. PLoS Biol. 2011;9(5):e1001059.

[18] Banerjee S., Marchetti M.C. Contractile stresses in cohesive cell layers on finite-thickness substrates. Physical Review Letters. 2012;109(10): 108101.

[19] Sedef M., Samur E., Basdogan C. Real-time finite-element simulation of linear viscoelastic tissue behavior based on experimental data. Computer Graphics and Applications. 2006;26(6): 58–68.

[20] Holle A.W., Engler A.J. More than a feeling: discovering, understanding, and influencing mechanosensing pathways. Current Opinion in Biotechnology. 2011;22(5):648–654.

[21] DiMilla P., Barbee K., Lauffenburger D. Mathematical model for the effects of adhesion and mechanics on cell migration speed. Biophysical Journal. 1991;60(1):15.

[22] Li Y., Bhimalapuram P., Dinner A.R. Model for how retrograde actin flow regulates adhesion traction stresses. Journal of Physics: Condensed Matter. 2010;22(19):194113.

[23] Yeh Y.C., Ling J.Y., Chen W.C., Lin H.H., Tang M.J. Mechanotransduction of matrix stiffness in regulation of focal adhesion size and number: reciprocal regulation of caveolin-1 and β1 integrin. Scientific reports. 2017;7(1):15008.

[24] Stéphanou A., Mylona E., Chaplain M., Tracqui P. A computational model of cell migration coupling the growth of focal adhesions with oscillatory cell protrusions. Journal of Theoretical Biology. 2008;253(4):701–716.

[25] Wu Y.I., Frey D., Lungu O.I., Jaehrig A., Schlichting I., Kuhlman B., Hahn K.M. A genetically encoded photoactivatable rac controls the motility of living cells. Nature. 2009;461 (7260):104–108.

[26] Das T., Safferling K., Rausch S., Grabe N., Boehm H., Spatz J.P. A molecular mechanotransduction pathway regulates collective migration of epithelial cells. Nature Cell Biology. 2015;17(3):276.

[27] Watt F.M., Green H. Involucrin synthesis is correlated with cell size in human epidermal cultures. The Journal of Cell Biology. 1981;90:738–742.

[28] Hu S., Wang R., Tsang C.M., Tsao S. W., Sun D., et al. Revealing elasticity of largely deformed cells flowing along confining microchannels. RSC Advances. 2018;8:1030–1038.

[29] Zielinski R., Mihai C., Kniss D., Ghadiali S.N. Finite element analysis of traction force microscopy: Influence of cell mechanics, adhesion, and morphology. Journal of Biomechanical Engineering. 2013;135:071009.

[30] Changede R., Sheetz M. Integrin and cadherin clusters: A robust way to organize adhesions for cell mechanics. BioEssays. 2017;39:1–12.

[31] Nematbakhsh A., Sun W., Brodskiy P.A., Amiri A., Narciso C., et al. Multi-scale computational study of the mechanical regulation of cell mitotic rounding in epithelia. PLoS Computational Biology. 2017;13: e1005533.

[32] Rappel W.J., Edelstein-Keshet L. Mechanisms of cell polarization. Current Opinion in Systems Biology. 2017;3:43–53.

[33] Ansardamavandi A., Tafazzoli-Shadpour M., Shokrgozar M.A. Behavioral remodeling of normal and cancerous epithelial cell lines with differing invasion potential induced by

substrate elastic modulus. Cell Adhesion and Migration. 2018;12(5):472–488.

[34] Cai P., Layani M., Leow W.R., Amini S., Liu Z., et al. Bio-inspired mechanotactic hybrids for orchestrating traction-mediated epithelial migration. Advanced Materials. 2016;28:3102–3110.

Chapter 7

Nuclear Reactor Thermal Expansion Reactivity Effect Determination Using Finite Element Analysis Coupled with Monte Carlo Neutron Transport Analysis

Chad Pope and Edward Lum

Abstract

The energy released from the nuclear fission process drives thermal expansion and mechanical interactions in nuclear reactors. These phenomena cause changes in the neutron chain reaction which results in further changes in thermal expansion and mechanical interactions. Coupling finite element analysis with Monte Carlo neutron transport analysis provides a pathway to simulate the thermal expansion and mechanical interaction to determine a fundamental parameter, namely, thermal expansion temperature coefficient of reactivity. Knowing the coefficient value allows predictions of how a reactor will behave under transient conditions. Using the coupling of finite element analysis and Monte Carlo neutron transport analysis, the thermal expansion temperature coefficient of reactivity was determined for the Godiva-IV reactor (−2E−05 Δk/k/°C) and the Experimental Breeder Reactor-II (EBR-II) (−1.4E−03 $/°C). The Godiva-IV result is within 3% of the measured result. The thermal expansion and mechanical interactions within EBR-II are sufficiently complex that experimentally measuring the isolated coefficient of reactivity was not possible. However, the calculated result fits well with the integral EBR-II reactivity coefficient measurements. Coupling finite element analysis with Monte Carlo neutron transport analysis provides a powerful technique that gives reactor operators and designers greater confidence in reactor operating characteristics and safety margins.

Keywords: FEA, Monte Carlo, reactivity, temperature coefficient, reactor

1. Introduction

Nuclear reactors exhibit remarkably complicated behavior ultimately originating from the energy released through the nuclear fission process. The complicated behavior involves many phenomena including nuclear, thermal, and mechanical. Individually, these phenomena involve processes that are challenging to quantify, measure, and model. When interactions between these phenomena are considered, the quantification, measurement, and modeling challenges become daunting. This chapter describes finite element analysis (FEA) coupled with Monte Carlo analysis as a methodology for quantification of a particularly important nuclear parameter which is primarily influenced by thermal and mechanical phenomena present in nuclear reactors.

1.1 Background

The multiplication factor, k, is used to quantify the fission chain reaction in nuclear reactors. Numerous definitions exist for k, with each definition applying to a particular situation. A simple definition of k is that it represents the ratio of the number of fissions in one generation to the number of fissions in the preceding generation. Through this definition, one can see that if k is less than unity, the number of fissions declines over time, and if k is greater than unity, the number of fissions increases over time. A unique situation exists when k is exactly equal to one. In that case, the number of fissions remains constant over time and is referred to as critical.

A companion parameter to k is reactivity, ρ. Reactivity represents the deviation from the critical state, as shown in Eq. (1).

$$\rho = \frac{(k-1)}{k} \tag{1}$$

The decimal form of reactivity can be converted to units of $ by dividing the decimal value by the fraction of delayed neutrons resulting from the fission process. Delayed neutrons are those neutrons emitted during the decay of select radioactive fission products rather than being emitted at the moment of fission. For uranium-235, the delayed neutron fraction is 0.0065.

When operating a nuclear reactor, frequently, one is interested in knowing the change in reactivity resulting from various activities such as control rod movements. Other changes resulting from thermal and mechanical phenomena can produce reactivity changes. Frequently, these reactivity changes are quantified in terms of the change in reactor temperature. The result is known as a temperature coefficient of reactivity defined by Eq. (2).

$$\alpha_T = \frac{\Delta\rho}{\Delta T} \tag{2}$$

The temperature coefficient of reactivity can be further subdivided into explicit subjects such as coolant temperature, fuel temperature, and even thermally driven reactor geometry changes.

From a reactor safety perspective, a negative temperature coefficient is indicative of inherently stability. If a reactor transient was initiated that results in a temperature increase, the resulting change in reactivity will necessarily be negative, which means the multiplication factor will be reduced. Eventually, the temperature increase will produce a sufficient reduction in k such that the reactor will shut down. Contrarily, a positive reactivity temperature coefficient is indicative of inherent instability. With a positive coefficient, a transient resulting in a reactor temperature increase will result in a positive reactivity change and a resulting increase in the multiplication factor. The increased multiplication factor will be accompanied by an increase in the number of fissions and resulting heat release and corresponding temperature increase which will subsequently produce an additional positive change in reactivity. The reactor will continue on this path until it is acted upon by a more dominate negative action or the reactor will ultimately be damaged or even destroyed.

Reactivity coefficients can be determined for numerous phenomena. For example, reactivity coefficients can be established for changes in reactor power. Thus, as the reactor power is increased, the reactivity change needed to compensate for the power change can be identified. Another interesting phenomenon that has a significant reactivity effect centers on bubble or void formation as the result of coolant boiling. Reactor designers must pay particular attention to the reactivity effect associated with coolant bubble formation because it can have a significant safety impact.

In the case of reactors that use water as a coolant, the water has a significant effect on the overall neutron energy spectrum in the reactor. Neutrons tend to be born at high energies on the order of several million electron volts. As the neutrons collide with various nuclei in the reactor, they tend to lose energy in a process called moderation. As the neutrons lose energy, they become more likely to be absorbed in uranium-235, which can then fission and release additional neutrons. Similar to the three different regimes for k, there are three regimes for moderation: under-moderation, optimum-moderation, and over-moderation. If a reactor is designed with under-moderation, the loss of coolant through bubble formation will result in a reactivity decrease because fewer neutrons will be slowed to energies where they are more likely to be absorbed in uranium-235. In the case of a reactor designed with over-moderation, the formation of bubbles will tend to result in a reactivity increase because more neutrons will be slowed to the point where they will be absorbed by uranium-235 causing an increase in the number of fissions.

The most dramatic and tragic demonstration of positive reactivity due to bubble formation was seen in the 1986 Chernobyl accident. When operated at low power, the Chernobyl reactor had a positive void reactivity coefficient. Thus, if the reactor coolant began to boil, the bubbles created by the coolant boiling led to a positive reactivity change thereby driving an increase in the multiplication factor and a corresponding increase in the number of fissions occurring in the reactor. The heat released from the additional fissions led to additional coolant boiling which drove a very rapid power increase and subsequent steam explosion and reactor destruction.

Reactivity coefficients tied to geometry changes are of interest in certain situations because they are typically fast acting and can have important safety implications. In many cases, thermally driven geometry changes are coupled with resulting mechanical interactions that severely complicate quantification and modeling approaches. While the change in geometry causes the reactivity change, typically temperature is used to quantify the reactivity coefficient since it is a change in temperature that causes thermal expansion and mechanical interaction. Thus, a reactor may have a thermal expansion temperature coefficient of reactivity or even more specifically a thermal expansion/mechanical interaction temperature coefficient of reactivity. The thermal expansion temperature coefficient of reactivity in two reactors is described below followed by a demonstration of using finite element analysis to model the thermally driven geometric changes followed by use of a Monte Carlo simulation to determine the corresponding multiplication factor value.

1.2 Godiva-IV and Experimental Breeder Reactor-II

Two reactors serve as the test bed for evaluating the analysis approach described in this chapter. One reactor a uses comparatively simple design and the other is significantly more complicated. The simple reactor design is called Godiva-IV. The Godiva-IV reactor, see **Figure 1**, is unique in that it is designed to provide a burst

Figure 1.
Godiva-IV reactor [1].

of neutrons rather than being designed for extended steady state power production. The Godiva-IV reactor is very compact with a simple cylindrical shape with a 178 mm diameter and a 156 mm height. The reactor design uses a solid construction of approximately 66 kg of 93% enriched uranium alloyed with 1.5 wt.% molybdenum. No active cooling arrangement is used. The reactor construction is somewhat more complicated than a monolithic cylinder of enriched uranium. The Godiva-IV reactor uses six ostensibly equal rings. Three stacked cylinders of differing heights are located within the six rings to complete the overall cylindrical shape. Three large C-clamps are attached to the outer radius for the reactor to restrain the fuel movement during burst operations.

When the Godiva-IV reactor is operated, a large power pulse occurs and heat from the fission process is deposited in the uranium alloy. The heat causes a temperature increase and subsequent thermal expansion. As the individual components of the reactor expand, they mechanically interact. As the reactor components expand, neutron leakage from the reactor increases which leads to a decrease in the multiplication factor and subsequent termination of the reactor power pulse. Thus, Godiva-IV has a negative reactivity temperature coefficient. That is, as the reactor temperature increases, the resulting reactivity change is negative which provides an inherent shutdown mechanism.

The other reactor used to evaluate the analysis approach described in this chapter is the Experimental Breeder Reactor-II (EBR-II) [2]. The EBR-II design is significantly more complicated than Godiva-IV, see **Figure 2**. EBR-II uses liquid sodium metal as the coolant. The fuel is 67% enriched uranium metal alloyed with a collection of various metals totaling 5 wt.%. The fuel is formed into individual 3.3-mm diameter pins along with stainless steel cladding. The fuel portion of the pins is 343 mm long while the cladding portion is 638 mm long. The additional length of the cladding allows for the containment of fission product gasses. A collection of 91 fuel pins are arranged into a hexagonal configuration which is commonly referred to an assembly. The 91 fuel pins in each assembly are contained within a stainless-steel hexagonal duct. The EBR-II reactor core consists of an arrangement of 637 assemblies. The core is fundamentally divided into two regions, a driver region containing the fissile material, and a blanket region containing depleted uranium. Within the driver region there are approximately 100 assemblies including control rods, experimental assemblies, stainless steel dummy assemblies, stainless steel reflector assemblies, and assemblies that use reduced fuel content. Surrounding the driver region is a collection of approximately 500 assemblies constructed of depleted uranium. The depleted uranium assemblies absorb neutrons that leak for the driver region to transmute depleted uranium to plutonium to breed new reactor fuel.

DOI: http://dx.doi.org/10.5772/intechopen.93762

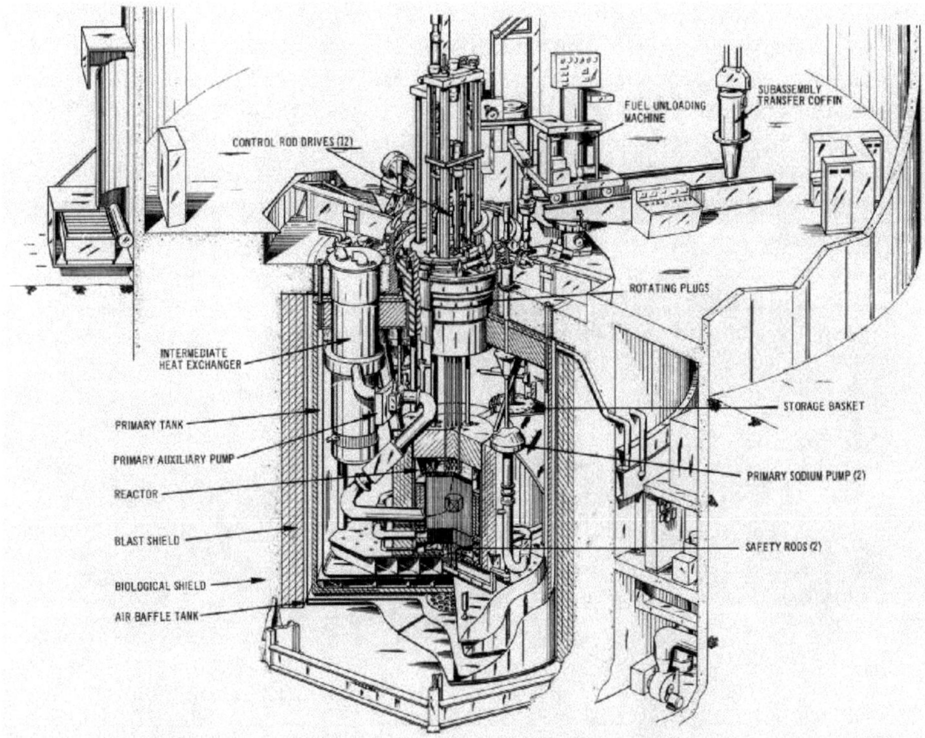

Figure 2.
EBR-II.

As EBR-II ascends to its operating power, heat from the fission process causes the fuel pins, hexagonal ducts, and all other components of the reactor to expand due to the temperature increase. The components undergo a complicated process involving thermal expansion and mechanical interaction. While the Godiva-IV thermal expansion process which leads to comparatively simple thermal expansion and mechanical interaction, the thermal expansion and mechanical interactions in EBR-II are significantly more complicated and must be subdivided into different areas. One area that is comparatively simple to understand and evaluate is the spacing of the assemblies in the hexagonal arrangement. As the reactor temperature increases during the reactor assent to power, the grid plate that holds the fuel assemblies thermally expands and the spacing between the fuel assemblies increases which results in increased neutron leakage and a decrease in the multiplication factor. A much more complicated process involves the thermal expansion and mechanical interaction of the stainless-steel hexagonal assembly ducts. Measuring and calculating the reactivity effect of the hexagonal duct thermal expansion and mechanical interaction is particularly challenging. The analysis method described in this chapter is used to evaluate the reactivity coefficient associated with the thermal expansion driven spacing of the assemblies along with the much more complicated reactivity coefficient associated with the thermal expansion and mechanical interaction of the fuel assembly hexagonal ducts.

2. Method

One of the difficulties with quantifying a geometric temperature coefficient is the complexity of the thermal expansion. Thermal expansion coefficients are nominally nonlinear, leading to different rates of expansion depending on how hot the geometry is in a given location. This leads to nonlinear thermal expansion. This is an important concept to understand because it drives the necessity for using more complex structural analysis techniques than first principles expansion. This is especially true when geometric expansion is mechanically restrained by other expanding materials.

The key to successfully quantifying a thermal expansion derived temperature coefficient is not the calculation of the coefficient itself, but more the mechanical model that is used to derive the geometry changes. To that end, finite element analysis is used to provide a high fidelity mechanical input into the Monte Carlo simulation [3]. **Figure 3** shows the generalized process for quantifying the temperature coefficient.

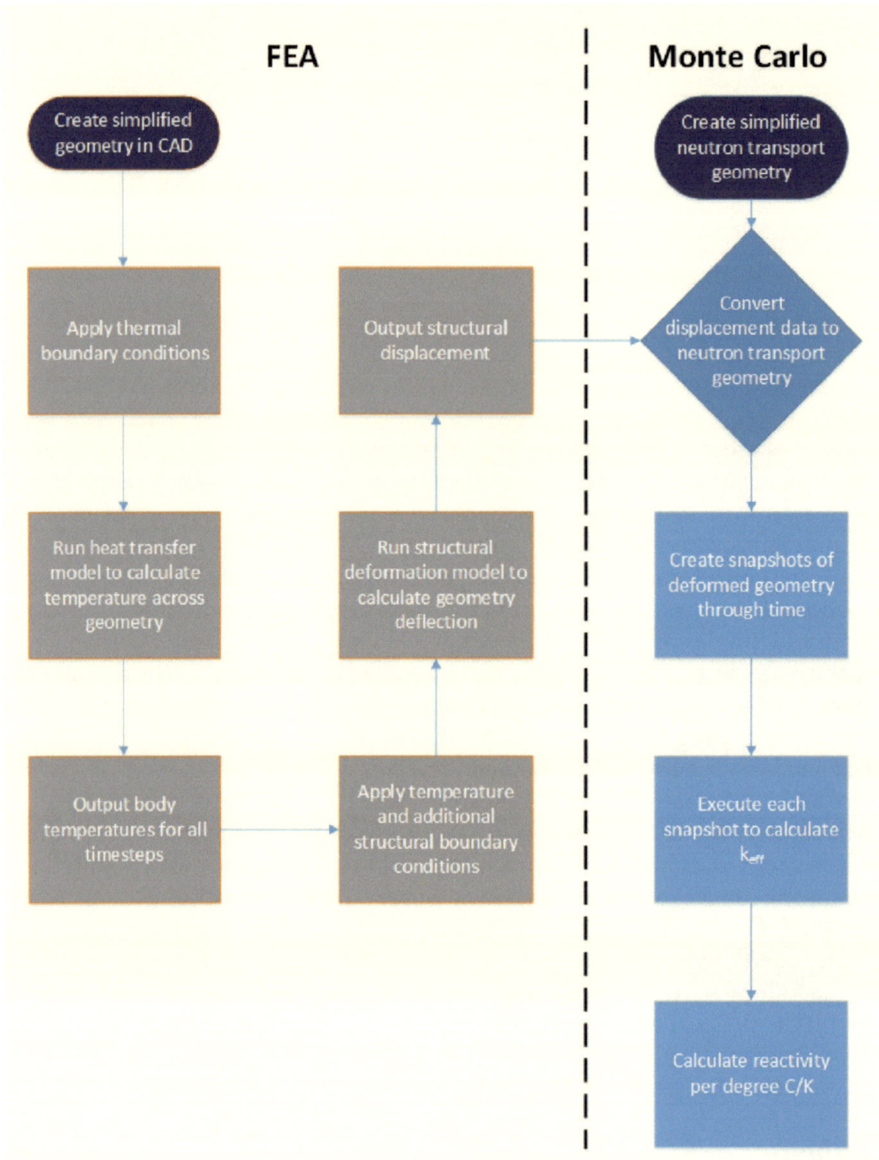

Figure 3.
General process flow.

2.1 Finite element analysis

Regardless of the source of the geometry information, whether an existing CAD model is defeatured or built from scratch, a simplified CAD geometry should be generated. The simplified geometry should contain enough information such that any complex expansion is captured, but simple enough to reduce the overall element count. A common example is removal of bolts and generally any small features from large geometries. FEA models in general run the risk of being too-large-to-compute without using resources unavailable to the typical engineer. Keeping total element count to a minimum is a driving factor when constructing an FEA model.

2.1.1 Mesh size

Exceeding 10 million nodes in a given model almost certainly means the model cannot be executed on a workstation in any reasonable amount of time. The reason for this is the sheer size of the data generated. A 10 million-node model requires 10 million positions (x, y, z), temperatures, and displacements (x, y, z) for one solution step. A double precision number requires eight bytes leading to each node requiring seven-, eight-byte numbers (56 bytes). 56 bytes per node applied to a 10 million-node mesh leads to 560 MB to store just the results of the model for 1 timestep or substep, not including the other required parameters for the solution, heat flux, power, boundary conditions, structural support, etc. Assuming the model requires several hundred timesteps and thousands of substeps, the total data requirement becomes multiple terabytes that needs to be loaded into memory. At the time of writing, several terabytes of memory was only available on very high-end workstations and was problematic for large HPC machines due to the memory allocation per CPU.

Given the difficulties noted above, simplifying the CAD model to reduce node count is critically important. The limited memory should prioritize the expansion effects not necessarily geometric fidelity. Even with these reductions, the model might take weeks of runtime to complete.

2.1.2 Boundary conditions

Boundary conditions are required as inputs into the FEA models. The boundary conditions are what simulate the reactor state that causes the structural change. They are divided into two types, thermal and structural.

The thermal boundary conditions consist of heat transfer coefficients and a thermal load. The thermal load will nominally be the fission source distribution based upon total power output. Determining the heat transfer coefficients can be done either using calculated values, measured values or a combination of both. Applying this method to a real system generally necessitates both. For example, a coolant flow rate is known for the entire reactor but is unknown on an assembly-by-assembly basis. The main goal of determining boundary conditions is to take what is known, and calculate what is needed for the FEA simulation. A similar problem exists with the thermal boundary conditions as with the mesh size, creating individual boundary conditions for every assembly and coolant channel can lead thousands of boundary conditions. It will be up to the user to determine if a thermal hydraulic simulation is required to calculate the heat transfer coefficients, or if hand calculations can suffice.

The structural boundary conditions in many ways are easier and less numerous. This is because the boundary conditions are only needed to simulate real restrains and conditions to aid in FEA convergence. An example of a real restraint is supporting the body in space such that it does not fall forever due to gravity. This is known as rigid body movement.

Convergence aids are sometimes required such that the simulation will converge. Convergence aids can be limiting body movement to a reasonable amount or declaring that a body cannot move during a particular substep. The primary pitfall with structural boundary conditions will be to overconstrain the model such that whatever subtle structural effect that drives the thermal coefficient is not nulled by the boundary conditions.

The boundary conditions need to be expansive enough to both simulate the reactor state and give enough information to allow the model to come to a solution. Additionally, the boundary conditions need to not overly constrain the model such that multiple solutions exist for a given state.

2.1.3 Timesteps

Selection of timestepping in FEA is another balance of model fidelity and analysis time. Finer timestepping leads to more information captured for a given effect, but can lead to longer computation times. For example, if the model has 100 steps that need a week of computation time to simulate 100 seconds of reactor time, then how are those steps distributed such that the necessary effects are captured? Are 90 steps used over 2 seconds of reactor time enough to capture the effect, with 10 steps used for 98 seconds of reactor time? Answering that question is highly problem dependent and takes multiple iterations to refine. Nominally, high fidelity stepping is required when the model is undergoing rapid geometry changes. For example, a pulse reactor can go from room temperature to several hundred degrees in a matter of milliseconds. During that time, many solution steps are required since the model is changing significantly between each solution step, whereas during cool-down of that same system, the model is changing slowly as conduction takes place.

2.2 Monte Carlo neutron transport

One of the problems stated in a previous section was the lack of modeling fidelity to capture subtleties in geometric changes. While the solution for the mechanical input utilizes unstructured mesh to define the geometry, Monte Carlo tools use more simplified geometry. Some Monte Carlo codes can take an unstructured mesh as an input to create their geometries, but nominally, some amount of translation will be needed to take the unstructured mesh and import it into a Monte Carlo code. Details that were required for the FEA may not be needed for the neutron transport.

Determining the multiplication factor using the Monte Carlo method requires four fundamental items, first, explicit geometric and material descriptions, second, detailed material nuclear property data, third, mathematical processes for sampling nuclear data, and fourth, a method for generating a string of numbers that satisfy rigorous tests for randomness. Using the four fundamental items, a simulation can be conducted for an individual neutron. The simulation begins by selecting the initial birth location for a neutron. The location is initially specified by the analyst but is later selected based on the location of fission locations from a prior generation. Once the birth location is known, the neutron energy can be randomly selected using numbers from the string of numbers that satisfy the randomness criteria and the mathematical distribution of possible neutron energies. The neutron direction can then be determined by randomly selecting an azimuthal and polar direction in the case of an isotropic direction assumption. With the neutron energy and direction being randomly selected, the distance the neutron travels before colliding with a nucleus can be randomly determined based on nuclear data associated with the probability of interaction, commonly referred to as a cross-section. Once the distance traveled is known, the type of interaction (e.g., scattering, absorption, and fission) can be randomly determined based on the ratio of cross-sections for the various interactions. It is also possible that the selected distance may result in the neutron leaking from the system and a new neutron must be generated. In the case of a scattering event, a new direction and neutron energy, based on collision mechanics and nuclear data, are randomly selected, and a new path length is selected. In the case of absorption, tracking of that neutron is discontinued, and a new neutron must be generated. If a fission event is selected, the location is recorded, and the number of neutrons produced by the fission event is randomly selected [4].

To accelerate the process, the analog simulation described above is modified with mathematically justified non-analog variance reduction techniques. These non-analog variance reduction techniques are selected based on the trade-off between computational time and a reduction in the statistical uncertainty of the result. For example, a process referred to as survival biasing is commonly applied where neutrons that are selected for absorption are only "partially" absorbed thereby allowing the remaining portion of the neutron to continue being tracked. The general idea is that it is more efficient to track a portion of a neutron than to track a neutron for an extended history only to have it eliminated in a meaningless reaction. As long as the variance reduction technique maintains a fair game, it can be used. The process, using analog and non-analog techniques, is repeated a great number of times, and then parameters of interest such as the multiplication faction can be inferred. The multiplication factor is inferred by the ratio of neutrons generated in one generation to the number of neutrons generated in the prior generation. A simulation can require more than 10^{12} random numbers, billions of neutrons, and thousands of generations to obtain sufficient statistical confidence in the result. For the work discussed in this chapter, the Monte Carlo code MCNP® was used for the multiplication factor calculations [5].

2.2.1 FEA interface to neutron transport code

Nominally, the results generated from structural FEA will be nodes that have been displaced in space. These nodes will need to be translated to the Monte Carlo neutron transport geometry definition. Even using unstructured mesh as an input will require some modification and additional geometry because not all of those parts were required for the FEA, but might need to be in place for the neutron transport. As stated previously, the mesh size will need to be kept to a minimum, hence some amount of defeaturing took place. Some of those features will need to be restored for the Monte Carlo neutron transport such that the particle population is simulated correctly. An example of geometry that would be removed for the FEA, would be explicit detail of the fuel pins in a nuclear fuel assembly. The FEA nominally would not require such detail, opting instead for bulk heating of the channel. Adding geometry after the FEA presents the problem of fitting un-deformed geometry into deformed spaces. Resolving this issue requires the use of custom computer codes to perform the geometry translation and geometry checking to make sure there are no overlaps. These codes nominally are custom to the particular reactor or nuclear system. In the following section "A Complex Example," EBR-II required a code called MICKA to perform the Monte Carlo geometry construction and translation [6].

2.2.2 Quasi-static snapshots

Nominally, thermal expansion reactivity coefficients that are nonlinear need to be analyzed through the whole power range of the reactor to capture all of the possible geometry states. The FEA will calculate the geometry displacement through the power-band and then export the data. That data will be exported and translated as a series of geometry snapshots with each snapshot representing the deformation of the reactor at particular power level.

2.2.3 Temperature coefficient calculations

The quasi-static snapshots are individually analyzed for their respective multiplication factor. Each individual snapshot is not intrinsically valuable because temperature coefficients in general represent a trend over a particular range. The important value to calculate over these snapshots is the change in multiplication factor, reactivity. Each reactivity point associated with a particular bulk temperature of the reactor is plotted. The slope of the linear fit of those reactivity points is the temperature coefficient. For nonlinear temperature coefficients, a set of linear fits are derived where each coefficient has an associated temperature band where the coefficient holds true. Before demonstrating the fitting process, change in multiplication from a noncritical state needs to be discussed.

Change in multiplication from critical was shown in Eq. (2). While change from critical does have a use, most real multiplying systems are never perfectly critical (k = 1), they are nominally slightly super or subcritical. This holds true for analyzed systems as well. Monte Carlo methods by definition have uncertainty associated with whatever values are calculated and rarely yield k = 1. A more common occurrence is k = 0.998 ± 0.004. The previous value would be considered critical in any real sense; however, from a calculation perspective it is noncritical. A modification to Eq. (2) is needed. Eq. (3) can be used for change in the reactivity.

$$\Delta\rho = \frac{(k_2 - k_1)}{k_2 k_1} \tag{3}$$

With an understanding of change in reactivity, a linear temperature coefficient can be determined from the Monte Carlo analysis. A linear regression is applied to the temperature dependent reactivity. This yields Eq. (4).

$$\mathrm{y} = mx + \mathrm{b} \tag{4}$$

The slope of the previous equation is the temperature coefficient. If the particular coefficient is nonlinear, multiple regressions will be required. The coefficient can be expressed as a nonlinear equation; however, temperature coefficients are traditionally expressed as linear quantities over temperature ranges.

3. A simple example

The following sections demonstrate how finite element analysis can be applied to nuclear systems to calculate extremely complex phenomena. The tools used in these works were ANSYS, for the finite element analysis, and MCNP® was used for the neutron transport [5, 7].

3.1 Godiva-IV

Confirmation of the methodology described above begins with a relatively simple geometry reactor. While certainly not a homogeneous single component bare cylindrical reactor, the Godiva-IV reactor provides an excellent case to apply the methodology described above and compare the modeling results to measured results. In particular, the Godiva-IV reactor temperature coefficient of reactivity is dominated by thermal expansion with limited mechanical interaction effects. Furthermore, the Godiva-IV reactor has been thoroughly characterized and detailed descriptions are available to allow construction of both the FEA model as well as the Monte Carlo model. Finally, detailed temperature measurements and the corresponding reactivity have been recorded which allow for comparison with the modeling results.

3.1.1 Thermal analysis

The thermal analysis required several boundary conditions as inputs into the model. The most important was the temperature data taken from an experiment on the Godiva-IV. The temperature data provided the pulse shape and total magnitude of the temperature rise. It also conveyed the time dependency of the FEA model. Experimental measurements are important to creating accurate models. They provide a more accurate input, depending on the quality of the measurement, than necessarily calculating and input from first principles. The experiment input data were for a 1.029$ reactivity insertion with a 68°C temperature rise over 300 seconds.

The second type of boundary conditions applied were the heat transfer coefficients, primarily the convection coefficients. These were hand calculated from heat and mass transfer equations. Hand-calculating these coefficients is normally required because these values are not normally measured for these facilities. For Godiva-IV specifically, capturing the thermal expansion temperature coefficient means modeling the thermal conduction paths as well as the convection into the room. The temperature differential of all of the components as they heat up and subsequently cool due to convection to the room is the primary driver to the complexity of the expansion. **Figure 4** shows the thermal FEA results and the temperature differentials. For more information specifically on the boundary conditions, see the reference [8].

Given the rapid structural response of the Godiva-IV pulse, the analysis type chosen was transient. The solution steps were focused on the pulse. Of the 400 solution steps, 300 steps surrounded the pulse that consisted of 10s, with the other 100 steps covering 290 seconds. The reason for this was that the model was rapidly changing during the initial nuclear heating, while the rest of the steps only contained relatively slow thermal conduction. After the temperature analysis model completed, the temperature data were exported to the structural analysis model.

3.1.2 Structural analysis

As stated previously, the structural boundary conditions are less numerous but can be more difficult to determine. For Godiva-IV, the primary structural boundary conditions were, support of the safety block, control rods, and providing a fixed support for the back side of the clamps. These boundary conditions were more straightforward than for typical models, weak springs were not required, and fixed supports were the only type of boundary conditions that were necessary. **Figure 5** shows exaggerated displacement at the end of the temperature input data. The exaggeration was required because the structural displacement on average was 0.2 mm and imperceptible to the human eye.

The structural analysis had similar timestepping as the thermal analysis. The data generated from the FEA were a set of averaged displacements on particular

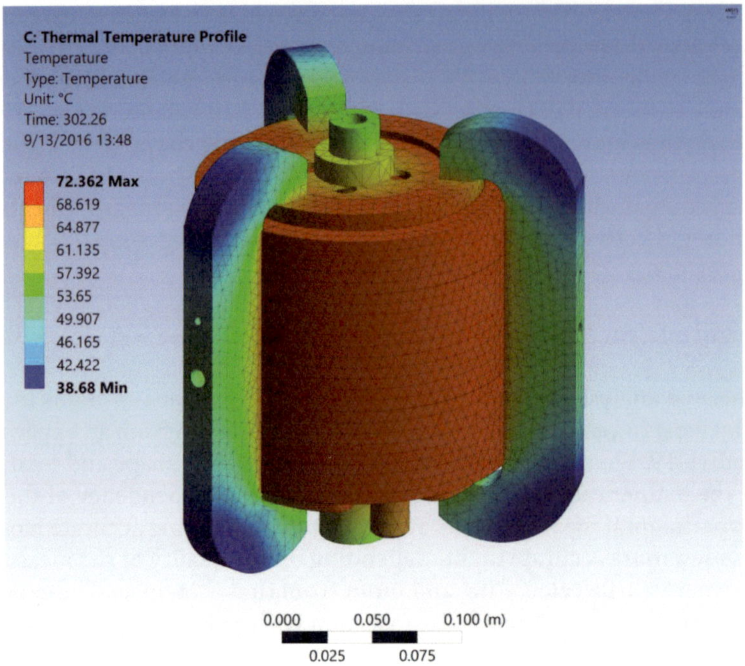

Figure 4.
GODIVA-IV thermal analysis results.

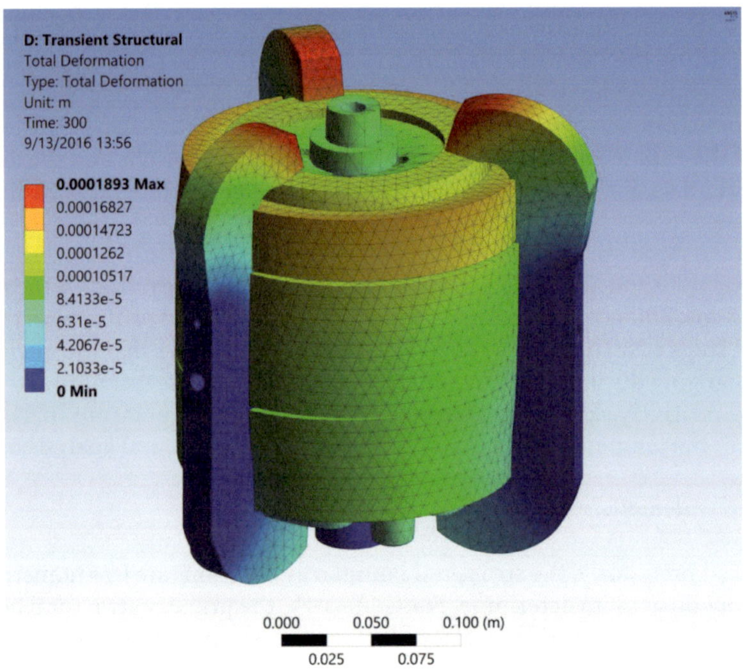

Figure 5.
Exaggerated structural displacement.

surfaces. These surfaces surrounded the curved faces of the fuel rings. The fuel rings were the focus because only the fuel movement and expansion matters to the neutronics of the reactor.

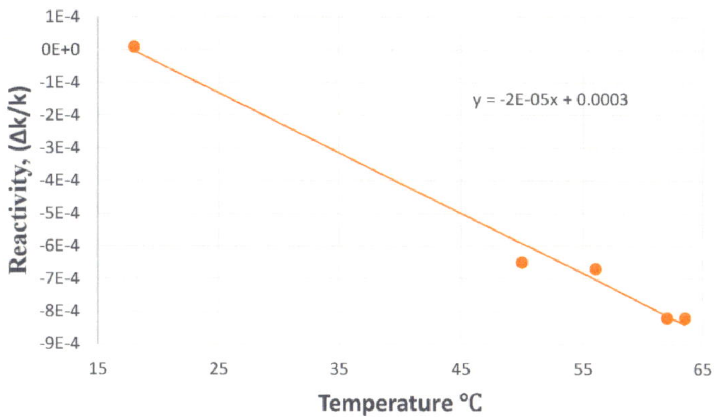

Figure 6.
Godiva-IV temperature coefficient result.

Source	Temperature coefficient (Δk/k/°C)
Measured	−1.95E−05
Calculated	−2E−05

Table 1.
Measured and calculated Godiva-IV temperature coefficient.

3.1.3 Neutron transport

The exported data were applied to the neutron transport model where a series of models were created, each with a different set of displacements per an average temperature. These results are shown in **Figure 6**. The slope of the linear regression is −2E−05 Δk/k/°C. The comparison to the measured value is shown in **Table 1**. This is the temperature coefficient of the Godiva-IV reactor. The results demonstrate that a coupling method of FEA and Monte Carlo Neutron Transport has to be potential to accurately predict the temperature coefficient.

4. A complex example

4.1 EBR-II

With a comparatively simple application providing excellent comparison results, a more challenging application is warranted. As noted above, the EBR-II design includes numerous fuel assemblies, molten sodium coolant, and a complicated thermal expansion and mechanical interaction process. Detailed characterization of the reactor components and materials along with measurements of control rod critical positions and corresponding bulk coolant temperatures are available [9]. These measured data allow confirmation of the methodology for certain aspects of the reactivity coefficients present in EBR-II such as the thermal expansion of the reactor grid plate. Extrapolation of the methodology can then occur for the more complicated thermal expansion and mechanical interaction of the assembly hexagonal flow ducts.

While no reliable method of measuring the reactivity coefficient associated with the hexagonal duct expansion and mechanical interactions is known to exist, the methodology described here can be applied and a reliable estimate of the reactivity coefficients can be obtained.

4.1.1 Thermal analysis

EBR-II required more extensive thermal boundary conditions than Godiva-IV which was considered the simple system because the heating was simple conduction through the materials and the ultimate heat sink was convection into the room air. EBR-II was more complicated because there was forced convection using liquid sodium that flowed over the fuel elements. The ultimate heat sink was a series of heat exchangers that cooled the sodium. Heating was also not symmetric from assembly to assembly. Modeling this complex behavior would require a complex thermal-hydraulic model to simulate the various coolant channels. Creating this model would have substantially complicated the thermal FEA analysis and additional would require input information that was not measured at EBR-II. Instead, a simple cooling model was developed for each assembly. The cooling model stated

Figure 7.
EBR-II simplified simulation model.

that the sodium inside of the duct entered the bottom of the coolant channel as cold sodium, and over the part of the channel where the fuel was located, the sodium was heated such that the outlet temperature match measurements taken at the EBR-II. Each fuel assembly type had a different cooling profile. **Figure 7** shows the different assembly types in the EBR-II FEA model.

Simplifying the coolant channels in this manner was sufficient because previous work done on the EBR-II suggested that duct-bowing was entirely driven by the temperature profile of the duct material and not by the internal structures. Thus, only the duct needed to be heated correctly.

The power input for EBR-II was derived from a linear interpolation of the ascent to power. All of the heat generation inputs were linearly scaled over timesteps. The timing did not match the real ascent to power, but that was not necessary since the model would be in thermal equilibrium for each calculated step. The more important aspect was that the thermal model would be a series of steps, each step corresponding to a different power level.

4.1.2 Structural analysis

The structural analysis required a simple boundary condition to hold the model in place, as well as a boundary condition to fix the center duct. Fixing the center duct meant that it was not allowed to thermally expand and was considered a rigid body. This was necessary to achieve convergence. Without fixing the center duct, the model could not resolve the contact overlap that existed between the ducts on the first solution step.

The structural FEA required significantly more time to solve than Godiva-IV due to the sheer size of the model (~5 million nodes) and the complexity of the thermal expansion. **Figure 8** shows an exaggerated displacement of the ducts. The southeast quadrant shows how the differences in the assembly, types, and powers can impact thermal expansion. Additionally, it demonstrates why FEA was necessary to capture all of the geometric detail of the duct-bowing temperature coefficient.

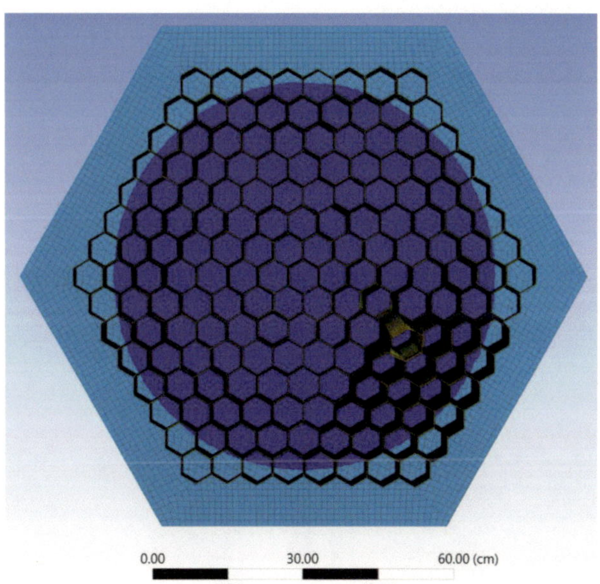

Figure 8.
EBR-II exaggerated structural displacement.

4.1.3 Neutron transport

Similar to the Godiva-IV model, the displacement data were exported out of ANSYS and imported into a series of MCNP® models [7]. The major difference was in the translation method. The Godiva-IV translation was averaging nodal thermal expansion and manually applying the change in radii and heights to the MCNP® input files. That approach was prohibitive for EBR-II because the resulting data exceeded 1 TB. A custom code called MCNP® Input Card and KCODE Architect (MICKA) was written to perform the node translation and MCNP® input construction. The MCNP® model for the EBR-II was itself expansive and required special data handling. More inf0rmation can be found in the reference [6].

One additional difficulty with the translation was that MCNP® cannot model a bowed-duct, only a straight hexagonal duct. To overcome this geometry limitation,

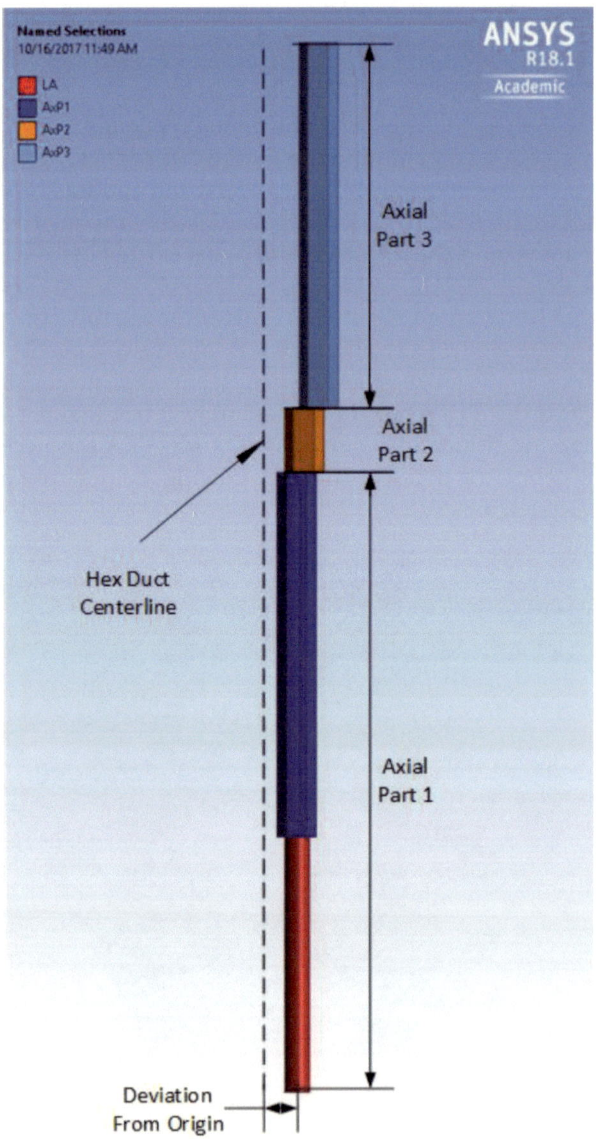

Figure 9.
Axial sections to simulate a bowed-duct in MCNP®.

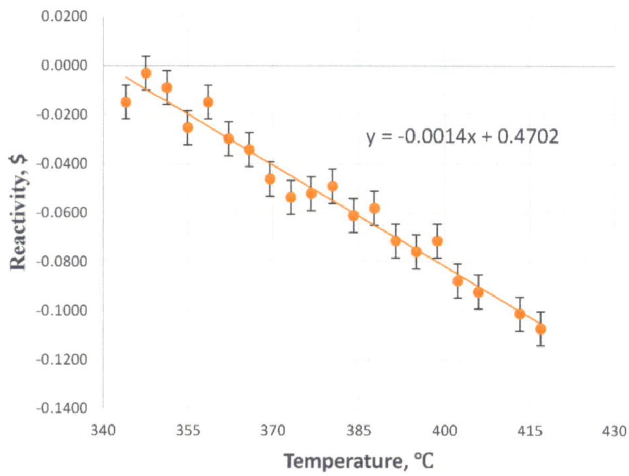

Figure 10.
Temperature coefficient results for duct-bowing coefficient.

the straight duct was divided into axial sections. Each axial section was moved in space to approximate a bowed-duct. **Figure 9** shows an example of the axial slices in an assembly.

After the translation of the nodal data from the FEA to MCNP®, the analysis process was similar to that of Godiva-IV. A series of snapshots at various bulk temperatures were taken and a linear regression was performed to calculate the slope of the points. **Figure 10** shows the results of the reactivity change per degree. The coefficient was calculated to be −1.4E−03 $/°C. While the data had a clear linear trend, some nonlinearity existed in sets of data points at lower bulk temperatures. This was consistent with historical measurements at EBR-II where lower powers exhibited a nonlinear trend in the reactivity change.

5. Conclusions

The energy released from the nuclear fission process drives complicated thermal expansion and mechanical interactions in nuclear reactors. These expansions and interactions subsequently cause changes in the neutron chain reaction balance within a reactor which results in further changes in thermal expansion and mechanical interactions. Measurement of these coupled phenomena occurring within a reactor has proven to be elusive. However, coupling finite element analysis with Monte Carlo neutron transport analysis provides a pathway to simulate the thermal expansion and mechanical interaction driven by the energy released in the neutron-induced fission process and then to subsequently determine fundamental nuclear parameters, namely, thermal expansion temperature coefficient of reactivity.

There are important safety implications associated with the thermal expansion temperature coefficient of reactivity and its relation to other temperature coefficients of reactivity. Knowing both the sign and magnitude of individual coefficients allows reactor designers to predict how a reactor will behave under transient conditions.

Using the coupling of finite element analysis and Monte Carlo neutron transport analysis, the thermal expansion temperature coefficient of reactivity was determined for the Godiva-IV reactor and found to be within 3% of the experimentally measured value.

The coupling technique was also used to determine the thermal expansion temperature coefficient of reactivity for EBR-II. The thermal expansion and mechanical interactions within EBR-II are sufficiently complex that experimentally measuring the isolated thermal expansion temperature coefficient of reactivity was not possible. However, using the coupling technique, a calculated value of −1.4E−03 $/°C was determined for the thermal expansion temperature coefficient of reactivity. This result fits well with integral EBR-II reactivity coefficient measurements.

With the Godiva-IV comparison results and the EBR-II results, it can be concluded that coupling finite element analysis with Monte Carlo neutron transport analysis provides a powerful technique for determining important reactor safety parameters. The technique can be applied to existing reactors and reactors proceeding through the design process which gives reactor operators and designers greater confidence in reactor operating characteristics and safety margins.

Author details

Chad Pope[1*] and Edward Lum[2]

1 Idaho State University, Pocatello ID, USA

2 Los Alamos National Laboratory, Los Alamos NM, USA

*Address all correspondence to: popechad@isu.edu

References

[1] Goda J, Bounds J, Hayes D, Sanchez R. Godiva IV Startup at NCERC: Delayed Critical through Prompt Critical. Los Alamos: Los Alamos National Laboratory, LA-UR 14-24398; 2014

[2] Lum E. Simulating the Katana Effect Monte Carlo Neutron Transport Combined with Finite Element Analysis to Calculate Negative Reactivity Due to Duct-Bowling. Pocatello: Idaho State University; 2017

[3] Finck PJ. A Technique for Computing the Reactivity Feedback Dueto Core-Assembly Bowing in LMFBR's. Argonne: Argonne National Laboratory; 1987

[4] L. L. Carter and E. D. Cashwell, Particle Transport Simulation with the Monte Carlo Method, TID-26607. New Mexico: Los Alamos National Laboratory; 1975.

[5] Goorley T. MCNP6.1.1-Beta Release Notes. New Mexico: Los Alamos National Laboratory; LA-UR-14-24680; 2014

[6] Stewart R, Lum E, Pope C. Design of an experimental breeder reactor run 138B reactor physics benchmark evaluation management application. Journal of Nuclear Science and Technology. 2019;**57**(3):323-334

[7] ANSYS, Inc. ANSYS Mechanical, Release 18.1 [Internet]. 2017. Available: https://www.ansys.com/products/structures/ansys-mechanical

[8] Lum E, Pope C. GODIVA-IV reactivity temperature coefficient calculation using finite element and Monte Carlo techniques. Nuclear Engineering and Design. 2018;**331**:116-124

[9] Lum ES, Pope CL, Stewart R, Byambadorj B, Beaulieu Q. Evaluation of run 138B at experimental breeder reactor II, a prototypic liquid metal fast breeder reactor (EBR2-LMFR-RESR-001-CRIT). In: Nuclear Energy Agency International Reactor Physics Experiment Evaluation Project. Paris, France: Nuclear Energy Agency; 2018

Chapter 8

Finite Element Analysis and Its Applications in Dentistry

Vinod Bandela and Saraswathi Kanaparthi

Abstract

Finite Element Analysis or Finite Element Method is based on the principle of dividing a structure into a finite number of small elements. It is a sophisticated engineering tool, which has been used extensively in design optimization and structural analysis first originated in the aerospace industry to study stress in complex airframe structures. This method is a way of getting a numerical solution to a specific problem, used to analyze stresses and strains in complex mechanical systems. It enables the mathematical conversion and analysis of mechanical properties of a geometric object with wide range of applications in dental and oral health science. It is useful for specifying predominantly the mechanical aspects of biomaterials and human tissues that cannot be measured *in vivo*. It has various advantages, can be compared with studies on real models, and the tests are repeatable, with accuracy and without ethical concerns.

Keywords: finite element analysis, finite element method, stress, dentistry, implants

1. Introduction

Dentistry is the fastest growing branch of medical field, deals with the study of diagnosis, prevention, and treatment of diseases, disorders, and conditions of the oral cavity. Although primarily associated with teeth, the field of dentistry is not limited to teeth but includes other aspects of the craniofacial complex including the temporomandibular joint (TMJ) and other supporting, muscular, lymphatic, nervous, vascular, and anatomical structures.

Virtually, every phenomenon in nature; whether biological, geological or mechanical, can be described with the aid of law of physics, in terms of algebraic, differential or integral equations relating various quantities of interest. Finite Element Analysis (FEA) or Finite Element Method (FEM) is a computer-based numerical method to analyze the structure based on the principle of dividing a structure into a finite number of small elements that are connected with each other at the corner points called nodes. For each element, its mechanical behaviour can be written as the function of displacement of the nodes. These nodes when subjected to certain loading conditions results in behaviour of the model similar to the structure it represents. When a computer analysis is performed on this, a system of simultaneous equations can be solved to relate all forces and displacement of the nodes. From this, stress and strain can be established in each element and the whole structure can be evaluated [1].

There were many articles published before on FEA and their uses, this chapter mainly focus on the brief application of FEA in dentistry, apart from the historical perspective, planning of analysis, workflow of FE study, merits, shortcomings, and future of FEA.

2. Historical perspective

The first researcher who developed this technique was Richard Courant, a mathematician with the main goal of minimizing the calculative procedures in gaining absolute solution to bio-mechanical system in early 1940's. Turner et al., in 1956 attempted to describe this method by developing broader definition of these numeric analyses in aeronautical engineering. Ioannis Argyris and R.W Clough coined the term 'Finite Element' in 1960. Weinstein et al., in 1976 used this technique in implant dentistry to evaluate various loads of occlusion on implant and adjacent bone. Since then, evolution of this technique has been observed in a very rapid and sophisticated scale in micro-computer as well as analysis of large-scale structural system [1, 2].

3. Planning of analysis

3.1 Pre-processor

In this stage, the material properties are assigned (**Figure 1**) [1, 2].

3.1.1 Specifying the title

It is specifying the name of the problem. This is optional but very useful, especially if a number of design iterations to be completed on the same base model.

3.1.2 Setting the type of analysis

In this, the type of analysis that is going to be used is done. Eg: structural, fluid, thermal or electromagnetic etc.

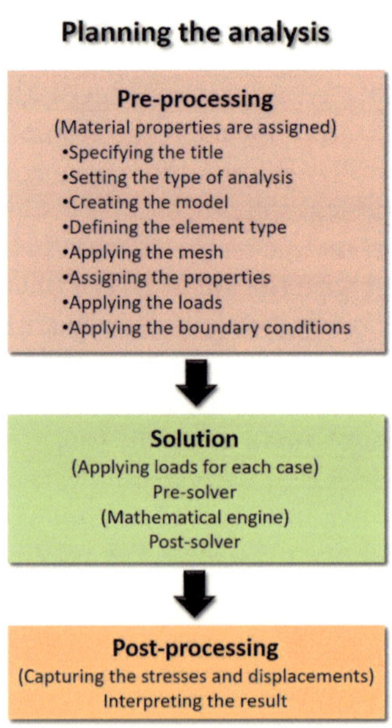

Figure 1.
Planning of analysis.

3.1.3 Creating the model

The model is drawn in 1-D (dimensional), 2-D, or 3-D space in the appropriate units (M, mm, inch etc.).

3.1.4 Defining the element type

This may be 1-D, 2-D, or 3-D.

3.1.5 Applying a mesh

Mesh generation is the process of dividing the analysis continuum into a number of discrete parts or finite elements. The finer the mesh, the better is the result but longer the analysis time.

3.1.6 Assigning properties

Material properties (Young's modulus, Poisson's ratio, density and if applicable coefficient of expansion, friction, thermal conductivity, damping effect, specific heat etc.) have to be defined in this step. In addition, element properties may need to be set.

3.1.7 Applying loads

Usually, some type of load is applied to the analysis model. The loading may be in the form of a point load, a pressure or a displacement in a stress (displacement) analysis. The loads may be applied to a point, an edge, a surface or even a complete body.

3.1.8 Applying boundary conditions

When applying a load to the model, in order to stop accelerating infinitely through the computer's virtual ether, at least one constraint or boundary condition must be applied. A boundary condition may be specified to act in all directions - axes (x, y, z) or in certain directions only. They can be placed on nodes, key points, areas or on lines.

3.2 Solution

This part is fully automatic and it can be logically divided into three main parts: the pre-solver, the mathematical engine and the post-solver. The pre-solver reads the model created by the pre-processor and formulates the mathematical representation of the model. The results are returned to the solver and the post-solver is used to calculate strains, stresses, etc., for each node within the component or continuum.

3.3 Post-processor

Here the results of the analysis are read and interpreted. They can be presented in the form of a contour plot, a table, deformed shape of the component or the mode shapes and natural frequencies if frequency analysis is involved. Most post-processors provide an animation service, which produces an animation and brings the model to life. All post-processors now include the calculation of stress and strains in any of the x, y or z directions or indeed in a direction at an angle to the co-ordinate axes. The principal stresses and strains may also be plotted or if required the yield stresses and strains according to the main theories of failure.

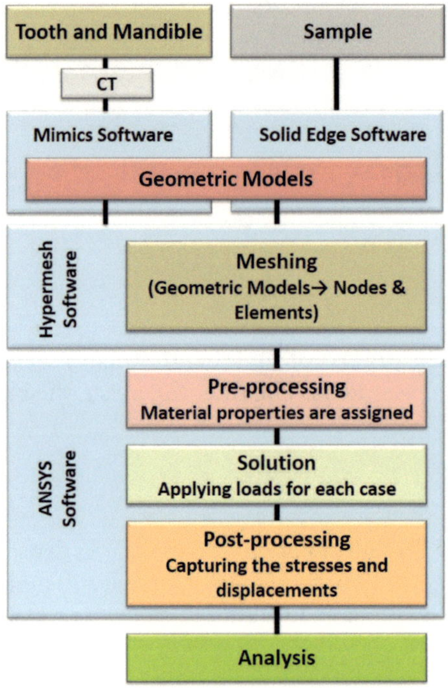

Figure 2.
Workflow of FE analysis.

In brief, the FE is a mathematical method for solving differential equations. It has the ability to solve complex problems that can be represented in differential equation form that occur naturally, in virtually all fields of the physical sciences. Accurate modeling is essential to ensure the relevance of the result for the corresponding FEA. The results solely depend on the model that has been created. Workflow of the entire finite element study is shown in **Figure 2**.

4. Application of FEA in oral radiology

Oral and maxillofacial radiology is the specialty of dentistry concerned with performance and interpretation of diagnostic imaging used in examining the dental, craniofacial, and adjacent structures. Use of FEA in this specialty helps for proper diagnosis and possibility of knowing iatrogenic effects.

Szücs et al., in 2010 analyzed the effect of removing various amounts of bone around an impacted mandibular third molar and predicted the possibility of iatrogenic fracture. FEA was used to generate 3-D models of a human mandible with impacted third molars. They found highest stress occurred during normal clenching if the surgical procedure involved the external oblique ridge. The peak stress occurred at the site of removal of the third molar, during contralateral loading of the mandible. They concluded that with FEA they could able to identify the accumulation of stress and strain at specific parts of the mandible and predicted the responses of bone to mechanical activity. FEA could prove to predict the likelihood of iatrogenic fracture of the jaws after surgical removal of mandibular bone, such as occurring during the extraction of third molar. This allow the dentists to change/modify their approach to tooth removal in certain cases [3].

Oenning et al., in 2018 simulated functional forces in a mandible model by means of FEA and then assessed the biomechanical response produced by impacted third molars on the roots of the second molar. They found areas of high-energy dissipation and compression stress in the second molar root, independently of the inclination of the impacted third molar. They concluded that, impacted third molars in close proximity with the adjacent tooth can generate areas of compression concentrated at the site of contact, suggesting an involvement of mechanical factors in triggering of resorption lesions [4].

Kihara et al., in 2019 evaluated the longitudinal change quantitatively in mandibular volume and configuration in a patient with craniofacial fibrous dysplasia (FD). The 3-D models were analyzed morphologically and volumetrically using FEA. They found FD lesion in the mandible enlarged non-uniformly and had site specificity. They suggested that compression stress induced by the occlusal force through the denture may have influence on the direction of enlargement in FD [5].

5. Application of FEA in restorative dentistry

Restorative dentistry refers to the diagnosis and integrated management of diseases of the teeth and their supporting structures and rehabilitation of the dentition for functional and esthetic requirements of an individual. Restorative dentistry. It is a broader term encompasses the dental specialties of endodontics, prosthodontics, and periodontics.

Many newer materials have been developed owing to the increasing interest in the field of esthetic dental restorations. In order to minimize the stress concentration of the restorative materials and to decrease the incidence of restorative failure; physical properties like modulus of elasticity should be near or equal to that of the natural dental tissue. Due to the lack of proper understanding on the biomechanical principles of the materials involved in restorative procedure, lead too many detrimental effects causing a restorative failure. Therefore, in order to know the behaviour of materials and dental tissue, biomechanical studies are very crucial [6, 7].

Goel et al., in 1991 investigated stress variation in the enamel and dentin adjacent to the Dentinoenamel Junction (DEJ) on FEM of maxillary first premolar. The results suggested that, because of mechanical interlocking between enamel and dentin in the cervical region is weaker than in other regions of the DEJ, enamel in this region may be susceptible to belated cracking that could eventually contribute to the development of cervical caries than other areas of tooth [8].

Rees in 2002 examined the effect of varying position of an occlusal load on the stress contour in the cervical region of a lower second premolar using a 2-D plane strain FEM. A 500 N load was applied vertically to either of the cusp tips or in various positions along the cuspal inclines. He found that, loads applied to the inner aspects of the buccal or the lingual cuspal inclines produced maximum principal stress values of up to 358 MPa, which is exceeding the known failure stresses for enamel [9].

Ausiello et al., in 2002 conducted a 3-D FEA study to identify the thickness and flexibility of the teeth adhesively restored with resin-based material. No difference in the stress relief between the application of a thin layer of more flexible adhesive with low elastic modulus and thick layers of less flexible adhesive of high-elastic modulus was found. They observed a relatively small cuspal deformation in all the models with increased cusp-stabilizing effect of ceramic inlays compared to composite restorations [10].

Ausiello et al., in 2004 investigated the composite inlay restored class-II MOD cavities the effect of differences in the resin-cement elastic modulus on stress-transmission to ceramic or resin-based during vertical occlusal loading. They found better stress dissipation in indirect composite resin-inlays. Glass ceramic inlays transferred stresses to the resin cement and adhesive layer [11].

Magne et al., in 2006 described a rapid method of generating FE models of dental structures and restorations. They evaluated five models: natural tooth, mesial-occlusal (MO), and mesial-occlusal-distal (MOD) cavities, MO, and MOD endodontic access preparations and found a progressive loss of cuspal stiffness in MO to MOD to endodontic access, as there is loss of tooth structure with these type of restorations. The natural tooth and the tooth with the MOD ceramic inlay retained 100% cuspal stiffness [7].

Ichim et al., in 2007 investigated the influence of the elastic modulus (E) on the failure of cervical restorative materials (Glass ionomer cement (GIC) and composite) and identified an E value that minimizes the mechanical failure under clinically realistic loading conditions. They found that the materials used in non-carious cervical lesions are unsuitable for restorations as they are less resistance to fracture and suggested that the elastic modulus of a restorative material to be in the range of 1 GPa [2].

Asmussen et al., in 2008 analyzed the stresses generated in tooth and restoration by occlusal loading of Class-I and Class-II restorations restored with resin composite; suggested that the occlusal restorations of resin composite should have a high modulus of elasticity in order to reduce the risk of marginal deterioration [12].

Coelho et al., in 2008 conducted a study to test the hypothesis that micro-tensile bond strength values are inversely proportional to dentin-to-composite adhesive layer thickness through laboratory mechanical testing and FEA. They found micro-tensile bond for Single Bond as increased adhesive layer thickness did not reduce Clear fil SE Bond strength [13].

Magne and Oganesyan in 2009 measured cuspal flexure of intact and restored maxillary premolars with MOD porcelain, and composite-inlay restorations and occlusal contacts (in enamel, at restoration margin, or in restorative material). They found a relatively small cuspal deformation in all the models and an increased cusp-stabilizing effect of ceramic inlays compared with composite ones [9].

5.1 Dental composites

Composites are the resin restorative materials developed to overcome the disadvantages of amalgam restorations, which are unaesthetic and toxic. Composites are filled resins, exhibit high compressive strength, abrasion resistance, ease of application, and high translucency. FEA has been in use to analyze stresses generated in teeth and restorations. It is a proven useful tool in understanding biomechanics of tooth and the biomimetic approach in restorative dentistry [14].

Lee et al., in 2007 conducted a study to measure the cusp deflection by polymerization shrinkage during composite restoration for MOD cavities in premolars, and examined the influence of cavity dimension, C-factor, and restoration method on the cusp deflection. They found that, the cusp deflection increased with increasing cavity dimension and C-factor and suggested the use of an incremental filling technique or an indirect composite inlay restoration to reduce the cuspal strain [15].

Choi et al., in 2011 analyzed the disintegration of a dental composite restoration around the margin due to contraction stress by measuring the circumferential strain on the outer surface of a ring-type dental substrate. They found increase in the marginal gap size representing the increase in the number of cracking's along the margin due to polymerization contraction [16].

Jongsma et al., in 2011 studied to find out the rationale of using whether 60% increase in push-out strength with a two-step cementation procedure of fiber posts is equivalent to the layering technique of composite restorations or not. They found two-step cementation of fiber posts lead to a decrease in internal stresses in the restoration, resulted in higher failure loads and less microleakage [17].

5.2 Dental ceramics

Dental ceramics are in-organic, non-metallic, and brittle restorative materials producing dental prosthesis that are used to replace missing or damaged dental structures which has high compressive strength and low tensile strength. FEM provides a mathematic analysis to predict strength values without the potential for errors in dental ceramics [18].

Tensile stresses tend to be more critical than compressive stresses for ceramic materials. The strength of ceramic restorations is significantly affected by the presence of flaws or other microscopic defects. Tensile stress concentration at cementation surface of the ceramic layer suggested as the predominant factor controlling ceramic failure [6].

Belli et al. in 2005 evaluated the effect of hybrid layer on distribution and amount of stress formed under occlusal loading in a premolar tooth restored with composite or ceramic inlay. They concluded that the hybrid layer has an effect on stress distribution under loading in restored premolar tooth model with composite or ceramic inlay [19].

Rezaei et al., in 2011 determined the effect of buccolingual increase of the connector width on the stress distribution in posterior FPDs made of IPS Empress. Three models of three-unit bridges replacing the first molar were prepared with the buccolingual connector width varied from 3.0 to 5.0 mm. They were loaded vertically with 600 N at one point on the central fossa of the pontic and a load of 225 N at 45° angle from the lingual side. They concluded that, increasing the connector width decreases the failure probability when a vertical or angled load is applied [20].

Thompson et al., in 2011 compared the inlay supported all-ceramic bridge with that of traditional full crown supported all-ceramic bridge. They demonstrated peak stresses in the inlay bridge around 20% higher than in the full crown supported bridge. They suggested the use of an ideal inlay preparation form and an optimized bridge design emphasizing on broadening of the gingival embrasure, so that the forces derived from mastication can be distributed adequately to a level that are within the fracture strength [21].

Matson et al., in 2012 compared the stress distribution generated in a veneer restoration of an upper central incisor to intact teeth by applying a 10 N lingual buccal load at the incisal edge. Veneers used in restorative rehabilitations for anterior teeth are retained by the adhesive systems and resin cements. These restorations are mechanically not strong, because they are made of brittle materials, but they have good retention due to the resin-dentine bonding. They recommended the use of veneers to replace enamel for rehabilitation [22].

6. Application of FEA in endodontics

Endodontology/Endodontics is the branch of dental sciences concerned with the form, function, health, injuries to and the diseases of the dental pulp and periradicular region, and their relationship with systemic health and well-being. Endodontic therapy involves either root canal filling techniques by conventional methods; or endodontic surgery with the use of biocompatible restorative materials, instruments, and techniques performed. The objective of endodontic instrumentation is to produce a tapered continuous preparation that should preserve the anatomy of root canal and maintain a good apical seal and foramen as small as possible, without any deviation from the original canal curvature [23].

During canal instrumentation, pressure is generated against the dentinal walls that may lead to inappropriate canal preparation or microcracks. These microcracks may lead to vertical fracture - one of the cause for tooth loss. During instrumentation, nickel-titanium (NiTi) are the commonly used for shaping the root canal. So, in order to perform well and avoid instrument breakage inside the canal, the material used and the technique performed should be followed meticulously. FEA helps to analyze and predict the treatment outcome [24].

Satappan et al., in 2000 analyzed the type and frequency of defects in NiTi rotary endodontic files after routine clinical use and reasons for their failure. They found torsional failure by using too much apical force during instrumentation as the more frequent cause than flexural fatigue, which resulted from the use in curved canals [25].

Hong et al., in 2003 analyzed the stress variations by vertical and lateral condensation on mandibular first molar mesio-buccal root canal by step-back technique. They found vertical condensation technique generating high stresses and the reason for vertical root fracture was due to over-force and improper operation [2].

Subramaniam et al., in 2007 compared the torsional and bending stresses in two simulated models of Ni-Ti rotary instruments, ProTaper and ProFile. They found the distribution of stresses was uniform in ProTaper model and stiffer by 30% than ProFile model, which shows ProFile is more flexible than ProTaper [26].

Kim et al., in 2008 compared the stress distribution during root canal shaping and estimated the residual stress in three brands of Ni-Ti rotary instruments: ProFile, ProTaper, and ProTaper Universal (Dentsply Maillefer). They found that the original ProTaper design showed greatest pull in the apical direction and highest reaction torque from the root canal wall while, ProFile showed the least. The residual stress was highest in ProTaper followed by ProTaper Universal and ProFile. In ProTaper, stresses were concentrated at the cutting edge [27].

Lee et al., in 2011 investigated on cyclic fatigue resistance of various Ni-Ti rotary files in different root canal curvatures by correlating cyclic fatigue fracture tests. They concluded that stiffer instrument had the highest stress concentration and the least number of rotations until fracture in the cyclic fatigue test. Increased curvature of the root canal generated higher stresses and shortened the lifetime of Ni-Ti files [28].

Belli et al., in 2011 evaluated the effect of interfaces on stress distribution in incisor models of primary, secondary, and tertiary monoblocks generated either by adhesive resin sealers in combination with a bondable root filling material or by different adhesive posts. The concept of creating mechanically homogenous units within the root dentine is theoretically excellent, but accomplishing in the canal space is challenging because bonding is compromised by volumetric changes in resin-based materials to dentine, debris on canal walls, configuration factors, and differences in bond strengths. They found stresses within roots increased with an increase in the number of the adhesive interfaces [29].

6.1 Application of FEA in post and core

A considerable amount of tooth structure lost due to caries, endodontic therapy, and placement of previous restorations will compromise the tooth structure to resume its full function to serve satisfactorily. The type of the tooth restoring and the amount of remaining coronal tooth structure are the two factors that influence the choice of technique. The second factor is probably the key important indicator in determining the prognosis a tooth that is restored. If a substantial amount of coronal structure is missing, a cast post and core is indicated [30].

The method of restoring a structurally weakened tooth is post and core system, which is most common and widely used. This system can be categorized into two; custom cast metal posts and cores that are single piece, and a two component design comprising a prefabricated post to which other core materials is subsequently adapted. While fabricating a custom post and core, the difference in the elastic modulus of dentine and post material may be a source for root structure because of stress and debonding of posts due to stress contraction of the cement. Design of the post also effects the stress distribution, which was found as the most common mode of failure. Ferrule preparation creates a positive effect in reducing the stress concentration in an endodontically treated tooth. FEM can be used in various types of materials like carbon, metal, glass fiber, and zirconia ceramic and different configurations of dowel like smooth and serrated on the stress distribution of the teeth [6, 7].

Studies have showed that the increase in elastic modulus of post material cause decrease in the stress in dentin. However, Boschian et al., in 2006 have reported that higher the elastic modulus of post material than dentin can cause a dangerous, non-homogenous stress in root dentin. Also Silva et al., in 2009 reported that the stress distribution is more related to endodontically treated teeth restored with a post than the post's external configuration. Therefore, whenever the clinician is planning to use a post he has to choose a post material, which has the stiffness similar to dentin. They evaluated the stress distribution in maxillary central incisor, which is endodontically treated and restored with fiberglass and metallic prefabricated posts [7].

Necchi et al., in 2008 conducted a study on rotary endodontic instruments to demonstrate the usefulness of the FEM in improving the knowledge of the mechanical behaviour of Ni-Ti and stainless steel ProTaper F1 instrument during root canal preparation. The results found the radius and position of the canal curvature as the most critical parameters in determining the stress whilst high stress levels are produced by decrease in the radius and instrumenting apical to the mid root position. They advised to discard the instrument after its use in those type of root canals [31].

The use of glass fiber dowels showed less stress than the metal, carbon, and ceramic posts which few researchers found. However, there are some differences in the material properties, boundaries and loading conditions. A study by Eraslan et al., in 2009 showed a reduction in VM stress in an endodontically treated tooth restored with all-ceramic post and core than with zirconium oxide ceramic post and fiber post at the dentin wall and within the post [32].

In a study by Zhou et al., in 2009 a mandibular second premolar was used to evaluate the stress distribution restored with fiber post and core with various shapes and diameter in axial and non-axial loads. They found no significant change with the increase in post diameter irrespective of the shape. They recommended trapezium and cone fiber posts as the ideal design for restoring the crown and root portion as they produced least maximum stress in non-axial loads than in axial load [33].

For fixation of post and core to the remaining tooth structure cements like zinc-phosphate, glass ionomer, resin-modified glass ionomer, and resin cement are used. The difference in elastic modulus of these cements, post materials and dentin results in stress concentration under function. In 2010, Soares et al., found zinc-phosphate and conventional glass ionomer cement producing high stress concentrations at dentin-cement interface. They also demonstrated that resin cement recorded higher fracture resistance values than other cements, which was in accordance with the study done by Suzuki et al., in 2008 [7].

A systematic review in 2010 by Al-Omiri et al., discussed the importance of ferrule and emphasized the use of adhesive resin-fiber posts and composite cores as the best luting technique with respect to the biomechanical behaviour and tooth fracture resistance [34].

Al-Omiri et al., in 2011 conducted a study on 3-D FEM of maxillary second premolar restored with an all-ceramic crown supported by a titanium post and a resin-based composite core to analyze the stress concentration areas. They found higher incidence of deep root fractures in teeth restored with post-retained crowns below the level of crestal bone due to the increase in intracanal stresses with horizontal loads generating more dentinal stress than vertical loads. Though endodontic posts provide retention for coronal restoration, the dentinal stress value was higher than those without posts were. Smaller diameter posts with modulus of elasticity similar to dentine were associated with better stress distribution. More the amount of radicular dentin around the post better/reduced dentinal stress concentration within the root [35].

7. Application of FEA in prosthodontics and implantology

The branch of dentistry pertaining to the restoration and maintenance of oral function, comfort, appearance, and health of the patient by the restoration of natural teeth and/or the replacement of missing teeth and craniofacial tissues with artificial substitutes. FEA helps in studying the stress patterns and their distribution between the tooth and the material used in restoring the natural or missing tooth/teeth structure and predicting the favorable outcome with least chance of failure.

Zarone et al., in 2005 conducted a study on maxillary central incisor, the influence of tooth preparation design restored with alumina porcelain veneer on the stress distribution under functional load. They suggested the use of chamfer with palatal overlap design when restoring with porcelain veneers as it restored the natural distribution of stress than window technique [9].

FEA has been extensively used in implant dentistry to predict the biomechanical behaviour of various dental implant designs, as well as the effect of clinical factors for predicting the clinical success. Stress patterns in implant components and surrounding bone are well studied. The achievement of any FE study depends on the accuracy of simulating structures used. They are the material properties of implant and bone, surface characteristics and geometry of the implant and its components, loading method and support conditions, and the biomechanical behaviour of implant-bone interface. The prime difficulty in simulating the living tissues and the responses to the applied load can be successfully achieved with the use of advanced imaging techniques [36].

FEA gives an in-depth idea about the patterns of stress in the implant and more importantly in the peri-implant bone and this helps in the betterment of the implant design and implant insertion techniques. Several studies had been put forward on the effect of material properties of implant, implant number, size (length and diameter), thread profile, and on the quality and quantity of surrounding bone on stress distribution. The stresses of various kinds such as von Mises stress, maximum shear stress, maximum and minimum principal stress are used to assess the mechanical stress on the bone, implant, and bone-implant interface. Amongst, von Mises stress is most frequently and mainly used scalar-valued stress invariant to evaluate the yielding, and or failure behavior of dental materials. While minimum principal stress gives an idea on the compressive stress, maximum principal stress gives on tensile stress. Principal stress is used to study both ductile and brittle properties of a bone [36].

Siegele and Soltesz in 1989 conducted a study using implants of various shapes to evaluate the patterns of stress generation in the jawbone found that different shapes produced different stress patterns and conical implant showed higher stress than screw shaped and cylindrical implants [2].

Mailath et al., in 1989 evaluated the stress values at the level of bone while placing implants with different designs and shapes (cylindrical and conical). They found more desirable stress patterns in the cylindrical implants than conical implants, large implant diameters provides more favorable stress distributions and implant materials should have a modulus of elasticity of at least 110,000 N/mm^2. Slipping between implants and cortical bone is desirable [37].

Geng et al., in 2001 did literature review on the application of FEA in implant dentistry. They advised the use of advanced digital imaging technique for preparing the models with high accuracy, considering anisotropic and non-homogenous material and simulating the exact boundary conditions and mimicking the implant and its components [7].

Chun et al., in 2002 found that the square thread shape filleted with a small radius was more effective in stress distribution than other dental implants used in the analyses also maximum effective stress decreased not only as screw pitch decreased gradually but also as implant length increased [38].

Himmlova et al., in 2004 conducted a study by taking implants of various lengths and diameters to evaluate the stress values produced at implant-bone interface. They found maximum stress at the collar of the implant and an increase in the implant diameter decreased the maximum von Mises equivalent stress around the implant neck more than an increase in the implant length [39].

Ding et al., in 2009 conducted a study on immediate loading implants showed that the masticatory force around the implant neck was decreased with increased diameter of an implant. Several studies found higher risk of bone resorption occurring in the implant neck region. By using FEM, authors could able to compare the elastic modulus and deformation with different types of bone, and implant materials which helps clinicians to better understand the process of bone remodeling, and for further improvements in surgical techniques [40].

Eraslan et al., in 2009 evaluated the effects of different implant thread designs on stress distribution characteristics at supporting structures. Four different thread-form configurations for a solid screw implant was prepared with supporting bone structure. V-thread, buttress, reverse buttress, and square thread designs with a 100-N static axial occlusal load applied to occlusal surface of abutment to calculate the stress distribution. They found that the implant thread forms has no effect on von Mises stress distribution in the supporting bone, but produced dissimilar compressive stress intensities in the bone [7].

Dos Santos et al., in 2011 conducted a study to evaluate the influence of height of healing caps and the use of soft liner materials on the stress distribution in peri-implant bone during masticatory function in conventional complete dentures during the healing period in submerged and non-submerged implants. They found non-submerged implants with higher values of stress concentration and soft liner materials gave better results. They stated that use of soft liners with submerged implants to be the most suitable method to use during the period of osseointegration [41].

Demenko et al., in 2011 emphasized that, selecting an implant size is one of the important factor in determining the load bearing capacity. The most common reason of mandibular implant supported overdenture failure was peri-implantitis due to the loss of osseointegration without any sign of infection [42].

The increase risk of mechanical failure can occur with the increase in crown to implant ratio, which was substantiated by many FE studies. A study by Verri et al., in 2014 found an oblique loading induced high stress on the abutment screw when the crown:implant ratio was 1.5:1 which is in agreement with the study done by Urdaneta et al., in 2010 on correlation between screw loosening, fracture of prosthetic abutments, and crown to implant height [36].

7.1 Prosthesis for maxillectomy or hemi-mandiblectomy

FEA is important in predicting the success of implant supported prosthetic rehabilitation of maxillectomy patients. In case of maxillary or partial mandibular resection patients, FE models can be used to simulate the resection areas and biomechanics of maxillary obturator or mandibular partial or implant supported prosthesis can be studied. de Sousa and Mattos in 2014 conducted a study to evaluate the stability and functional stress caused by implanted-supported obturator prostheses in simulated maxillary resections of an edentulous maxilla corresponding to Okay Classes Ib, II, and III, with no surgical reconstruction. They found that the implant-supported obturator prostheses tended to rotate toward the surgical resection site, the region where there is no osseous support. As the osseous support and the numbers of implants and clips diminished, the tensile and compressive stresses in the gingival mucosa and in the cortical bone increased. They concluded that the osseous tensile and compressive stresses resulting from the bar-clip retention system for Okay Classes Ib, II, and III maxillectomy may not be favorable to the survival rate of implants [36].

8. Application of FEA in trauma and fractures

Oral and maxillofacial surgery is one branch of dentistry, which has always been associated with biomechanics. Trauma surgery, orthognathic surgery, reconstructive surgery are the subdivisions where understanding the mechanism of fractures and its biological response to the biomechanical change are worth knowing for optimal treatment method and outcome [43].

When present technology was not available in the past, cadaveric studies were the only way of information and it is not possible to carry out designing and executing which at present times have ethical issues often challenging to have valid and reliable results. Furthermore, post mortem alterations and the age do not match in a typical facial trauma cadaver. One such example was René Le Fort, a French army surgeon, conducted a series of thorough experiments on the heads of cadavers. His work gave rise to a system of classifying facial fractures, now known as Le Fort types I, II and III [36, 43].

Since the maxillofacial region has vital anatomical structures, intervention in this region needs precise work to be carried out in restoring function and esthetics of the tissues in obtaining predictable and favorable long-term outcomes. In the field of trauma surgery, to identify the craniofacial region that are potential prone to fracture, FEA enables precise mapping of the maxillofacial region to know the biomechanics and stress pattern distribution of trauma that helps in evaluation of patient and optimizing the surgical protocol for treating the fractures [43].

Today, with the help of FEA mechanical properties of facial hard and soft tissues, osteosynthesis materials, implant components for fixing the fractured parts, and various biological and synthetic bone substitutes can be easily generated and determined due to the advancement in the computing and virtual analysis. It allows the testing of various fixation system to prevent the future failure due to its improper selection or inappropriate positioning. It made us possible to know the impact in biomechanical behaviour of testing materials on the biological responses of the bone tested as well as adjacent anatomical structures more accurate, repeatable, time saving, and cost-effective way regardless of their complexity [43].

Isolated orbital floor fracture (IOFF), zygomatic bone fracture are the examples of more complex traumas occurring frequently in contact sports and their pathomechanism were also studied with the aid of FEA. In relatively rare facial traumas like in case of blast or gunshot wounds, FEA helps in exploring, analyzing and determining the mechanism of anatomical structures damaged and ways in reconstructing them. The pathomechanism underlying the type and method of fracture is exceptionally important as it may help in designing the helmets, other protecting devices. Rigid fixation is one of the key element in determining the long-term success for osseointegration. Inappropriate selection of an osteosynthesis component for the biological tissues can cause complication in fusion of bone. Therefore, FEA helps in determining and designing various fixation systems and methods [44, 45].

Osteosynthesis of condylar fracture and fixing the element is a challenging aspect for a maxillofacial surgeon due to its specific anatomy and surgical access. Through FEA, it has become possible for the researchers to find the better way and an exceptionally handy, easy mountable and durable element for optimal stabilizing and fixing the fractured fragments. A new type of "A-shape condylar plate" was designed for all levels of neck fractures and it can be used for stabilization of existed coronoid process fracture. FEA has proved to be a useful tool in investing and thorough evaluation of newer materials and solutions, which are more optimized, durable and light weight components before they can be used in the clinical situations [46].

Bujtár et al., in 2010 analyzed the stress distribution in detailed models of human mandibles at 3 different stages of life (12, 20, and 67 years) with simulation of supra normal chewing forces at static conditions. They found higher elasticity in younger models in all regions of the mandible whereas higher levels of stress in a 67 year old at the mandibular neck region of edentulous mandible [47].

Huempfner-Hierl et al., in 2014 showed a pattern of von Mises stresses beyond the yield point of bone that corresponded with fractures commonly seen clinically. They found Naso-orbitoethmoid fractures account for 5% of all facial fractures. They concluded that, FEM can be used to simulate the injuries occurring to the human skull that provides information about the pathogenesis of different types of fracture [48].

Murakami et al., in 2014 evaluated the strength of mandible after removal of a lesion to illustrate the theoretical efficacy of preventive measures against pathologic fracture. They found plate application is effective to decrease the stress on the mandible after surgical removal of a cyst including third molar [43].

Santos et al., in 2015 analyzed the stress distributions on the symphyseal, parasymphyseal, and mandibular body regions in the elderly edentulous mandible under applied traumatic loads, which enabled precise mapping of the stress distribution in a human elderly edentulous mandible (neck and mandibular angle) [49].

9. Application of FEA in orthodontics and dentofacial orthopedics

Orthodontics is a specialty of dentistry, which deals with the diagnosis, prevention and correction of malpositioned teeth and jaws. It also focuses on determining and modifying the facial growth, known as dentofacial orthopedics. Abnormal alignment of the teeth and jaws is common. In the field of Orthodontics and Dentofacial Orthopedics, FEM has proved to be a reliable and valid procedure in evaluating the applied orthodontic forces.

Tanne et al., in 1995 did a 3-D FE study to investigate the location of nasomaxillary complex centre of resistance (CRe). 9·8 N of force directed anteriorly and inferiorly were applied at five different levels, parallel and perpendicular to the occlusal plane. When a horizontal force was applied at a point in the horizontal plane, passing through the superior ridge of the pterygomaxillary fissure, the complex exhibited a translatory displacement of 1·0 μm approximately in forward direction. Whereas, clockwise or counter clockwise rotation when the forces were applied at the remaining levels suggesting that CRe of the nasomaxillary complex is located on the postero-superior ridge of the pterygomaxillary fissure, registered on the median sagittal plane [2].

Many researchers have developed various FE models in order to understand the interaction between tooth mobility and periodontal ligament. Jones et al., in 2001 validated an FE model and found PDL as the main mediator for orthodontic tooth movement and the material properties of PDL are difficult to quantify [7].

The use of the lingual orthodontic technique has increased over time, as adults dislike the visibility of orthodontic appliances. Sung et al., in 2003 evaluated the effect of compensating curves on canine retraction between the lingual and the labial orthodontic techniques. The compensating curve was increased on the .016-in stainless steel labial or lingual archwire, and a 150-g force was applied distally on the canine. The pattern of tooth movement (with or without a compensating curve) was found to be different between labial and lingual techniques. As the amount of compensating curve increased (0, 2, and 4 mm) in the archwire, the rotation and the distal tipping of the canine was reduced. The anti-tip and anti-rotation action of compensating curve on the canine retraction was greater in the labial archwire than in the lingual archwire [50].

Cattaneo et al., in 2009 studied on Orthodontic tooth movement (OTM) which occurs when an orthodontic force is applied to the brackets. The modeling and remodeling process of the supporting structures occurs by alteration in the distribution of stress/strain in the periodontium. As per the classical OTM theories, symmetric zones of compression and tension are present in the periodontium. However, they did not consider the complex mechanical properties of the PDL, the morphology of alveolar structures', and magnitude of the applied force. The authors could not confirm the classical ideal of symmetrical compressive and tensile areas in periodontium as per the OTM scenarios. They found light continuous orthodontics forces will be perceived as intermittent by the periodontium. They expressed that, as the roots and alveolar bone morphology are patient-specific, FEA should not be based on general models [51].

Lingual orthodontics has developed rapidly in recent years; however, research on torque control variance of the maxillary incisors in both lingual and labial orthodontics is still limited. Liang et al., in 2009 generated maxilla and maxillary incisors models to evaluate the torque control during retraction in labial and lingual orthodontic technique for maxillary incisors. They found loads of the same magnitude produced translation of the maxillary incisor in labial orthodontics but lingual crown tipping in lingual orthodontics. This suggested the loss of torque control during retraction of the maxillary incisors in extraction patients is more likely in lingual orthodontic treatment [52].

Field et al., in 2009 investigated the stress–strain responses of teeth to orthodontic loading. Two cases were analyzed, consisting of a single-tooth system with a mandibular canine, and a multi-tooth system with mandibular incisor, canine, and first premolar that are subjected to orthodontic tipping forces. They found stress levels greater in the multi-tooth system than in the single-tooth system also, elevated distortion strain energies at the alveolar crest area and tensile and compressive stresses at the apical sites clinically associated with root resorption [22].

9.1 Orthognathic surgery

Orthognathic surgery also known as corrective jaw surgery or simply jaw surgery is aimed to correct the conditions of jaw and face. They relate to correct the structure, growth modification, disorders of TMJ, sleep apnea, malocclusion problems owing to skeletal disharmonies, or other orthodontic problems that cannot be treated with orthodontic braces. It involves the surgical manipulation of the structures of the facial skeleton in restoring the suitable anatomy and their functional relationship with dentofacial skeletal abnormalities for the patient's sense of self and well-being. Successful outcome depend on meticulous preoperative planning until finalization of occlusion. Virtual planning promotes a more accurate analysis of dentofacial deformity and preoperative planning with the help of computer-based technique like FEA, an invaluable tool in providing comprehensive patient education. Today's orthognathic treatment consists of standard orthognathic procedure in correcting jaw deformities like maxillary and mandibular prognathism, open bite, difficulty in chewing and swallowing, TMJ dysfunction pain, excessive wear of the teeth, and receding chins. It includes adjunctive procedures like genioplasty, septorhinoplasty, and lipectomy of the neck to improve hard and soft tissue contours [53].

Chabanas et al., in 2002 presented their study on the treatment protocol – a computer aided maxillofacial sequence for orthognathic surgery in the patients with large gnathic defects because the treatment protocol is difficult and time consuming [43].

Erkmen et al., in 2005 conducted a study and found that the use of 2.0 mm lag screws placed in a triangular configuration provided most sufficient stability and lesser stress fields at the osteotomy site compared to other rigid fixation methods [54].

For successful outcome in any orthognathic surgeries, selection of an appropriate bridging element is a key determinant, corrective mandibular surgery like bilateral sagittal split osteotomy (BSSO) is not an exception to stabilize the bony segments with different fixing elements and FEA is an important tool [43].

Stróżyk et al., in 2011 compared three types of fixation during BSSO using 3-D FE model divided into 3 segments with 5 mm gap in between according to BSSO line. Three fixation systems were bridged to the osteotomized fragments, a 20 N and 80 N force applied at the incisor and molar area respectively. They concluded that the most stable bridging after BSSO can be obtained with bicortical screw fixation [55].

Surgically Assisted Palatal Expansion (SARPE) is an orthognathic surgical procedure that is performed frequently in the patients with narrower maxilla. De Assis et al., in 2014 investigated the stress distribution in maxillae that underwent SARPE. They constructed five maxillary models with no osteotomy, Le Fort I osteotomy with a step in the zygomaticomaxillary buttress, Le Fort I osteotomy with a step in the zygomaticomaxillary buttress and the pterygomaxillary disjunction, Le Fort I osteotomy without a step, and Le Fort I osteotomy with pterygomaxillary disjunction and no step. The distribution of tensions in maxillae that underwent SARPE was simulated by the FEM and they revealed that the steps in the zygomaticomaxillary buttress and the pterygomaxillary disjunction seems to be important in decreasing the harmful dissipation of tensions during SARPE [56].

A more complex surgery involving correction of deformation of both the jaws simulating the maxillary and mandibular jaw osteotomy using FEA was also executed. Fujii et al., in 2017 conducted a study to determine whether non-linear 3D-FEA can be applied to simulate pterygomaxillary dysjunction during Le Fort I osteotomy (LFI) not involving a curved osteotome (LFI-non COSep), and to predict potential changes in the fracture pattern associated with extending the cutting line. In their study, the rate of agreement between the predicted pterygomaxillary dysjunction patterns and those observed in the postoperative 3D-CT images was 87.0%. The predicted incidence of pterygoid process fracture was higher for cutting lines that extended to the pterygomaxillary junction than for conventional cutting lines. They also added that, 3-D FEA can be a useful tool in predicting pterygomaxillary dysjunction patterns and provides useful information in selecting safe procedures during LFI-non-COSep [57].

Knoops et al., in 2019 conducted a study to compare the soft tissue prediction accuracy of several available computer programmes like Dolphin, ProPlan CMF, and Probabilistic Finite Element Method (PFEM) in patients with Le Fort I osteotomy. They concluded that patient or population-specific material properties can be defined in PFEM, while no soft tissue parameters are adjustable in ProPlan. Therefore, PFEM provides accurate soft tissue prediction and can be a useful tool in preoperative patient communication [58].

10. Application of FEA in reconstructive surgery

The FEM technique can also be used in oncosurgeries and reconstructive surgery where an extensive resection is needed and reconstruction of jawbones are done. The crucial parameter form the postoperative point of view is the amount of bone segment removed from the surgical site, which includes size, shape, and location. The aim of reconstructing the bone defect should result in restoration of the integrity, its anatomy and the functionality of stomatognathic system. With the aid of digital technology; modeling, simulation and analysis, it is possible to know and compare the stress levels and distribution on and at the bone-graft interface and predictable behaviour of the reconstructed site to identify the most suitable transplant for a given clinical situation and to find the appropriate bone fusion under favorable conditions in the reconstructed area [43].

Moiduddin et al., in 2017 studied to present an integrated framework model in designing and analyzing customized porous reconstruction plate based on the selection of implant design techniques. Reconstruction of large mandibular defects often leads to complications while using reconstruction plates. Studies proved that implants with porous structures can effectively enhances the biological fixation to the bone but, no study reported on the design and analysis of the customized porous mandibular reconstruction. In their study, two customized implant design techniques; mirroring and anatomical were compared. They recommended the use of mirror design reconstruction technique in mandibular bone repair, which not only improves the stability but also the flexibility of mandibular reconstruction under chewing conditions [59].

Hu et al., in 2019 performed a study to characterize the mechanical behaviour of 3-D printed anisotropic scaffolds as bone analogs by fused deposition modeling (FDM). Using topological optimization and 3-D printing technology, designing and manufacturing of a customized graft with porous scaffold structure is necessary in repairing large mandibular defects. They used CBCT images of an edentulous 50-year-old patient. The topological optimized graft provided the best mechanical properties. They highlighted the use of numerical simulations and 3-D printing technology in designing and manufacturing the artificial porous graft [60].

11. Application of FEA in periodontics

PDL is a highly specialized soft connective tissue that is present between the tooth root and the alveolar bone. The primary function is to support the tooth and is the most important component of periodontium. Various studies included and investigated on its biomechanics and stress distribution under normal, masticatory, and traumatic loads. PDL is the crucial aspect in designing as it influences the properties of a 3-D model, though it is difficult in modeling and not a concern for the study. Ignoring the PDL may result in inaccurate values of stress and strain distribution [36].

Tuna et al., in 2014 conducted a study and pointed out the advantage of simulating as a contact model at the interface of tooth root and alveolar process instead of a solid meshed FE model with poor geometric morphology or very dense mesh to save the time. They reinforced the use of PDL in designing the tooth model and its associated structures that increases the accuracy and contribution to the smoothness of interface stress distributions [61].

12. Merits of finite element method

- FEA is a non-invasive method [1, 2, 9, 62].
- Results can be easily interpreted in physical terms as well as it has a strong mathematical base.
- Non-homogenous structures also can be dealt by merely assigning different properties to different elements.
- It is even possible to vary the properties to different elements and within an element according to the polynomial applied.
- It minimizes the requirement for laboratory testing, but not replaces entirely.
- Applicable to linear and non-linear as well as solid and fluid structural interactions.
- Any problems can be split into smaller number of problems.
- It is very easy to simulate any biological condition in pre-operative, intra-operative, and post-operative stages for more accurate and reliable results.
- Reproducibility of the results does not affect the physical properties of the materials involved.
- It can replace stereo lithographic models for pre-surgical planning.
- With FEA, static and dynamic analysis is possible.
- It is less time consuming even with the complex structures.
- No extensive instrumentation is required.
- The study can be repeated as many times as the operator wants.
- The systematic generality of finite element procedure makes it a powerful and versatile tool for a wide range of problems.

13. Shortcomings of FEM

- The solution obtained from FEM can be realistic if and only if the material properties are known precisely [1, 2, 9, 62].
- The major drawback is sensitivity of the solution on the geometry of the element such as type, size, number, shape and orientation of element used.
- FEM programs yield a large amount of numerical data as results and it is very difficult to separate out the required results from the pile of numbers.
- Inability to simulate the biological dynamics of the tooth and its supporting structure accurately. For example, in non-carious cervical lesions, due to the exposure to oral environment the structure of dentin (tertiary or reparative dentin) undergoes variable amount of changes such as attrition, erosion or abrasion, which has formed as a response to stimulus.
- Misguided results due to inaccurate data or information or interpretation.
- Due to their complex anatomy and lack of complete knowledge about the mechanical behaviour, modeling of human structures are extremely difficult.
- The results depend on the personnel involved in the process due to assumptions.
- Until well-defined physical properties of enamel, dentin, PDL, cancellous, and cortical bone are available, the progress and the process in the FEA will be limited.

14. Advances in FEM

- Early FE models had the difficulty in allocating physical characteristics to the different constituent parts of the tooth, as they were considered as isotropic which in real are not [1, 2, 9, 62].
- The non-linear simulation and dynamic behaviour of PDL and other soft tissue properties has become an increasingly powerful approach that provides precision and reliability in calculating stress and strain with a wide range of tooth movements.
- The transient and residual stresses in dental materials are also included in non-linear FEM calculations also include. Residual stresses in ceramic and metal restorations, contraction stresses in composites, and permanent deformation prediction of materials are some to mention for non-linear application to be applied and investigated.
- The phenomena of sliding and friction critically affect the stress and strain created on the contact surfaces between teeth that play a major role in the mechanical behaviour. This non-linear property can be solved by contact analysis depend various factors like region of contact, load, material, and environment that are highly unpredictable. The frictional response depends on the pair of surfaces in contact, temperature, and humidity.

- Research is also going on polyhedral meshing and mesh-less (or mesh-free) analysis for reducing the meshing time. Advantages of polyhedral meshing being; less meshing time, high accuracy, and too less number of degrees of freedom (DOF).
- Hybrid meshing (hex-pyram-tetra) is a very special option but not all software supports its application.

15. Conclusion

The power of the Finite Element Method is its versatility. It is a well-established numerical analysis used not only in aerospace, automotive industry and civil engineering, but also in health care. It addresses the biomedical problems that are challenging due to structural complexity. The structure analyzed may have arbitrary shape, arbitrary support, and arbitrary loads therefore; it is ideally suited for the analysis of bibliographical structures, which are non-homogeneous. The modeling and simulation of the structures and or materials saves time and money in conducting the experiment. Therefore, this tool has been successfully employed in various areas of dentistry.

A finite element analysis does not produce formula as a solution, nor does it solve a class of problems. This method is a way of getting a numerical solution to a specific problem. Finite element analysis is an accurate tool in assessing stress distribution, only of the given set of values are effective. However, it varies from person to person as the situation and biomechanical properties of living structures interpretation differs. Hence, the obvious shortcomings should be kept in mind before any decision making procedure in experimental as well as clinical dentistry. The experiments done are repeatable with no ethical concern and study designs can be modified as per the requirement. Certain limitations of FEA do exist. Keeping in mind the limitations, FEA research should be accompanied with clinical evaluation.

Conflict of interest

The authors declare no conflict of interest.

Author details

Vinod Bandela[1*] and Saraswathi Kanaparthi[2]

1 Fixed Division, Prosthetic Dental Sciences, College of Dentistry, Jouf University, Kingdom of Saudi Arabia

2 Department of Pedodontics and Preventive Dentistry, St. Joseph Dental College and Hospital, Andhra Pradesh, India

*Address all correspondence to: vinod.bandela@gmail.com

References

[1] Geng JP, Tan KB, Liu GR. Application of finite element analysis in implant dentistry: a review of the literature. J Prosthet dent. 2001;**85**(6):585-598

[2] Mohammed SD, Desai H. Basic concepts of finite element analysis and its applications in dentistry: An overview. Journal of Oral Hygiene & Health. 2014 Aug 14:1-5

[3] Szücs A, Bujtár P, Sándor GK, Barabás J. Finite element analysis of the human mandible to assess the effect of removing an impacted third molar. J Can Dent Assoc. 2010;**76**(1):72

[4] Oenning AC, Freire AR, Rossi AC, Prado FB, Caria PH, Correr-Sobrinho L, et al. Resorptive potential of impacted mandibular third molars: 3D simulation by finite element analysis. Clin. Oral Investig. 2018;**22**(9):3195-3203

[5] Kihara T, Ikawa T, Shigeta Y, Shigemoto S, Nakaoka K, Hamada Y, et al. Mandibular 3-dimensional finite element analysis for a patient with an aggressive form of craniofacial fibrous dysplasia. Oral Surg Oral Med Oral Pathol Oral Radiol. 2019;**128**(6):e214-e222

[6] Borcic J, Braut A. Finite element analysis in dental medicine. Finite Element Analysis: New Trends and Developments. 2012 Oct;**10**:1

[7] Piccioni MA, Campos EA, Saad JR, Andrade MF, Galvão MR, Rached AA. Application of the finite element method in Dentistry. RSBO Revista Sul-Brasileira de. Odontologia. 2013;**10**(4):369-377

[8] Goel VK, Khera SC, Ralson JL, Chang KH. Stresses at the dentino-enamel junction of human teeth: A finite element investigation. J Prosthet Dent. 1991;**66**:451-459

[9] Srirekha A, Bashetty K. Infinite to finite: an overview of finite element analysis. Indian J Dent Res. 2010;**21**(3):425-432

[10] Ausiello P, Apicella A, Davidson CL. Effect of adhesive layer properties on stress distribution in composite restorations: A 3D finite element analysis. Dent Mater. 2002;**18**:295-303

[11] Ausiello P, Rengo S, Davidson CL, Watts DC. Stress distributions in adhesively cemented ceramic and resin-composite class II inlay restorations: A 3D-FEA study. Dent Mater. 2004;**20**:862-872

[12] Asmussen E, Peutzfeldt A. Class I and Class II restorations of resin composite: an FE analysis of the influence of modulus of elasticity on stresses generated by occlusal loading. Dent Mater. 2008;**24**(5):600-605

[13] Coelho PG, Calamia C, Harsono M, Thompson VP, Silva NR. Laboratory and FEA evaluation of dentin-to-composite bonding as a function adhesive layer thickness. Dent Mater. 2008;**24**:1297-1303

[14] Ravi RK, Alla RK, Shammas M, Devarhubli A. Dental Composites-A Versatile Restorative Material: An Overview. Indian J Dent Sci. 2013;5(5):111-115.

[15] Lee MR, Cho BH, Son HH, Um CM, Lee IB. Influence of cavity dimension and restoration methods on the cusp deflection of premolars in composite restoration. Dent Mater. 2007;**23**(3):288-295

[16] Choi NS, Gu JU, Arakawa K. Acoustic emission characterization of the marginal disintegration of dental composite restoration. Composites Part A: Appl Sci Manuf. 2011;**42**(6):604-611

[17] Jongsma LA, de Jager Ir N, Kleverlaan CJ, Feilzer AJ. Reduced contraction stress formation obtained by a two-step cementation procedure for fiber posts. Dent Mater. 2011;27(7):670-676.

[18] Shenoy A, Shenoy N. Dental ceramics: An update. J Conserv Dent. 2010;**13**(4):195-203

[19] Belli S, Eskitascioglu G, Eraslan O, Senawongse P, Tagami J. Effect of hybrid layer on stress distribution in a premolar tooth restored with composite or ceramic inlay: an FEM study. J Biomed Mater Res Part B: Appl Biomater. 2005;**74**(2):665-668

[20] Rezaei SMM, Heidarifar H, Arezodar FF, Azary A, Mokhtarykhoee S. Influence of Connector Width on the Stress Distribution of Posterior Bridges under Loading. J Dent. 2011;**8**:67-74

[21] Thompson MC, Field CJ, Swain MV. The all-ceramic, inlay supported fixed partial denture. Part 2. Fixed partial denture design: a finite element analysis. Aust Dent J. 2011;**56**:301-311

[22] Matson MR, Lewgoy HR, Barros Filho DA, Amore R, Anido-Anido A, Alonso RC, et al. Finite element analysis of stress distribution in intact and porcelain veneer restored teeth. Comput Methods Biomec. 2012;**15**(8):795-800

[23] San Chong B. Harty's Endodontics in Clinical Practice E-Book. Elsevier Health Sciences; 2016 Jul 28.

[24] Carpegna G, Alovisi M, Paolino DS, Marchetti A, Gibello U, Scotti N, et al. Evaluation of Pressure Distribution against Root Canal Walls of NiTi Rotary Instruments by Finite Element Analysis. Appl Sci. 2020;**10**(8):2981

[25] Sattapan B, Nervo GJ. palamara JE, Messer HH. Defects in rotary nickel-titanium files after clinical use. J Endod. 2000;**26**:161-165

[26] Subramaniam V, Indira R, Srinivasan MR, Shankar P. Stress distribution in rotary nickel titanium instruments-a finite element analysis. J Conserv Dent. 2007;**10**(4):112-118

[27] Kim HC, Cheung GS, Lee CJ, Kim BM, Park JK, Kang SI. Comparison of forces generated during root canal shaping and residual stresses of three nickel–titanium rotary files by using a three-dimensional finite-element analysis. J Endod. 2008;**34**(6):743-747

[28] Lee MH, Versluis A, Kim BM, Lee CJ, Hur B, Kim HC. Correlation between experimental cyclic fatigue resistance and numerical stress analysis for nickel-titanium rotary files. J Endod. 2011;**37**(8):1152-1157

[29] Belli S, Eraslan O, Eskitascioglu G, Karbhari V. Monoblocks in root canals: a finite elemental stress analysis study. Int Endod J. 2011;**44**:817-826

[30] Rosenstiel SF, Land MF. Contemporary Fixed Prosthodontics-E-Book. Elsevier Health Sciences; 2015 Jul 28.

[31] Necchi S, Taschieri S, Petrini L, Migliavacca F. Mechanical behavior of nickel-titanium rotary endodontic instruments in simulated clinical conditions: A computational study. Int Endod J. 2008;**41**:939-949

[32] Eraslan O, Aykent F, Yücel MT, Akman S. The finite element analysis of the effect of ferrule height on stress distribution at post-and-core-restored all-ceramic anterior crowns. Clin Oral Invest. 2009;**13**(2):223-227

[33] Zhou LY, Shen QP, Han DW. Stress analysis of mandibular second premolar restored with fiber post-core with different shapes and

diameters. Shanghai Kou Qiang Yi Xue. 2009;**18**:324-328

[34] Al-Omiri MK, Mahmoud AA, Rayyan MR, Abu-Hammad O. Fracture resistance of teeth restored with post-retained restorations: an overview. J Endod. 2010;**36**(9):1439-1449

[35] Al-Omiri MK, Rayyan MR, Abu-Hammad O. Stress analysis of endodontically treated teeth restored with post-retained crowns: A finite element analysis study. J Am Dent Assoc. 2011;**142**(3):289-300

[36] Trivedi S. Finite element analysis: A boon to dentistry. J Oral Biol Craniofac Res. 2014;**4**(3):200-203

[37] Mailath G, Stoiber B, Watzek G, Matejka M. Bone resorption at the entry of osseointegrated implants--a biomechanical phenomenon. Finite element study. Z Stomatol. 1989;**86**:207-216

[38] Chun HJ, Cheong SY, Han JH, Heo SJ, Chung JP, Rhyu IC, et al. Evaluation of design parameters of osseointegrated dental implants using finite element analysis. J Oral Rehabil. 2002;**29**:565-574

[39] Himmlova L, Kácovský A, Konvičková S. Influence of implant length and diameter on stress distribution: a finite element analysis. J Prosthet Dent. 2004;91(1):202-205.

[40] Ding X, Zhu XH, Liao SH, Zhang XH, Chen H. Implant–bone interface stress distribution in immediately loaded implants of different diameters: a three-dimensional finite element analysis. J Prosthodont. 2009;**18**(5):393-402

[41] Dos Santos MB, Da Silva Neto JP, Consani RL, Mesquita MF. Three-dimensional finite element analysis of stress distribution in peri-implant bone with relined dentures and different heights of healing caps. J Oral Rehabil. 2011;**38**(9):691-696

[42] Demenko V, Linetskiy I, Nesvit K, Shevchenko A. Ultimate masticatory force as a criterion in implant selection. J Dent Res. 2011;**90**:1211-1215

[43] Lisiak-Myszke M, Marciniak D, Bieliński M, Sobczak H, Garbacewicz Ł, Drogoszewska B. Application of Finite Element Analysis in Oral and Maxillofacial Surgery-A Literature Review. Materials. 2020;**13**(14):3063

[44] Wanyura H, Kowalczyk P, Smolczyk-Wanyura D, Stopa Z, Bossak M. Finite element analysis of external loads resulting in isolated orbital floor fractures. J. Stoma. 2011;**64**:476-489

[45] Schaller A, Voigt C, Huempfner-Hierl H, Hemprich A, Hierl T. Transient finite element analysis of a traumatic fracture of the zygomatic bone caused by a head collision. Int. J. Oral Maxillofac. Surg. 2012;**41**:66-73

[46] Kozakiewicz M, Swiniarski J. "A" shape plate for open rigid internal fixation of mandible condyle neck fracture. J. Craniomaxillofac. Surg. 2014;**42**:730-737

[47] Bujt_ar P, S_andor GKB, Bojtos A, Szucs A, Barab_as J. Finite element analysis of the human mandible in 3 different stages of life. Oral Surg Oral Med Oral Pathol Oral Radiol Endod. 2010;**110**:301-309

[48] Huempfner-Hierl H, Schaller A, Hemprich A, Hier T. Biomechanical investigation of naso-orbitoethmoid traumaby finite element analysis. Brit J Oral Maxillofac Surg. 2014;**52**:850-853

[49] Santos LS, Rossi AC, Freire AR, Matoso RI, Caria PH, Prado FB. Finite-element analysis of 3 situations of trauma in the human edentulous

mandible. J. Oral Maxillofac. Surg. 2015;**73**:683-691

[50] Sung SJ, Baik HS, Moon YS, Yu HS. Cho YS. A comparative evaluation of different compensating curves in the lingual and labial techniques using 3D FEM. Am J Orthod Dentofacial Orthop. 2003;**123**(4):441-450

[51] Cattaneo PM, Dalstra M, Melsen B. Strains in periodontal ligament and alveolar bone associated with orthodontic tooth movement analyzed by finite element. Orthod Craniofac Res. 2009;**12**(2):120-128

[52] Field C, Ichim I, Swain MV, Chan E, Darendeliler MA, Li W, et al. Mechanical responses to orthodontic loading: a 3-dimensional finite element multi-tooth model. Am J Orthod Dentofacial Orthop. 2009;**135**(2):174-181

[53] Khechoyan DY. Orthognathic surgery: general considerations. In Seminars in plastic surgery 2013 Aug (Vol. 27, No. 3, p. 133). Thieme Medical Publishers.

[54] Erkmen E, Simsek B, Yucel E, Kurt A. Comparison of different fixation methods following sagittal split ramus osteotomies using three-dimensional finite elements analysis. Part 1: Advancement surgery-posterior loading. Int. J. Oral Maxillofac. Surg. 2005;**34**:551-558

[55] Stró˙zyk, P.; Nowak, R. Finite elements method analysis of fixation for bilateral sagittal split osteotomy. Dent. Med. Probl. 2011;48:157-164.

[56] De Assis DS, Xavier TA, Noritomi PY, Goncales ES. Finite element analysis of bone after SARPE. J. Oral Maxillofac. Surg. 2014;**72**:167

[57] Fujii H, Kuroyanagi N, Kanazawa T, Yamamoto S, Miyachi H, Shimozato K. Three-dimensional finite element model to predict patterns of pterygomaxillary dysjunction during Le Fort I osteotomy. Int. J. Oral Maxillofac. Surg. 2017;**46**:564-571

[58] Knoops PGM, Borghi A, Breakey RWF, Ong J, Jeelani NUO, Bruun R, et al. Three-dimensional soft tissue prediction in orthognathic surgery: A clinical comparison of Dolphin, ProPlan CMF, and probabilistic finite element modelling. Int. J. Oral Maxillofac. Surg. 2019;**48**:511-518

[59] Moiduddin K, Anwar S, Ahmed N, Ashfaq M, Al-Ahmari A. Computed assisted design and analysis of customized porous plate for mandibular reconstruction. IRBM. 2017;**38**:78-89

[60] Hu J, Wang JH, Wang R, Yu XB, Liu Y, Baur DA. Analysis of biomechanical behavior of 3D printed mandibular graft with porous scaffold structure designed by topological optimization. 3D printing in medicine. 2019;**5**(1):5

[61] Tuna M, Sunbuloglu E, Bozdag E. Finite element simulation of the behavior of the periodontal ligament: a validated nonlinear contact model. J Biomech. 2014;**47**:2883-2890

[62] Chandrupatla TR, Belegundu AD, Ramesh T, Ray C. Introduction to finite elements in engineering. Vol. 10. Upper Saddle River, NJ: Prentice Hall; 2002 Jan.

Chapter 9

Rolling Resistance Estimation for PCR Tyre Design Using the Finite Element Method

Sutisna Nanang Ali

Abstract

This study presents rolling resistance estimation in the design process of passenger car radial (PCR) tyre by using finite element method. The rolling resistance coefficient of tyres has been becoming one of main requirements within the regulation in many countries as it is related to the level of allowable exhaust gas emission generated by vehicle. Therefore, the tyre being designed must be digitally simulated using finite element method before the tyre is manufactured to provide a high confident level and avoid unnecessary cost related to failure physical product testing. The simulation firstly computes the deformation of several alternative designs of tyres under certain loading, and then the value of deformation force in each tyre component during deformation took place is calculated. The total force of deformation is considered as energy loss or hysteresis loss resulted in tyre rolling resistance. The experiment was carried out on three different tyre designs: two grooves, three grooves, and four grooves. The four groove tyre design gave the smallest rolling resistance coefficient (RRC). Finally, the simulation was continued to compare different crown radius of the tyres and the result shows that the largest crown radius generates the lowest rolling resistance.

Keywords: rolling resistance, PCR tyre, design, hysteresis loss, finite element method

1. Introduction

Passenger car radial (PCR) tyre is one of the most widely used tyres and is designed to follow the international standard and the regulation in the country where the tyre is being used. The recent regulation mainly concerns with the reduction the source of pollution and safety, such as rolling resistance, rolling noise, and wet grip. This study discusses the finite element simulation of tyre in order to design PCR tyre having low rolling resistance coefficient that lead to a low energy consumption tyre.

The energy consumption of vehicle to some extent is contributed by tyres. According to International Council on Clean Transportation [1], improving tyre energy efficiency will reduce fuel consumption by 3 to 5% which will reduce greenhouse gas emission by more than 100 million metric ton annually. Therefore, a low rolling resistance tyre is highly required to reduce the gas emission produce by vehicles.

Tyre rolling resistance is defined as the energy consumed per unit travel distance when the tyre rolls under load [2]. Therefore, lower energy use of vehicle can be obtained by using low rolling resistance tyre.

Tyre design process includes conceptual design, benchmarking, detail design, and design review and analysis. During design review and analysis phase, a simulation was conducted to estimate the value of rolling resistance coefficient. In this simulation, a finite element model was built by using Abaqus software to simulate the tyre deformation and calculate a complete energy loss absorbed by the deformation of rolling tyre under certain load and speed. The hysteresis energy loss was calculated using a user defined subroutine written in Python and is used as Abaqus plug-in.

2. Rolling resistance

Tyre rolling resistance requirement was outlined in United Nation Economic Commission for Europe (UNECE) regulation No. 117 Revision 2, together with rolling sound emission and adhesion on wet surface (wet grip). The country applying this Regulation may refuse to allow the sale or entry into service of a PCR tyre (C1 Class) which does not meet the stage 1 rolling resistance requirements from 1 November 2014 and the stage 2 rolling resistance requirements from 1 November 2018 [3].

However, different countries have different policies regarding implementation standard, date and rating. European Union implements tyre labeling requirement since 2012, where the label states the rating of Rolling Resistance Coefficient, Rolling Sound Emission, and Wet Grip. Gulf Cooperation Council implement GSO standard tyre labeling starting 2014, mandatory for rolling resistance and wet grip.

2.1 Rolling resistance coefficient

UNECE Regulation No. 117–2 defines Rolling Resistance F_r as loss of energy (or energy consumed) per unit of distance traveled, and Rolling Resistance coefficient C_r as ratio of the rolling resistance to the load on the tyre (**Table 1**).

As stated by Tonachel [4], rolling resistance occurs as tyres deform during rotation. The load within the rubber material and rebar that construct the tyre are deformed and the loss of energy during these repeated deformations is then dissipated in the form of heat. The dissipation of energy in radial tyre occurs on crown is estimated about 70%, on sidewall 15%, and bead area 15% [5].

Standard rolling resistance		
	Stage 1	**Stage 2**
Tyre class	**Max value (N/kN)**	**Max value (N/kN)**
C1	12.0	10.5
C2	10.5	9.0
C3	8	6.5
For snow tyres, the limits shall be increased by 1 N/kN.		

Table 1.
Standard RR coefficient based on ECE R117–2.

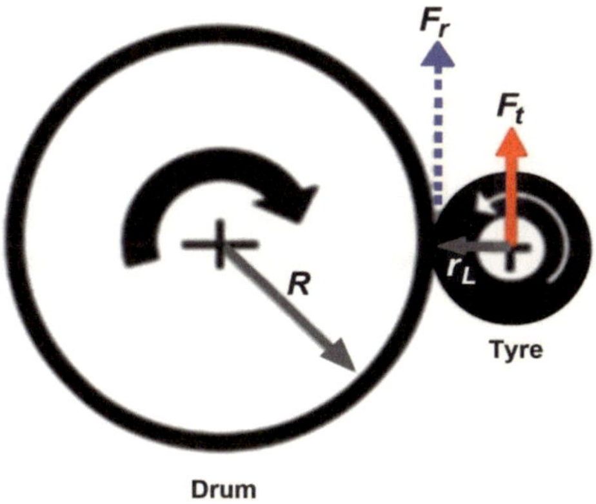

Figure 1.
Rolling resistance test method [6].

Therefore, the research was focused on the crown area. The simulation was done on crown initial radius and the stiffness of tread to study their effect on rolling resistance coefficient C_r.

2.2 Rolling resistance measurement

In this research, the rolling resistance test was conducted according to ISO 28580, i.e. using force measurement method (**Figure 1**). In this method, the tyre and drum wheel assembly is forced toward a drum wheel with the skim load F_{pl}, and the reaction force at the axle of tyre and drum wheel assembly F_t is then measured. The rolling resistance F_r at the contact of tyre and drum can be calculated using the following equation:

$$F_r = F_t \left(1 + (r_L/R) - F_{pl}\right. \quad (1)$$

where F_r Rolling Resistance (N)
F_t Measured force at the spindle (N)
r_L Tyre radius (m)
R Drum wheel radius (1.7 m)
F_{pl} Skim load (N)

3. Design methodology

Tyre design consists of several phases, including conceptual design, benchmarking, detail design, design review and analysis. Design review and analysis phase is important to ensure that the final product will be in accordance with the required performance as designed. One of the processes in this phase is doing simulation by using finite element method with the following steps [7]:

1. Define target performance
2. Tyre Simulation using FEA
3. Validation of FEA simulation result

3.1 Target Performance

The tyre being designed is PCR tyre with size of 175/65 R14, should have maximum Rolling Resistance Coefficient of 8.5 N/kN and good cornering stability.

3.2 Tyre simulation using FEA

The Finite Element Method is used to analyze the rolling resistance of PCR tyre, consisting of two steps:

1. Tyre simulation was performed using commercial finite element software Abaqus. The simulation consists of several steps as follows:

    a. FE tyre modeling, using axisymmetric modeling method.

    b. Define material properties and material modeling.

    c. Static footprint simulation and Radial stiffness.

2. Energy dissipation and rolling resistance were evaluated by using internally developed python code. The code extracts the strain energy results of the model and the same is post processed with viscous material data. The dissipation energy is calculated based on the strain energy function of Yeoh's model by taking the product of elastic strain energy and the loss tangent of materials. Computation of Tyre rolling resistance with its respective compounds developed for their applications, performed by considering different crown radius and radial stiffness.

3.2.1 Axisymmetric modeling

To model a tyre in Abaqus we use a cross section area of the tyre drawing and, imported as IGES file, this modeling technique is well known as axisymmetric modeling. All tyre component and their material properties are defined in this step. The tyre components that construct the tyre includes tread, base, wing, inner liner, side wall, apex, rim cushion, bead, JLB (join less belt), belt, and ply, are shown **Figure 2** and made up from four different types of materials and these are rubber compounds, textile fabrics, steel cords and bead wire.

The tyre model in Abaqus consists of two part partition: Carcass and Cord. Carcass and Cord partition were meshed separately, which are modeled in half axisymmetric model and then mirrored, become a complete assembly. In case the tread need to be included in the simulation, for instance to evaluate footprint, the

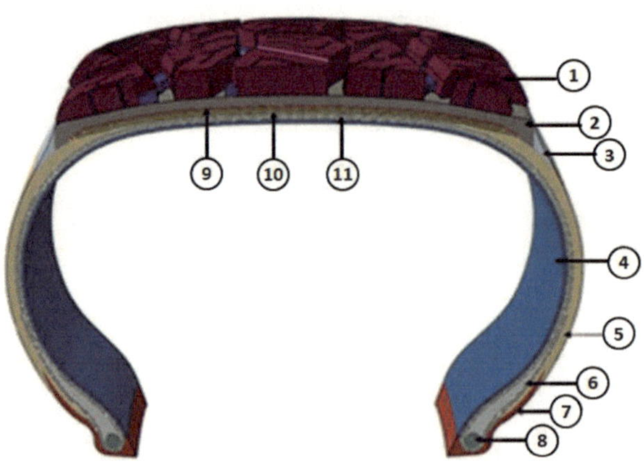

NO	ITEM	MATERIAL
1	TREAD	T61
2	BASE	T61
3	WING	S70
4	INNER LINER	I40
5	SIDE WALL	S70
6	APEX	A02
7	RIM CUSHION	R50
8	BEAD	BW08
9	JLB	N20
10	BELT	Z80
11	PLY	C32

Figure 2.
Sample of Tyre components [7].

DOI: http://dx.doi.org/10.5772/intechopen.94144

tread is meshed separately in addition to Carcass and Cord. Tread meshing need to be carefully done so that the nodes on tread and carcass will be matched perfectly. Later on the rim part is included in the assembly. The axisymmetric model of the tyre after meshing is illustrated in **Figure 3**.

The next steps are defining the mounting, creating constraints, defining boundary conditions, and loading (pressure) prior to running axisymetry function in Abaqus to form a full tyre. **Figure 4** illustrate the axisymmetric tyre with pressure and a full round of the axisymmetric tyre model.

3.2.2 Material properties and modeling

Material properties need to be input into Abaqus during this simulation step, each component should have the following material property data including hardness, density, stress, strain, Young's modulus, mu, Kappa, C10, and D1. **Table 2** exhibits the full material properties of all tyre components under study.

A model that represents the stress–strain relationship of the material is needed in finite element analyses of rubber components. There are several material models available In Abaqus to describe the mechanical behavior of rubber. The model to be

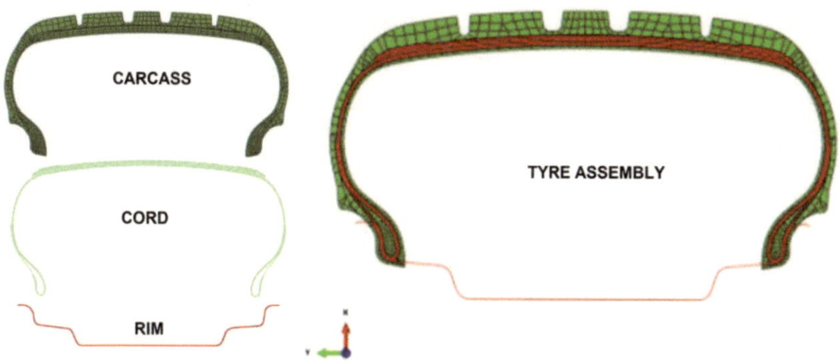

Figure 3.
Tyre meshing of axisymmetric model in Abaqus [7].

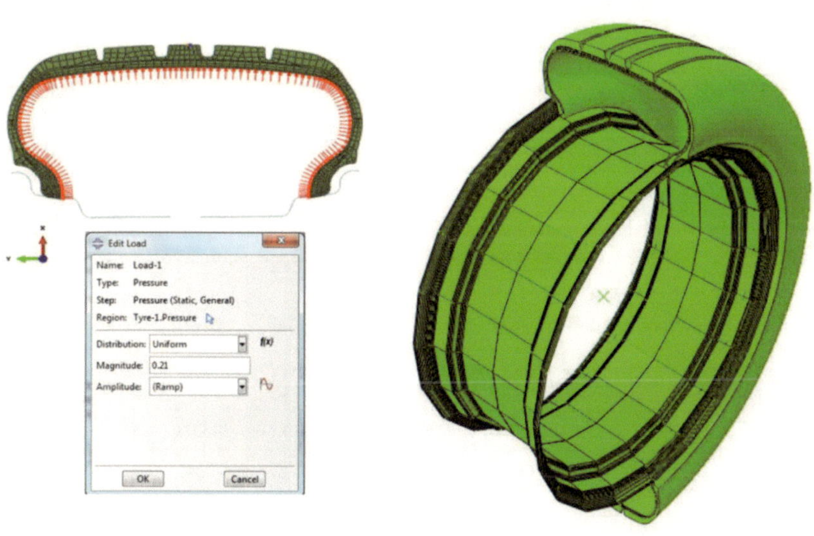

Figure 4.
Pressurized axisymmetric (left) and full axisymmetric Tyre model (right).

used in the analyses depends on several factors such as availability of experimental data, strain range, and complexity of loading.

Each tyre component shows different deformation response under external loading. Rubber exhibits non linear deformation and almost incompressible response, while fabric cords and steel wire withstand most both tension and compressions loads and therefore produce small strain. For rubber, hyper-elastic material models are used to describe high deformation. In this study, Yeoh's model was chosen to define hyperelastic property of rubber materials and Marlow model for reinforcements such as fabric and steel cords. Bead was modeled as an elastic material.

The Yeoh material model had a cubic form with only *I1* dependence and is applicable to purely incompressible materials. The strain energy density for Yeoh model is written as

$$W = \sum_{i=1}^{3} C_i (I_1 - 3)^i \tag{2}$$

where C_i are material constants. C_i quantity is 0.5 of the initial shear modulus.

The reason for using the Yeoh's model in the rubber material model, despite the fact that Abaqus supports other material models like Neo-Hookean and Mooney-Rivlin, because it is capable of predicting different deformation modes using data from a simple deformation mode like uni-axial tension test. A review by Wei et al. [8] found that most of material models are determined based on the polynomial expression of strain energy function. Although Mooney-Rivlin energy density function has been widely applied for tyre dynamic properties analysis, the function has a limitation that it could not be accurately applied to large deformation problems of the rubber material. Neo-Hookean material model also has a limitation that the coefficients derived from uni-axial deformation tests are not suitable to describe other deformation modes. In order to determine the parameters of rubber hyperelastic property, most of the material models need to combine three deformation tests (uni-axial, biaxial tension and pure shear), which is recognized as a complex and time consuming procedure.

No	Properties	Tread	Under tread	Wing	Inner liner	Side wall	Apex	Rim cushion	Bead	Belt	Ply
1	Hardness	74.00	74.00	62.67	67	59	89	73	82	71	69
2	Density	1.16	1.16	1.09	1.221	1.088	1.163	1.162	1.289	1.182	1.114
3	Stress, Mpa	0.95	0.95	0.49	0.78	0.50	2.00	1.09	1.53	0.84	0.96
4	Strain, %	0.16	0.16	0.16	0.17	0.15	0.17	0.16	0.16	0.16	0.17
5	Young's modulus	5.94	5.94	3.06	4.68	3.27	12.04	6.78	9.42	5.36	5.73
6	Poisson ratio	0.49	0.49	0.49	0.49	0.49	0.49	0.49	0.49	0.49	0.49
7	mu	1.99	1.99	1.03	1.57	1.10	4.04	2.28	3.16	1.80	1.92
8	Kappa	99.01	99.01	50.94	77.97	54.47	200.60	113.02	156.95	89.26	95.53
9	C10	1.00	1.00	0.51	0.78	0.55	2.02	1.14	1.58	0.90	0.96
10	D1	0.02	0.02	0.04	0.03	0.04	0.01	0.02	0.01	0.02	0.02

Table 2.
Material properties.

3.2.3 Footprint and radial stiffness analysis

After completing the axisymmetric tyre modeling, the next step is the simulation of tyre under static loading. From this simulation there are two analyses can be further performed: footprint analysis and radial stiffness analysis. For footprint analysis, the load needs to be applied on the tyre to represent the normal load according to the specified load index of the tyre. **Figure 5** shows the tyre under static loading and its respective footprint result.

For designing a new PCR tyre, there are three different tyre were taken for benchmark. The tyre being simulated is of the size 175/65 R14 and were inflated at 2.1 bar (30.5 psi) with various loads of 100 kg, 150 kg, and 200 kg using three types of tyre, called tyre A, tyre B, and tyre C. Tyre A has two grooves, tyre B has three grooves, and tyre C has four grooves.

To obtain more accurate footprint result, the full tyre with tread was modeled so that the contact pressure distribution on the tread which in contact with the road can be evaluated. **Figure 6** exhibits the footprint comparison of these three tyres.

In Abaqus, footprint simulation is performed under static loading and needs several input files for defining geometry, boundary condition, sequence and load of tyre and rim. The result of Abaqus footprint analysis as it is shown in **Figure 6**, suggests that the tyre having two grooves shows the largest contact area at shoulder. Large contact area on shoulder indicates better cornering stability.

The second simulation result is about radial stiffness of the tyre. The radial stiffness mainly depends on sidewall stiffness and affects the transversal bending of tyre. This transversal bending causes the tyre to lose its height by certain value, from initial radius R becomes deflected radius R_{def}, as shown in **Figure** 7.

The R_{def} resulted from simulation of two groove tyre is the largest (see **Table 3**), that means that its radial stiffness is also the largest. Larger radial stiffness gives more cornering stability.

By looking at footprint and radial stiffness, the two groove tyre indicates a better cornering stability compared to the other tyre types.

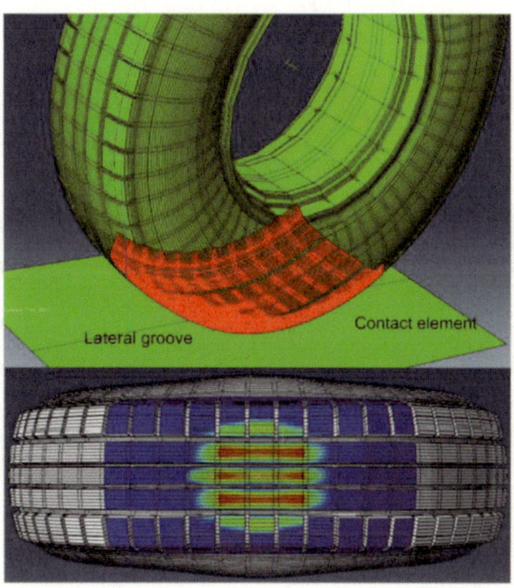

Figure 5.
Footprint simulation under static loading.

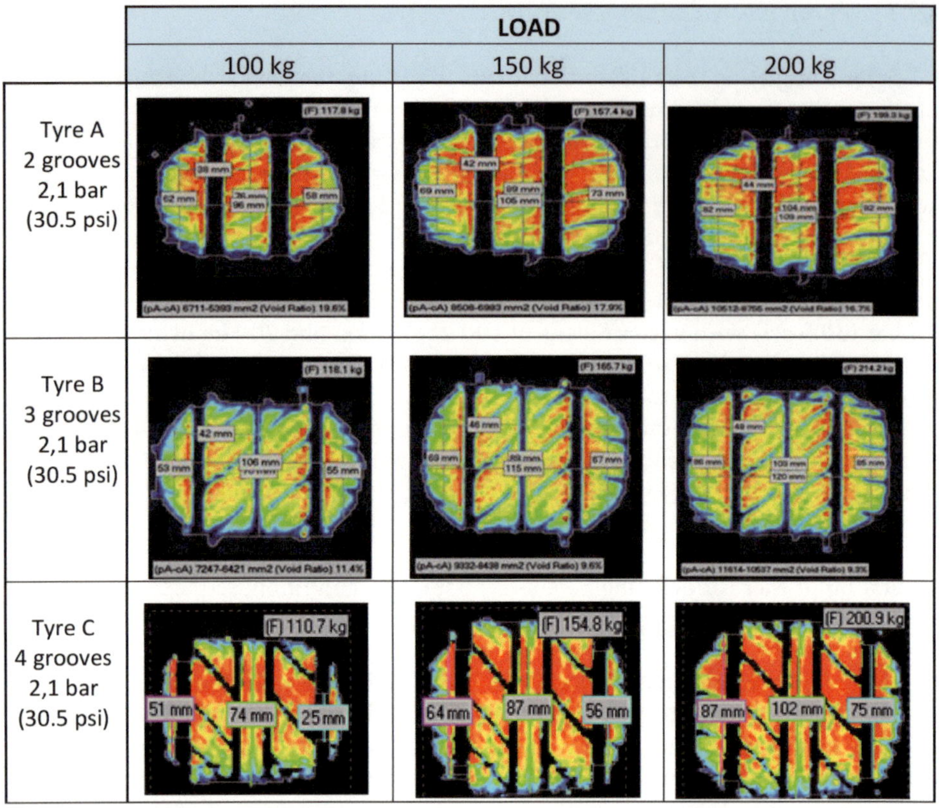

Figure 6.
Footprint comparison of benchmark tyres [7].

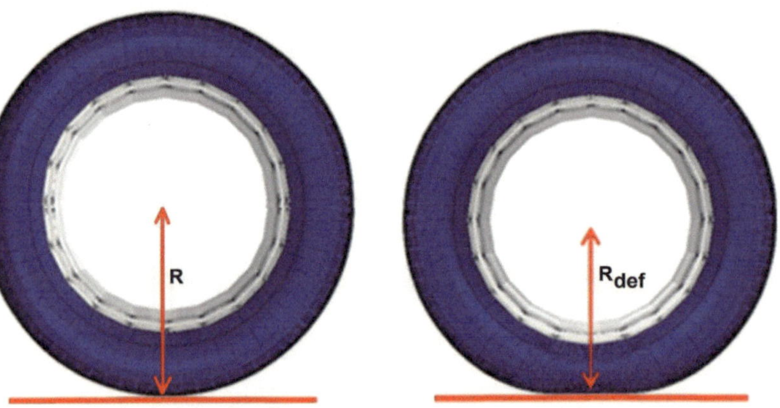

Figure 7.
Tyre deformation.

	2 Groove tyre	3 Groove tyre	4 Groove tyre
R deflection	269.06 mm	268.75 mm	268.54 mm
R initial 291.5 mm			

Table 3.
Value of Tyre radius during deflection.

3.3 Rolling resistance analysis

Rolling resistance force in tyre is mainly generated by friction force, drag force, and hysteresis loss. This study will only discuss rolling resistance force generated by hysteresis loss inside the rubber and cord. The analysis was performed in two main steps, those are static tyre simulation (footprint and radial stiffness) and calculation of strain energy loss to find rolling resistance force.

With review of a number of tyre rolling resistance simulations, it is found that rolling resistance calculation is based on the strain energy loss during a traveled distance. Aldhufairi et al. [9] used a script of Abaqus to extract the 3D tyre model data as input and an analytical rigid road drum with a straight and smooth surface was added to the model, equivalent to that used in the experiment, due to the limitation of the testing machine the travel speed was limited to 30 km/h. Ghosh et al. [10] suggested a method that implements a steady state rolling simulation using Abaqus software to obtained the strain energy and principal strains, together with the loss factors (Tan d) of the material obtained separately in the laboratory, are used to estimate the energy dissipation of a rolling tyre through post processing. The internal code was developed to perform such a task.

Lind [11] suggested three sequential steps for solving the rolling resistance model; inflation, footprint and rolling. The last rolling step was performed using a dynamic solver setting where the center node was moved in the x-direction with a prescribed acceleration up to a target speed. The rolling resistance was from the FE-simulation result computed in two different ways. The first method uses the contact forces from each node multiplied with its distance from the wheel centre; the second method uses the reaction forces from the constrained middle node and computes the rolling resistance. The result presented for the material model and for the rolling resistance does not aim toward representing any specific tyre rubber compound or tyre.

While the others used FE tyre model without tread, Cho et al. [12] Included the tread in FE tyre model. The hysteretic loss during one revolution was computed with the maximum principal value of the half-amplitudes of six strain components, and the temperature distribution of tyre was obtained by the steady-state heat transfer analysis. The static tyre deformation analysis is performed by ABAQUS/Standard in the deformation module and the strain and stress results are input into the in-house dissipation module where the hysteretic loss, rolling resistance and heat generation rate are computed.

In this study, the footprint analysis was carried out with patterned tyre model and for rolling resistance simulation used tyre model without pattern for the sake of computing time. However, the accuracy is a little sacrificed but still acceptable i.e. 6.2% error as describe in the Section 3.4.

The rolling resistance analysis was based on hysteresis of rubber and cord where the phase of stress lags behind the strain as it is shown in **Figure 8**. The hysteretic loss ΔW per unit volume during a period $T_c = 2\pi/\omega$ is:

$$\Delta W = \int_0^{T_c} \sigma(\tau)\frac{d\varepsilon(\tau)}{d(\tau)}d\tau = \int_0^{T_c} \sigma_0\varepsilon_0 \, sin\,(\omega\tau + \delta)\cos\,(\omega\tau)d\tau. = \pi\sigma_0\varepsilon_0\,\sin\,\delta \qquad (3)$$

where σ_0 and ε_0 being the stress and strain amplitudes and ω being the excitation frequency.

In engineering application, as suggested by Cho et.al [9] 3D viscoelastic bodies are subjected to more complicated multi-axial cyclic excitations, so the time histories of strains and stresses are neither one-dimensional nor sinusoidal.

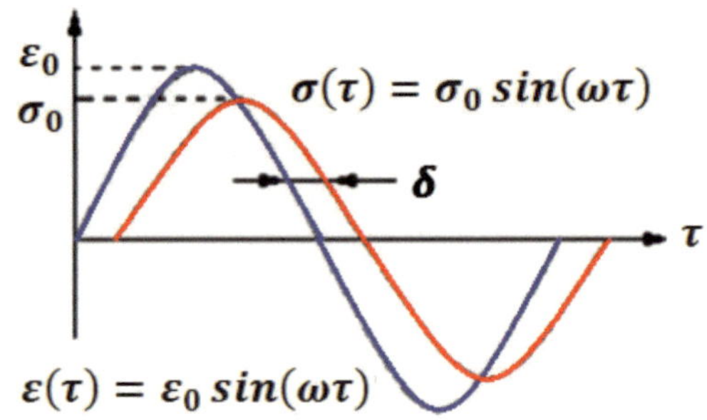

Figure 8.
Stress - strain phase.

Therefore, the hysteretic loss is expressed in a generalized form:

$$\Delta W = \int_0^{T_c} \sigma_{ij}(\tau) \frac{d\varepsilon_{ij}(\tau)}{d(\tau)} d\tau \tag{4}$$

The hysteretic loss can be converted to the heat generation, and the heat generation rate Q per unit volume during a cycle is:

$$Q = \frac{\Delta W}{T_c} = \frac{1}{T_c} \int_0^{T_c} \sigma_{ij}(\tau) \frac{d\varepsilon_{ij}(\tau)}{d(\tau)} d\tau \tag{5}$$

In order to calculate the energy loss during deformation, curve interpolation and FFT function were developed. Abaqus python contains NumPy which can do FFT. A python scripting is used to read signal curve, perform the FFT and create a new curve, i.e. amplitude vs. frequency, for plotting in Abaqus.

```
def interpolation(curve):
    myCurve = []
    i = 0
    n = len(curce)
    myCurve.append(0.0,curve[0])
    while i < n:
      myCurve.append(myAngle[i], curve[i])
      i = (i + 1)
    myCurve.append(360.0,curve[-1])
    i = 0
    n = len(myCurve)
    NewCurve = []
    while i < (n - 1):
      angle_A = (myAngle[i + 1][1] - myAngle[i][1]) /
                (myAngle[i + 1][0] - myAngle[i][0])
      yo = myAngle[i][1]
      xo = myAngle[i][0]
      j = myAngle[i][0]
      while j < myAngle[(i + 1)][0]:
    newAngle.append(yo + (angle_A * (j - xo)))
    j = (j + delta)
    i = (i + 1)
    if i == (n - 1) and newAngle.append(myAngle[i][1]):
            pass
```

```
        return newAngle
    def fourier(sigma, epsilon):
        FFT1 = 2 * abs(fft.fft(sigma)) / len(sigma)
        FFT2 = 2 * abs(fft.fft(epsilon)) / len(epsilon)
        k = 0
        total = 0
        while k < (len(FFT1) / 2):
          total = total + (FFT1[k] * FFT2[k]) * k
          k = k + 1
        return total
```

The input for sigma and epsilon are the interpolated stress and the interpolated strain respectively.

In a rolling tyre, the rubber compounds exhibit the complicated 3D dynamic viscoelastic deformation. The strains and stresses are constituted in terms of the complex modulus G* = G' + iG". In this case, G' is called the storage modulus and G" is the loss modulus. The complex modulus is a function of the strain amplitude $\boldsymbol{\varepsilon_o}$, frequency f, and temperature T. The correlation between storage and loss modulus in terms of the phase difference $\boldsymbol{\delta}$ as follows:

$$\text{Tan}\,\boldsymbol{\delta} = \frac{G''}{G'}\text{.and G''} = \text{G} * \sin\boldsymbol{\delta} \tag{6}$$

In Abaqus simulation the complex modulus G* can be obtained by extracting axisymmetric element data and therefore the heat dissipation from energy loss G" can be calculated by multiplying G and sin $\boldsymbol{\delta}$, and in terms of Python coding is written as follows:

$$\text{heat\_dissipation} = \text{energy} * \sin(\text{tand}) \tag{7}$$

Where energy is extracted from previous axisymmetric simulation element data and tand is from input data.

The rolling resistance force generated by the hysteretic loss is computed as the total hysteretic loss of the rolling tyre during one revolution divided by the traveling distance of tyre during the same period of time, hence:

$$\text{F}_{\text{RR}} = \frac{W}{2\pi r} \tag{8}$$

where $W = \int_{\Omega} \Delta W \, dV$

r = effective radius of tyre

Ω = material volume of tyre

Rolling resistance coefficient C_r is the indication of how large the rolling resistance is for a given load upon which it is rolling and is calculated by:

$$\text{C}_\text{r} - \frac{Total\ force\ (N)\ x\ 1000}{Load\ (N)}\ \text{N/kN} \tag{9}$$

Total force is meant the sum of force caused by hysteresis loss in each tyre component material.

To analyze the force produced by tyre component materials, a Python code was developed as plugin in Abaqus software. The process of analyzing the rolling resistance is describe in the following steps and is shown in **Figure 9**.

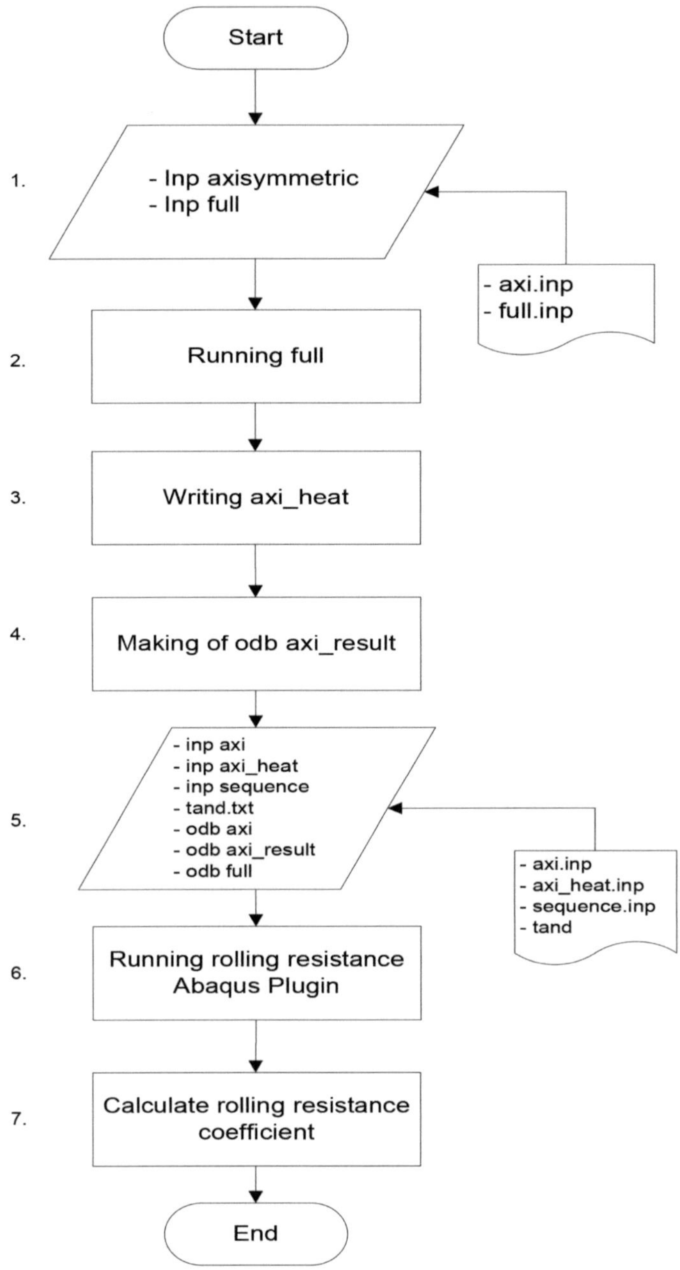

Figure 9.
Rolling resistance analysis process using Abaqus plugin.

1. Prepare input files, i.e.:

    - Axisymmetric input file (axi.inp)

    - Full tyre input file (full.inp)

2. Running full tyre model simulation in Abaqus using the input files in step 1 in command prompt with the following command:

    Abaqus job = full oldjob = axi cpus = 4

3. Writing axi_heat input file:

 - Copy axi, inp and rename it to axi_heat.inp
 - Change the tyre element type from cgax into dcax
 - Delete input of tyre_coord and rim
 - Delete all properties in each material and replace with:

   *conductivity: 0.2

 - Delete all existing steps and boundary conditions and replace with steps and boundary conditions necessary for rolling resistance simulation

4. Copy axi.odb and rename it into axi_result.odb

5. Input data needed for running rolling resistance simulation:

   a. Input files:

      - axi.inp
      - axi-heat.inp
      - sequence.inp

   b. Odb files:

      - axi. Odb
      - axi_result.odb
      - full.odb

   c. Tan delta data: tand.txt.

   Below is tan-δ example of tread compound.

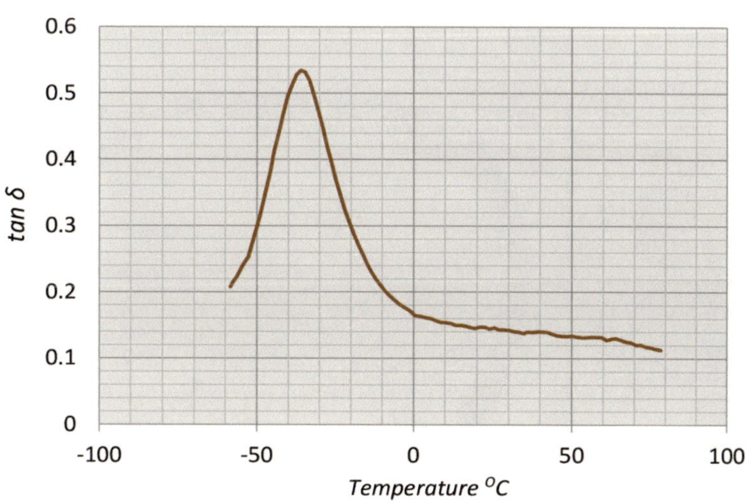

6. Running rolling resistance calculation using Abaqus plugin after specifying the required data as mentioned in step 5 in the pop up menu and other information needed for running the simulation such as:

    - Select how energy is interpolated from coordinate element to bulk elements
    - Define speed of tyre [km/h]
    - Define error limit for heat transfer [%]
    - Define interpolation parameter [deg]
    - Define parameter for tyre radius calculation.

7. After completing the calculation the output data will be presented in axi_RR_result file (see **Figure 10**), and the Rolling Resistance Coefficient (C_r) is then calculated using equation (9):

$$C_r = \frac{Total\ force\ (N)\ x\ 1000}{Load\ (N)}\ \mathrm{N/kN}$$

The example of the simulation result is shown below:
Results:
Force produced by material I40 is 2.08360116975 N
Force produced by material A02 is 1.29144915874 N
Force produced by material T61 is 18.8669933222 N
Force produced by material BW08 is 0.0 N
Force produced by material T61 is 1.67943517041 N
Force produced by material Z80 is 1.13411378677 N
Force produced by material S70 is 0.41902012456 N
Force produced by material S70 is 3.82462665603 N
Force produced by material R50 is 1.72655457713 N
Force produced by material N20 is 0.883588825178 N
Force produced by material C32 is 1.62535580812 N
Total force is 33.5347385989 N

Since load index of the tyre is 82, the maximum tyre load is equal to 475 kg. According to ETRTO standard, the tyre load for rolling resistance calculation is 80%

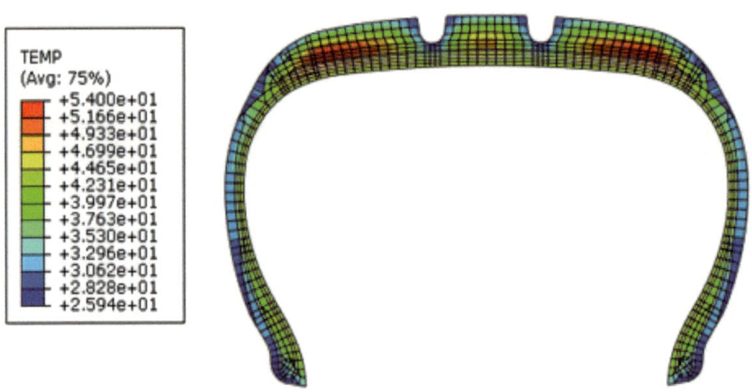

Figure 10.
Temperature distribution of 2 groove tyre.

of maximum load which is 380 kg or 3728 N, and then the rolling resistance coefficient is equal to:

$$C_r = \frac{33.5347\ x\ 1000}{3728} = 9\ \text{N/kN}$$

Using the same calculation for tyre B and C, we obtain the following result:
tyre A produces C_r = 9 N/kN
tyre B produces C_r = 8.77 N/kN
tyre C produces C_r = 8.4 N/kN

3.4 Validation of rolling resistance simulation result

The rolling resistance simulation result obtained from an Abaqus plugin code need to be validated by comparing the result with the actual test result. The actual test has been carried out using 14 different tyres that has been tested on RR machine conducted by certified bodies, such as TUV, and the results are compared with the RR result from simulation, as shown in **Figure 11**.

In average, the simulation result is higher than the actual testing result by 0.46 or 6.2%.

3.5 Rolling resistance on different radial stiffness and crown radius

The radial stiffness of tyre significantly affects the rolling resistance. **Table 4** shows that smaller stiffness of sidewall (indicated by higher R deflection) resulted in higher rolling resistance. This phenomenon explain that to deform a higher stiffness material needs more energy, meaning that the energy loss is higher and eventually the rolling resistance is also higher.

Tyre tread contour has a great influence on rolling resistance. To study this, the simulation was performed on three tyres with different crown radiuses, i.e.:
Tyre A: R1 = 250 mm and R2 = 150 mm,
Tyre B: R1 = 550 mm and R2 = 300 mm, and
Tyre C: R1 = 900 mm and R2 = 300 mm, as it is shown in **Figure 12**.

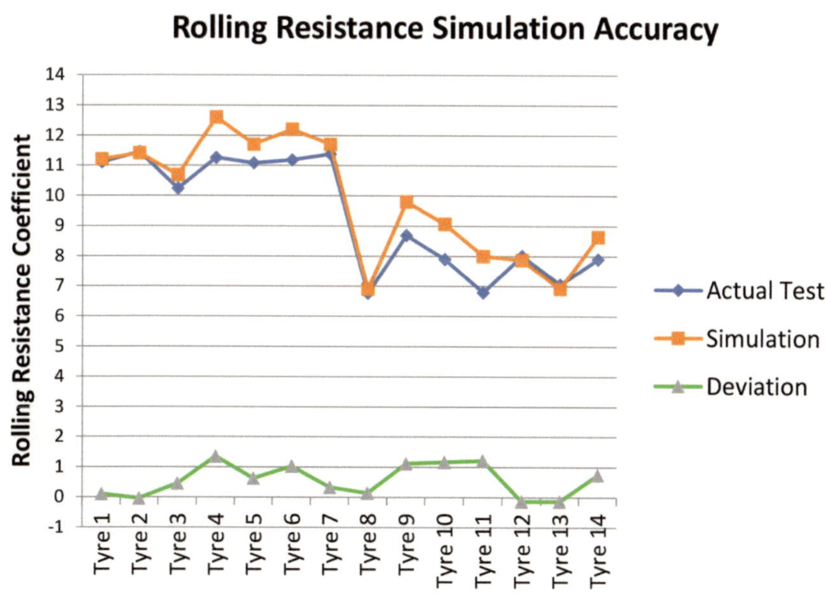

Figure 11.
Rolling resistance coefficient test result.

	2 groove tyre	3 groove tyre	4 groove tyre
R deflection	269.06 mm	268.75 mm	268.54 mm
Rolling resistance coef.	9 N/kN	8.77 N/kN	8.4 N/kN

Table 4.
Correlation between radial stiffness and rolling resistance.

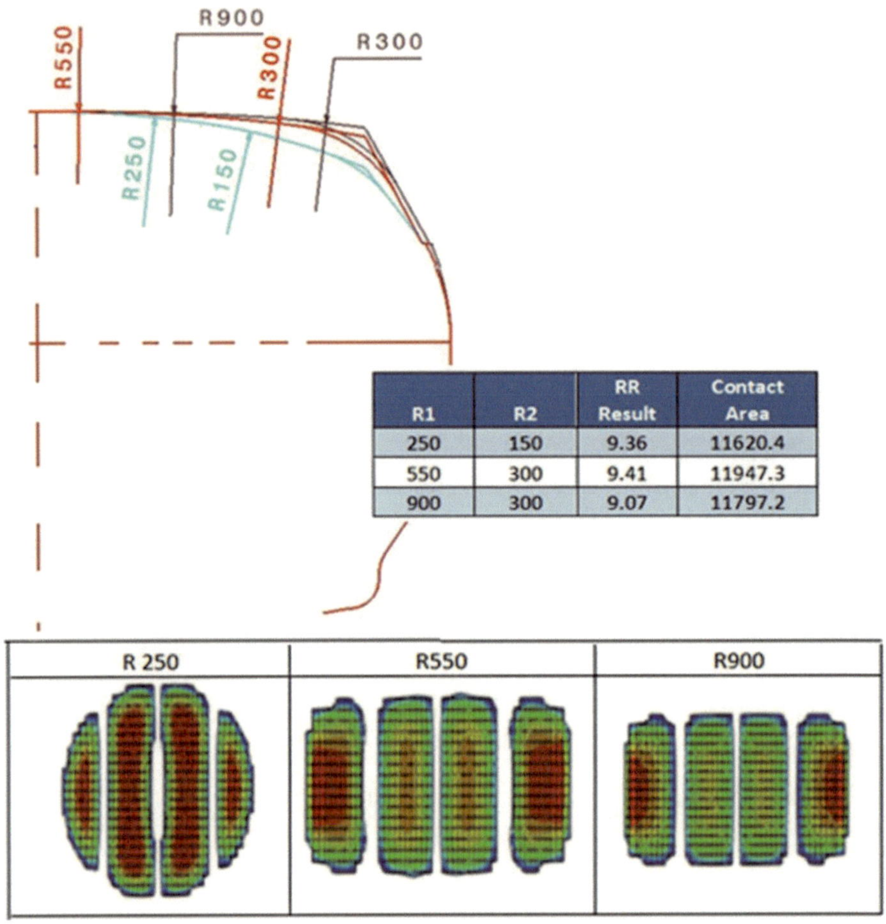

R1	R2	RR Result	Contact Area
250	150	9.36	11620.4
550	300	9.41	11947.3
900	300	9.07	11797.2

Figure 12.
Crown radius relation with footprint and rolling resistance [7].

4. Conclusion

During the PCR tyre design and development, there is several tyre performance parameters need to be considered, including rolling resistance, wet adhesion, noise, and cornering stability. In this study, a Finite Element simulation was carried out to perform prediction of rolling resistance and cornering stability.

The simulation was performed in two stages: steady state rolling simulation using Abaqus build in function and rolling resistance calculation using internally developed Python code as Abaqus plugin.

The validation was done by comparing the simulation result and actual test on RR machine and the average discrepancy of C_r is 0.46 or 6.2%. In addition, the RR plugin only need between 10 and 15 minutes to run, it is very short compared to pre processing time.

The simulation result suggests that the best estimated rolling resistance is four groove tyre with crown radiuses of R1 = 900 mm and R2 = 300 mm. However, two grooves tyre provides larger shoulder contact area which in turn gives better cornering stability, but has rolling resistance coefficient of 9 N/kN.

Considering that the rolling resistance coefficient (*Cr*) of two groove tyre is within the allowable value for stage 2 requirement of UNECE regulation No. 117–2 (the maximum *Cr* is 10 N/kN), so the suggested PCR tyre should have the following specification to meet the performance target: low rolling resistance and good cornering stability:

- Two grooves

- Crown radius R1 = 900 mm and R2 = 300 mm

Author details

Sutisna Nanang Ali
President University, Cikarang-Bekasi, Indonesia

*Address all correspondence to: nanang.ali@president.ac.id

References

[1] Pike Ed 2011, Opportunities to Improve Tire Energy Efficiency (Washington DC: International Council on Clean Transportation).

[2] Hall DE and Moreland JC 2001, Fundamentals of Rolling Resistance. Rubb. Chem.Technol. 74 No 3 525.

[3] United Nation Economic Commission for Europe (UNECE) regulation No. 117 Revision 2 Uniform Provisions concerning the Approval of Tyres with regard to Rolling Sound Emissions and to Adhesion on Wet Surfaces and/or to Rolling Resistance

[4] Tonachel L 2004, Fuel Efficient Replacement Tires (New York: National Resources Defence Council)

[5] LaClair TJ 2005, Rolling Resistance in The Pneumatic Tire (Washington DC: National Highway Traffic Safety Administration) p 483

[6] NHTSA 2009, Tire Fuel Efficiency Consumer Information Program Development: Phase 2 – Effects of Tire Rolling Resistance Levels on Traction, Treadwear, and Vehicle Fuel Economy.

[7] Sutisna, N.A and Gapsari, F. 2018, Rolling Resistance and Noise Estimation for Product Design and Development of Eco-Tyre using Finite Element and Numerical Method Proceeding of International Conference on Mechanical Engineering Research and Application 2018 IOP Conf. Ser.: Mater. Sci. Eng. 494 012021

[8] Wei, C, Olatunbosun, OA & Behroozi, M 2016, Simulation of tyre rolling resistance generated on uneven road, International Journal of Vehicle Design, vol. 70, no. 2, pp. 113–136. https://doi.org/10.1504/IJVD.2016. 074415

[9] Aldhufairi, H, Olatunbosun, R & Essa, K 2019, Determination of tire's rolling resistance using parallel rheological framework, SAE Technical Paper. https://doi.org/10.4271/2019-01-5069

[10] S. Ghosh, R.A. Sengupta, and G. Heinrich 2011, Investigations on Rolling Resistance of Nanocomposite Based Passenger Car Radial Tyre Tread Compounds Using Simulation Technique. Tire Science and Technology: September 2011, Vol. 39, No. 3, pp. 210–222.

[11] Lind, Petter. 2017, A finite element material modelling technique for the hysteretic behaviour of reinforced rubber. KTH Royal Institute Of Technology School Of Engineering Sciences.

[12] J.R. Cho, H.W. Lee, W.B. Jeong, K. M. Jeong, K.W. Kim, 2013 Numerical Estimation Of Rolling Resistance And Temperature Distribution Of 3-D Periodic Patterned Tire, International Journal of Solids and Structures, Volume 50, Issue 1, 2013, pp 86–96

Chapter 10

Finite Element Analysis in Nanotechnology Research

RameshBabu Chandran

Abstract

The Finite Element Analysis in the field of Nanotechnology is continually contributing to the areas ranging from electronics, micro computing, material science, quantum science, engineering, biotechnology, medicine, aerospace, and environment and in computational nanotechnology. The finite element method (FEM) is widely used for solving problems of traditional fields of engineering and Nano research where experimental analysis is unaffordable. This numerical technique can provide accurate solution to complex engineering problems. Over decades this method has become the noted research area for the mathematicians. The popularity of FEM is due to the advent of computer FEA software such as NASTRAN, ANSYS, ABAQUS, Matlab, OPEN Foam, Simscale and the like. With the development of nanoscience, the researchers found difficulties in spending funds for nano related projects. The FEA has evolved as the affordable methodology and offers solutions to all complicated systems of research.

Keywords: nanotechnology, FEM, FEA, research, nanoscience

1. Introduction

"To move precisely in nanoworld, you donot succeed by perfecting proven techniques".- Handelsblat. [1] . As stated, the nano research requires newer methodologies and techniques to be worked out to succeed. The microtechnology to nanotechnology needs a factor of thousand for size reduction. Different methodologies exist to club cooperation between macro, micro and nano robots and analytical based FEM for static, modal, harmonic and transient analysis of structures. Clubbed with multiparametric optimization and neural networks, FEM had developed as an optimal solution to all complicated problems of engineering, science, technology, medicine and research. The "bottom up" technology of late twentieth century promises the use of robotics for micro/nano manipulation processing [1]. The revolution of computers had led to development of closed form solutions which are extremely difficult to be obtained for any engineering problems [2]. This urge leads to adopting any one of the numerical techniques. FEA had been one of the options of researchers and choice of the method depends on the familiarity of the users. FEM exploits the research methodologies using direct approach, variational approach, direct approach, energy approach, weighted residual approach, Isoparametric formulation, static condensation and nonlinear analysis [2]. Numerical approximations can be reached by differential equations and PDE for various mathematical and nano technology problems of applied physics. FEM explores checking the validity of analytical studies of nanotechnology. Moreover the unaffordable experimental setup of nanoresearch could be replaced with FEA software as observed by engineers and scientists working on the field [2].

2. Applications of FEM in nano research

2.1 Electrospun nanofibrous Mats under biaxial tension

The nanofibrous mats can be used as freestanding electrodes in energy storage devices. These mats have non uniform material properties [3] with each and every single fiber having nonlinear characteristics. The research came up with two macroscopic continuum models with uniform or oriented nanofibre distribution to exactly replicate the nanofibrous mats under biaxial tension. The mechanical response of electrospun SF/PCL nanofibrous mats was explored. The model simulated by FEM exposed the deformation of nanofibrous mats and the gradual damage mechanism along with the microstructure.

2.2 Carbon nanotube reinforced composite's stress transfer

The improvement [4] in mechanical properties of hybrid composite system strongly depends on interfacial mechanical properties (interfacial stress transfer efficiency). The interfaces are zones of structural, compositional and property gradients. The width varies from single atom to micrometers. The properties of the composite depend on the surface of the single fiber and the resin used for bonding. The stress transfer of SWNT reinforced composite was studied using hybrid FEM approach. Three dimensional REV (representative elementary volume method) had been used along with Molecular Dynamics (MD) to exploit the stress transfer mechanism of CNT reinforced composite. The [4] effects of fiber volume fraction, interfacial stiffness and elastic modulus on the stress of the matrix were explored. The analytical models are difficult along with experiments in nanoscale as they are too expensive. Hence FEM simulated models were used to predict the stress transfer and found to be accurate on validation.

2.3 Formulation of 3D finite element implementation for adhesive contact at nanoscale

A research was carried out [5] for three dimensional nanoscale contact problems with strong adhesion. The contact description was based on Lennard Jonnes [5] description suitable to explain Vanderwaals attraction between interacting bodies. By incorporating the potential into nonlinear continuum mechanics, formulations had been arrived for surface force and body force. Based on these formulations, overall contact algorithm had been arrived using FEM. This model has an application in biomechanics as denoted by adhesion of gecko spatula [5]. Efficient FE formulation was arrived based on surface traction. The behavior of contact model described by SF formulation was more efficient than BF formulation for strong adhesive existence.

2.4 Finite element simulation of micromachining of nanosized silicon: carbide particle reinforced composite

The nanosized silicon – carbide – particle (SICp) reinforced aluminum matrix composite's micromachining process had been studied [6] using finite element method. The parameters of cohesive zone model had been found from stress – displacement curves of the molecular dynamics (MD) simulation. The model represented exactly the random properties of silicon – carbide particle distribution and the interfacial bonding between SICp and matrix [6]. The mechanism of machining was analyzed as per chip morphology, stress variation, temperature and cutting force. The FE simulation projected the fact that SICp caused non uniform interaction between the tool and reinforcement. The deformation mechanics [6] led to inhomogeneous stress variation and irregular cutting force.

2.5 FE modeling of double walled CNT based sensor

The Carbon Nanotubes (CNTs) are widely used for designing nano sensors, nano resonators and actuators [7]. The mass sensing characteristics of defective Double Walled Carbon nanotubes (DWNTs) were studied using FEM. Various finite element simulations covering chiral, zigzag and arm chair nanotubes with cantilever and bridged conditions using molecular [7] structural dynamics approach. The defects have been subdivided in to 6 missing atoms (A type) and 24 (B type) missing atoms on the outer wall of DWNT. COMBIN 14 element had been used for simulation of defective DWNTs with weak Vanderwaals force. The study revealed the fact that the frequency of defective DWNT reduced with increase in chiral angle. Also the frequency reduced with increase in pinhole defects.

2.6 Perspective on nanotips

Numerous methods have been fabricated for developing ultra-sharp tips for scanning probe microscopy and [8] electron microscopy. It has been observed that the sharp end terminates with very single atom in field ion microscope (FIM). The last atom had been intended to form atomic channel of electrons in field emission mode which would self-collimate a coherent electron beam with an outstanding brightness. Hence nanotips are found to behave as a source of self-collimated electron or ion beam. In this research [8] the distribution of electric field in the vicinity of nanotip apex that holds the topmost single atom had been studied analytically and numerically. The tip base was found to dominate nano protrusions which enhance electric field. The study revealed that nanotips with broad bases produce even less field than modest tips at the same voltage. This pronounced the fact that the tip base accounts for high voltages needed at imaging threshold field.

2.7 Axial vibration of embedded love – Bishop nanorods

In this research, nonlocal free vibration of axial rods embedded in elastic medium had been studied using Love – Bishop rod theory with FEM. Constitutive modeling for rod formulation using kinematic relations and dynamic equilibrium had been analyzed. Equation [9] of motion and boundary conditions were obtained by varying total potential of nano rod and were solved using separation of variation. Frequency equations of 4 types of nanorods were obtained. Size dependent FEM formulation was synthesized based on weighted residual method. The four parameters mode number, non-local parameter, rod length, and slenderness ratio were used to study the frequency parameters of nanorods. The free vibration frequencies of the simple axial rods had been found to be higher than that of Love bishop rods. These were evaluated by frequencies of two rod theories in higher modes.

2.8 Bending analysis of embedded nano plates using FEM

Eringen's nonlocal elasticity theory is capable to capture small length scale effect. Hence it is widely used to explore the mechanical behavior of nanostructures [10]. Instead of using differential form, the integral form should be used to avoid inconsistency in results. Arbitrary kernel functions are used for general form. The first order [10] shear deformation theory is used to model the nanoplates. The study evaluates the first order shear deformable embedded nanoplates for bending using Erignen's nonlocal theory. Using FEM approach the maximum deflection of the structure was evaluated. The results pronounced that the clamped or simply supported boundary conditions provided same trend for the effects of non-local parameter on the bending analysis of nano plate for Erignen's integral and differential formulation. Also the results proved that the elastic foundation increased the stiffness of the structure and decreased the influence of nonlocal parameter.

2.9 Stresses at bone: particle reinforced nano composite interface

The biomaterial should satisfy its bio functionality and biocompatibility. The tissue – implant interface plays a huge role in both the parameters. Nanobioceramics is a newer technology that had widened [11] range of biomedical and dental applications including increased bioactivity for tissue regeneration and engineering, drug and gene delivery, treatment of viral infections and implantable surface modified medical devices for [11] better hard and soft tissue attachments. FEA had been accepted for simulations in biomechanics for analyzing stresses and strains in dental implants and surrounding bone structures. The tissue engineering needs combining 3D scaffolds with living cells to deliver the much needed cells to damage sites in the human body. This scaffold should be capable of making cells to attach and multiply. Hence the design of scaffold is a challenging task which could be narrated by the finite element analysis. Also the nanotechnology had revolutionized nanobiomaterials, tissue engineering nano scaffold, nano – drug delivery and dental nanocomposites [11].

2.10 Elastic plastic analysis of ultrafine grained Si2N2O – Si3N4 composites

The development of micro/nanotechnology [12] had led to characterization of the mechanical properties at micro- and nano- scales. The nano indentation test has a diamond indenter to produce indentation load and the penetration depth from which load – penetration curve (P-h) is obtained. The P-h curve can be used to define mechanical properties including hardness, elastic modulus and toughness [12]. Only few studies had been reported on elastic–plastic property of brittle bulk ceramics. There are two methods to derive material properties from loading and unloading indentation curves. One of the curves involves the use of unloading curves and classical elastic solution of infinite half space [12]. This method is suitable for calculating the hardness and elastic modulus of materials. Another methodology involves producing loading and unloading response curves for various parameters through finite element modeling. Stress strain relations can be produced by using the nanoindentation experiment. The ultra-fine – grained Si2 N2O – Si3N4 had been produced by hot press sintering of amorphous nano – sized silicon nitride powders at 1600, 1650 and 1700 Deg Celsius with nanosized additives. After evaluating by nano identation through finite element formulation, the elastic modulus and P-h curve are obtained. A newer theoretical methodology for evaluating stress strain relation of brittle ceramic materials had been identified. Numerous coefficients in theoretical calculation formula had been found using calculation and simulation results.

2.11 Torsional statics and dynamics of circular nanostructures

Newer technologies for developing advanced materials [13] and structures are advancing towards a minute length scale (i.1., micro – or nano – scale). This is the root of nanotechnology. By reducing the size of the materials, the materials exhibit specific and interesting non classical mechanical, chemical and electrical properties. The classical continuum theories fail to replicate the minute length scale [13]. Hence explicitly new continuum mechanics /atomic dynamic simulation are required. Erignen developed Non local elasticity theory as one of the continuum models. In this research, the torsional static and dynamic nonlocal effects for circular nanostructures for concentrated and distributed torques were investigated based on nonlocal elasticity stress theory [13]. Variational energy principle is obtained to derive governing differential equation and strain energy and kinetic energy components are obtained. A new nonlocal finite element method (NL-FEM) had been developed to solve integral nonlocal equation. The statics and dynamics of nonlocal nanoshafts, nanorods, and nanotubes with various loads and boundary conditions revealed possible numerical solutions which were compared with analytical solutions.

2.12 Elastic properties of coiled CNT reinforced Nano composite

With the invention of Carbon nanotube with advanced material properties, various nano composites had been developed [14]. A new algorithmic representative volume element (RVE) and finite element method had been formulated to find the elastic properties of coiled Carbon NanoTubes (CCNT). The elastic properties had been explored with the respect to interphase, fiber volume fraction, orientation, number of coils, tube diameter, coil diameter and helix angle using FEM. The elastic moduli of the nanocomposites were found to decrease with increase in the number of coils. Also it had been found that SWNT offers better reinforcement when compared with CCNT reinforcement.

2.13 Elastic and fracture characteristics of graphene – silicon nanosheet composites

Graphene and its composites find application in various fields of aerospace, bioelectric sensors, bio engineering, electronics, energy technology, and lithium batteries due to appreciable electrical, mechanical and thermal properties. The single layer graphene sheets (SLG) needs an appropriate substrate which should not alter the properties of graphene. In this research, an efficient method was developed for evaluating nonlinear stress strain behavior and fracture strength of graphene – silicon nanosheet composites. Nonlinear finite element model [15] had been evolved to obtain constitutive model of the problem which are computed using molecular dynamics (MD) simulations. Graphene is modeled as multilinear elastic and silicon is simulated as isotropic material. Using this model nonlinear behavior of graphene, silicon and their stress strain curve including inflection point leading to failure had been arrived. The results of stress strain curves and elastic modulus and the critical stress of single layer graphene (SLG), silicon nanosheet and their composites with different thickness of silicon nanosheet agrees with that of the molecular dynamics [15].

2.14 Thermo electric bulk and nanostructured materials

Numerous analytical solutions had been formulated [16] for coupled nonlinear behavior of thermoelectric device in one dimension. These devices are used in refrigeration and energy harvesting. In this research, a nonlinear model of thermoelectricity was developed using finite element method. The simulated model takes in to account Seebeck, Peltier and Thomson effects [16]. The FEM is represented in potential variables i.e., voltage and temperature and solved using Newton method by formulating stiffness and Jacobian matrices. The results were verified with simulated one dimensional model. The FEM is then implemented to estimate energy conversion of nanostructured thermoelectric materials. Thus the advantages of nanostructured materials lie in the increased performance and miniaturization.

2.15 FEM of nano: indentation to characterize thin film coatings

Thin film coatings [17] are used in tribological, corrosion resistance of mechanical components, tooling, biomedical implants, electronics, microsystem packaging, cutting tool coatings and magnetic devices. The film coatings are explicitly used for reducing wear and tear. It is a current need to investigate the thin film coatings for critical loads that lead to ultimate fracture. Since nanoindentation is a nondestructive one, it is preferred and imperatively it could be simulated by finite element method (FEM). Using FEM, the hardness, elastic modulus, endurance loads, optimal thickness, optimal critical load, stress distribution and contact pressure between substrate and layer could be found.

2.16 Electric field gradients and bipolar electrochemistry effects

The effects of electric field in alternating (AC) and direct (DC)voltages had been explored in vivo and vitro with [18] electrodes in connection with tissues and implanted cells. The electro simulation through noncontact wireless settings by dipoles by bipolar chemistry is highly possible. FEM studies with same configuration that of experimental studies had proved that the voltage profiles are in qualitative agreement with known bipolar effects. There exist [18] a clear mapping of charge gradients at the material surface leading to growth of neurons. The insulating materials distort the electric space distribution while the dipole at the border of implanted conducting material extends along the material surface and much smoother in intercalation materials.

2.17 Elastic stability of curved nanobeam by finite element approach

The elastic stability of curved nanobeam had been investigated using Eringen's strain driven model [19] coupled with higher order shear deformation theory. The influence of different structural theories and analyses of nanobeam is taken into account while deducing the model. The governing differential equation is solved by finite element method using 3 –noded curved beam. The model had been validated using analytical/numerical solutions. The parameters such as thickness ratio, beam length, rise of curved beam, boundary conditions and size dependent [19] or nonlocal are analyzed based on buckling behavior of curved nanobeams. The results prove that the type of buckling mode corresponding to lowest critical value would be varying based on geometrical and internal material length scale parameter and boundary conditions [19].

2.18 Tensile modulus of CNT reinforced polypropylene composite

The reinforcing efficiency of carbon nanotubes (CNTs) in polymers had been found using finite element modeling. The probability distribution functions [20] of CNT diameter, orientation, dispersion and waviness had been incorporated in the finite element model to derive how the CNT characteristics affect the tensile modulus of CNT reinforced polypropylene composite. The scanning electron microscopy images of CNT/PP composites made by melt mixing and injection molding had been used by image analysis approach [20]. The predicted model had been found to be experimentally correct as per ASTM D638.

3. Conclusion

The FE methods had been used to study thermo-electrical–mechanical coupled model. The integrity of lumped element, distributed element and system level element for design, modeling and simulation of nano/micro mechanical systems (N/MEMS) had been achieved by FEM. The nanostructures, nanocomposites and CNTs and their composites had been modeled using FEA. Further FEM had been applied in nanomaterials and systems used in medicine, dental science, biotechnology and electric field in the form of electrospinning.

The investigation of material properties with 1 – 100 nm dimensions had been achieved by nanoscience and technology. Thanks to nanotechnology and FEM, one of the dimensional materials such as CNTs, silica carbide nanotube, nanowire, nanorod and nanobeam had been dream of innovation. The field effective transistors, gas sensors, nanoactuators, nanocantilevers are the live examples. It has been found that these structures have varied applications in nano/micro – electro – mechanical systems (NEMS /MEMS). Ultra capacitors had been found application in hybrid cars. This chapter elaborated the applications of finite element method in varied applications of nanotechnology including CNTs, nano beams, nanorods, nanobiomaterials, graphene coated materials, nanosensors, nanotips and curved nanobeams. Apart from these applications the nanotechnology extends hands to day to day applications such as self-cleansing walls, wall claddings, reinforcement to cement matrix. The FEM could be extended to these materials which had not been extensively covered. A correct mechanical model simulated by finite element modeling would replicate the exact experimental setup and provide solutions to constitutive modeling and all engineering problems. The best example is the usage of CNTs as reinforcements in composites and cementitious materials. The CNTs are costlier that almost 85% of the CNT reinforcement is studies using FEA software by the researchers instead of experimental studies. The last few decades had been dedicated to CNTs, sensors, diagnostic probes and multifunctional materials based on CNTs, electronic devices and energy storage devices. The sensors could be used to monitor all kind of structures including cracks in bridges and structural collapses of civil engineering structures. The replacement of silicon based sensors and transistors with CNT based printed transistors are future challenge and already researches are on using FEM and FEA software. With advent of computers and FEA software such as ANSYS, ABAQUS, NASTRAN the unaffordable experimental analysis of nanoscience had been replaced with analytical studies using FEM.

Acknowledgements

I thank AAA College of Engineering and Technology, Sivakasi management and faculty who directly or indirectly contributed to this work. I thank my PhD guide Dr. S. Prabavathy, Mepco Schlenk Engineering College Sivakasi who is my inspiration for whatever I do.

Author details

RameshBabu Chandran
AAA College of Engineering and Technology, Sivakasi 626005, India

*Address all correspondence to: rameshbabu_1979@rediffmail.com

References

[1] Sarhan, M. (n.d.). Computational Finite Element Methods in Nanotechnology. CRC Press, 2013 630p.

[2] Y.M. Desai, T.I. Eldo, A.H.Shah, Finite Element Method with Applications in Engineering, Pearson , 2011, 470p.

[3] Yin, Y., & Xiong, J. Finite element analysis of electrospun nanofibrous mats under biaxial tension. Nanomaterials, 2018 8(5), 1-19. https://doi.org/10.3390/nano8050348

[4] Spanos, K. N., Georgantzinos, S. K., & Anifantis, N. K. . Investigation of stress transfer in carbon nanotube reinforced composites using a multi-scale finite element approach. Composites Part B: Engineering, 2014, 63, 85-93. https://doi.org/10.1016/j.compositesb.2014.03.020

[5] Sauer, R. A., & Wriggers, P. Formulation and analysis of a three-dimensional finite element implementation for adhesive contact at the nanoscale. Computer Methods in Applied Mechanics and Engineering, 2009, 198(49-52), 3871-3883. https://doi.org/10.1016/j.cma.2009.08.019

[6] Pen, H., Guo, J., Cao, Z., Wang, X., & Wang, Z. Finite element simulation of the micromachining of nanosized-silicon-carbide-particle reinforced composite materials based on the cohesive zone model. Nami Jishu Yu Jingmi Gongcheng/Nanotechnology and Precision Engineering, 2018, 1(4), 242-247. https://doi.org/10.1016/j.npe.2018.12.003

[7] Patel, A. M., & Joshi, A. Y. Atomistic Finite Element Modeling and Analysis of pinholes in Double Walled Carbon Nanotube based mass sensor. Materials Today: Proceedings, 2016, 3(6) 1438-1443. doi:10.1016/j.matpr.2016.04.026

[8] Rezeq, M. Finite element simulation and analytical analysis for nano field emission sources that terminate with a single atom: A new perspective on nanotips. Applied Surface Science, 2011, 258(5), 1750-1755. https://doi.org/10.1016/j.apsusc.2011.10.034

[9] Civalek, Ö., & Numanoğlu, H. M. Nonlocal finite element analysis for axial vibration of embedded love–bishop nanorods. International Journal of Mechanical Sciences, 2020, 188. https://doi.org/10.1016/j.ijmecsci.2020.105939

[10] Ansari, R., Torabi, J., & Norouzzadeh, A. Bending analysis of embedded nanoplates based on the integral formulation of Eringen's nonlocal theory using the finite element method. Physica B: Condensed Matter, 2018, 534(January), 90-97. https://doi.org/10.1016/j.physb.2018.01.025

[11] Choi, A. H. Stress induced at the bone-particle-reinforced nanocomposite interface. In Interfaces in Particle and Fibre Reinforced Composites. Elsevier Ltd. 2020 https://doi.org/10.1016/b978-0-08-102665-6.00019-4

[12] Luo, J., Zhao, Z., Shen, J., & Zhang, C. Elastic-plastic analysis of ultrafine-grained Si2N 2O-Si3N4 composites by nanoindentation and finite element simulation. Ceramics International, 2014, 40(5), 7073-7080. https://doi.org/10.1016/j.ceramint.2013.12.039

[13] Lim, C. W., Islam, M. Z., & Zhang, G, A nonlocal finite element method for torsional statics and dynamics of circular nanostructures. International Journal of Mechanical Sciences, 2015, 94-95, 232-243. https://doi.org/10.1016/j.ijmecsci.2015.03.002

[14] Khani, N., Yildiz, M., & Koc, B., Elastic properties of coiled carbon nanotube reinforced nanocomposite:

A finite element study. Materials and Design, 2016, 109, 123-132. https://doi.org/10.1016/j.matdes.2016.06.126

[15] Gangele, A., & Pandey, A. K., Elastic and fracture characteristics of graphene-silicon nanosheet composites using nonlinear finite element method. International Journal of Mechanical Sciences, 2018, 142-143(May), 491-501. https://doi.org/10.1016/j.ijmecsci.2018.05.012

[16] Potirniche, G. P., & Barannyk, L. L. A nonlinear finite element model for the performance of thermoelectric bulk and nanostructured materials., Energy, 2019, 185, 262-273. https://doi.org/10.1016/j.energy.2019.07.040

[17] Alaboodi, A. S., & Hussain, Z. Finite element modeling of nano-indentation technique to characterize thin film coatings. Journal of King Saud University - Engineering Sciences, 2019, 31(1), 61-69. https://doi.org/10.1016/j.jksues.2017.02.001

[18] Abad, L., Rajnicek, A. M., & Casañ-Pastor, N. Electric field gradients and bipolar electrochemistry effects on neural growth: A finite element study on immersed electroactive conducting electrode materials. Electrochimica Acta, 2019, 317, 102-111. https://doi.org/10.1016/j.electacta.2019.05.149

[19] Polit, O., Merzouki, T., & Ganapathi, M. Elastic stability of curved nanobeam based on higher-order shear deformation theory and nonlocal analysis by finite element approach. Finite Elements in Analysis and Design, 2018 *146*(April), 1-15. https://doi.org/10.1016/j.finel.2018.04.002

[20] Bhuiyan, M. A., Pucha, R. V., Worthy, J., Karevan, M., & Kalaitzidou, K. Understanding the effect of CNT characteristics on the tensile modulus of CNT reinforced polypropylene using finite element analysis. Computational Materials Science, 2013, 79, 368-376. https://doi.org/10.1016/j.commatsci.2013.06.046

Chapter 11

Finite Element Method for Ship Composite-Based on Aluminum

Prantasi Harmi Tjahjanti and Septia Hardy Sujiatanti

Abstract

The structure and construction of ships made of aluminum alloy, generally of the type of wrought aluminum alloy, when experiencing fatigue failure caused by cracking of the ship structure, is a serious problem. Judging from the 'weaknesses' of aluminum material for ships, this chapter will explain the use of alternative materials for ship building, namely aluminum-based composite material which is an aluminum alloy AlSi10Mg (b) ship building material based on the European Nation (EN) Aluminum Casting (AC) - 43,100, with silicon carbide (SiC) reinforcement which has been treated with an optimum composition of 15%, so that the composite material is written with EN AC-43100 (AlSi10Mg (b) + SiC * / 15p. Composite ship model using ANSYS (ANalysis SYStem) software to determine the distribution of stress. The overall result of the voltage distribution has a value that does not exceed the allowable stress (sigma 0.2) and has a factor of safety above the minimum allowable limit, so it is safe to use. The reduction in plate thickness on the EN AC-43100 (AlSi10Mg (b)) + SiC * /15p composite vessel is significant enough to reduce the ship's weight, so it will increase the speed of the ship.

Keywords: ship composite based on aluminum, EN AC-43100 (AlSi10Mg (b) + SiC*/15p, software ANSYS, stress distribution

1. Introduction

The choice of material for ship building is carried out with several considerations, including physical properties, mechanical properties, material prices, and labor skills needed for the production process. Based on the material used to build ships, in fact it can be divided into two major parts, namely (a) steel ships and (b) non-steel ships. Non-steel ship materials include aluminum alloys which have been developing for more than 30 years and have replaced steel, namely in the use of commercial ships and on surface warships, especially for the deck and superstructure [1]. Even in Indonesia, on 18 December 2008, the Indonesian Navy (AL) launched its first warship (KRI), named KRI Krait-827 made of aluminum, with a speed of 25 knots. This warship is lighter than ships made of iron/steel (Jawa Pos, Desember 2008). The purpose of using aluminum/aluminum alloy is due to the density and modulus of aluminum 1/3 of the steel, thus significantly reducing the overall weight of the ship. The use of aluminum has become an alternative material used as a hull material in ship construction. Almost all of them use wrought aluminum with aluminum of marine grade: main alloy part magnesium (Mg) (alloys of marine grade), marine grade aluminum 5052 (used only for above water), marine grade a luminum 5083 (used for underwater hulls), marine grade aluminum 5086, and marine grade structure of aluminum 6061.

However, the structure of a ship made of aluminum alloy, if it experiences fatigue failure caused by cracks in the ship's structure, it is a serious problem. Cracking itself is usually caused by a combination of rotational stress (torque) and stress concentration interacting with areas of the weak material [2]. The rate of structural cracking in aluminum is 30 times faster than the crack rate in steel when tested at the same stress with the same crack size [3]. On the other hand, the wear resistance on aluminum is also low [4], because aluminum is classified as a "soft" material compared to other metals.

To 'fix' the aluminum material into a strong and hard material, namely adding/mixing it with a reinforcing material, which is a research to get a new material, called Composite Material is grouped in Metal Matrix Composite (MMC) [5–7]. If the method of mixing between the matrix and the reinforcement uses the casting method, it is called Metal Matrix Cast Composite (MMCC). Furthermore, if the metal used is aluminum-based, it is called Aluminum Metal Matrix Cast Composite (AMMCC).

Centered on the 'weaknesses' of ship aluminum sheet, this chapter offers an alternative sheet for shipbuilding, namely silicon carbide (SiC) reinforcement composite material based on aluminum. This aluminum alloy is made by casting aluminum alloy. Aluminum casting (AC) alloy is written: AlSi10Mg (b) in accordance with DIN EN (European Nation) 1706 expressed in chemical symbols written as EN AC-AlSi10Mg (b) and expressed in numeric, written EN AC-43100, so that the writing is combined to become EN AC-43100 (AlSi10Mg (b)). Reinforcement is SiC which has been treated with an optimum composition of 15% (written SiC*), so that the composite material is written with EN AC-43100 (AlSi10Mg (b) + SiC*/15p.

From the background above, this chapter will explain about making a numerical model of ships from the composite material EN AC-43100 (AlSi10Mg (b) + SiC */15p with the help of ANSYS ver.12.0 software to find out how the stresses are distributed. Wave input given is still water and dynamic waves (induced wave), not wave spectrum. Can be applied therein. From the results, it will be known which part of the ship building, the composite material AlSi10Mg (b) + SiC */15p can be applied therein.

2. Ship composite base on aluminum

2.1 Aluminum alloy EN AC-43100 (AlSi10Mg(b)) as matrix

EN AC43100 (AlSi10Mg(b)) alloy is an alloy of silicon aluminum which cannot be heat treated. It has strong flowability in a liquid state and almost no cracks occur in the freezing process [8]. This alloy is commonly used in the welding of aluminum alloys, both cast and wrought alloys as a welding medium or metal [Bergsma & Kassner, 1996]. The physical and mechanical properties of AlSi10Mg (b) can be seen in **Table 1**.

While the mechanical properties of aluminum casting EN AC-43100 (AlSi10Mg (b)) are summarized in **Table 2**.

2.2 Silicon carbide (SiC) ceramic particles as reinforcement

Silicon Carbide is a chemical compound composed of carbon and silicon alone. Created by electrochemical sand and carbon reactions at high temperature s. Silicon carbide has excellent abrasive properties, and has been developed and

Physical and mechanical properties of aluminum AlSi10Mg (b)	Grade
Density (gr/cm^3)	2703
Crystal lattice	FCC
Melting Temperature (°C)	660,22
Boiling Temperature (°C)	2500
Elasticity modulus (GPa)	70,000
Tensile Strengh (Rm) (MPa)	180
Yield Strength (Rp0,2) (MPa)	90
Elongation crack (%)	2,5
Hardness (Brinnel)	55

Source: MKB-material standards.

Table 1.
Physical and mechanical properties of aluminum AlSi10Mg (b) (casting material).

manufactured for over a hundred years into grinding wheels and other abrasive goods. High power, low heat expansion, high thermal conductivity, high hardnes, high elasticity modulus, excellent heat shock resistance and superior chemical inertness are the general properties of silicon carbide.In a crystal lattice, silicon carbide with a tetrahedral chemical structure of carbon and silicon atoms has a strong bond which results in a very hard and strong material. Silicon carbide prevents acids or alkaline salts to strike. In air, SiC forms a protective layer of silicon oxide at 1200° C, which can be used up to 1600°C. The high thermal conductivity combined with low thermal expansion and high strength gives this material exceptional resistance to heat shock.

Nowadays, silicon carbide has grown into a high technological quality ceramic with outstanding mechanical properties. Applications are commonly used in abrasive materials, refractories, electrical conductors and have resistance heating, ignition, and electronic component applications. The engineering properties of silicone carbide are shown in **Table 3**.

In fact, numerical modeling of wrought aluminum vessels has never been possible. It existed until recently, because small ships are already set and included. So it is necessary to decide if the material can be used for shipbuilding. Ship composite EN ACAlSi10Mg(b) + SiC*/15p must be numerically rendered Ship Modeling. Analysis of numerical computation using ANSYS software version 12.00 for seeing the stress distribution that occurs does not surpass the stress permits (0.2 sigma with that obtained from the tensile test), and also if it is safe for factor protection. The provided wave input is still induced by water and wave (the quasi-static one).

3. Numerical modeling ship

3.1 Type and sizes ship

Type of composite boats (EN AC-AlSi10Mg (b) + SiC*/15p) to be modeled numerically using software ANSYS version 12.0 is Fast Patrol Boats with length over all (LOA) is 42.0 meters. Ship size is as follows:

Material designation	Alloy description (DIN EN 1706)		Material short symbol	Material number	Material state*	Tensile strength	Yield strength	Elongation at break	Brinell hardness
	chem. Symbol	numeric	(DIN 1725–2)	(DIN 1725–2)		R_m (Mpa)	$R_{p0,2}$ (Mpa)	A_5%	HBS
GK AlSI12	EN AC-Al Si 11	EN AC-44000	G-AlSi11	3.2211	F	170	80	7	45
	EN AC-AlSi 12(b)	EN AC-44100			F	170	80	5	55
	EN AC-AlSi 12 (a)	EN AC-44200	G AlSi12	3.2581	F	170	80	6	55
GK-AlSI1DMg	EN AC-AlSi10Mg(a)	EN AC-43000	G-AlSi10Mg	3.2581	F	180	90	2,5	55
					T6	260	220	1	90
					T64	240	200	2	eo
	EN AC-AlSi10Mg(b)	EN AC-4310			F	180	90	2,5	55
					T6	260	220	1	90
					T64	240	200	2	80
	EN AC-AlSi10Mg(Cu)	EN AC-43200	G-AlSi10Mg(Cu)	3.2383	F	180	90	1	55
					T6	240	200	1	90
GK-AlS19Mg	EN AC-AlSi9Mg	EN-AC-43300	G-AlSi9Mg	3.2373	T6	290	210	4	90
					T64	250	180	6	80
GK-AlS17Mg	EN AC-AlSi7Mg	EN AC-42000			F	170	90	2,5	55
					T6	260	220	1	90
					T64	240	200	2	80
	EN AC-AlSi7Mg0,3	EN AC-42100	G-AlSi7Mg	3.2371	T6	290	210	4	90
					T64	250	180	8	80
	EN AC-AlSiMg0, 6	EN AC-42200			T6	320	240	3	100
					T64	290	210	6	90
GK-AlZn10SI8Mg						220	200	1	90

Material state:* *F: Casting state; T6: Solution annealed and completely temper-hardened; T64: Solution annealed and not completely temper-hardened*Source**: *Medjunarodna Klasifikacija Bolesti (MKB-material standards) [9].*

Table 2.
Mechanical properties of aluminum casting.

Properties of silicon carbide (SiC)			
Mechanical	**SI/Metric (Imperial)**	**SI/Metric**	**(Imperial)**
Density	gm/cc (lb/ft^3)	3.1	(193.5)
Porosity	% (%)	0	(0)
Color	—	black	—
Flexural Strength	MPa (lb/in^{2}x10^3)	550	(80)
Elastic Modulus	GPa (lb/in^{2}x10^6)	470	(64.5)
Shear Modulus	GPa (lb/in^{2}x10^6)	—	—
Bulk Modulus	GPa (lb/in^{2}x10^6)	—	—
Poisson's Ratio	—	0.14	(0.14)
Compressive Strength	MPa (lb/in^{2}x10^3)	3900	(566)
Hardness	Kg/mm^2	2800	—
Fracture Toughness KIC	MPa•m$^{1/2}$	4.6	—
Maximum Use Temperature (no load)	°C (°F)	1650	(3000)
Thermal			
Thermal Conductivity	W/m•°K (BTU•in/ft^2•hr.•°F)	120	(830)
Coefficient of Thermal	10^{-6}/°C (10^{-6}/°F)	4.0	(2.2)
Expansion			
Specific Heat	J/Kg•°K (Btu/lb.•°F)	750	(0.18)

Source: Silicon Carbide datasheet.

Table 3.
Technical properties of silicon carbide.

• Length over all (LOA)	= 42,00 m
• Length between perpendiculars (LBP)	= 39,00 m
• Breadth (b)	= 7,00 m
• Height (H)	= 4,00 m
• Draft (T)	= 1,8 m
• Maximum speed	= 24,0 knot
• Crew of ship	= 18 person

Shape hull of Fast Patrol Boat is known as V shaped hull, especially on the front (**Figure 1**). Planning regulations adapted to use the class from the Bureau Classification Indonesia (BKI) [10].

3.2 Ship model making

ANSYS modeling can be done in two ways, namely direct generation and solid modeling. In direct creation, element creation is done directly by defining the nodes required for an element. This method is best used if only a small number of elements are planned. But for complex shipbuilding with a large number of elements, this method was impractical. Whereas in solid modeling, the definition of the model is from the points (keypoint) serving a line. From these lines an area can be made and then the area can be formed by volume.

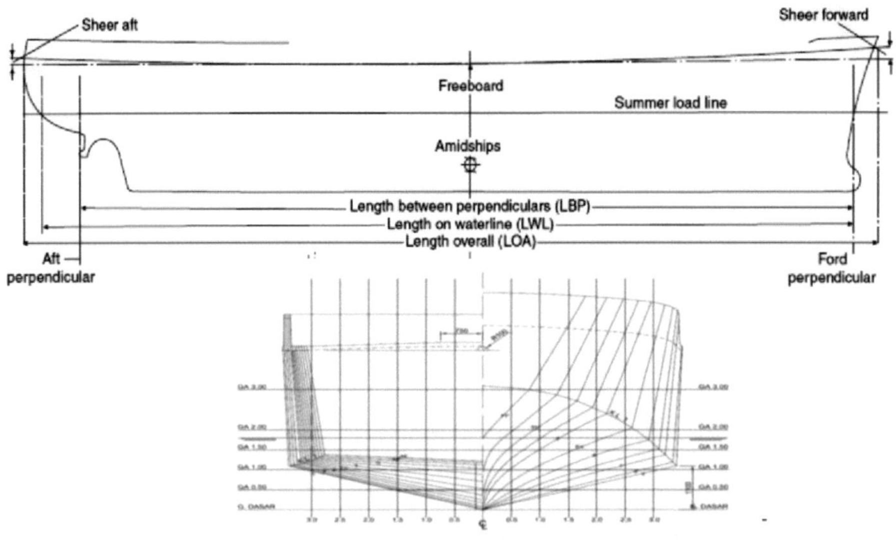

Figure 1.
Ship scheme. Source: (Model Fast Patrol Boat (FPB) 42m).

To make a ship model by means of solid modeling, the first thing to do is redrawing it. In this case, the drawing data obtained from AutoCAD is redrawn in ANSYS. This is done because it is difficult to make repairs if the drawing from AutoCAD is imported directly into ANSYS. In addition, this redrawing is done to avoid the possibility that there are parts that cannot be read in ANSYS during the model import process. Redrawing begins by entering the keypoint coordinates obtained from the AutoCAD drawing. The first keypoint coordinates entered are the lines plan coordinates followed by the accommodation deck coordinates. The keypoint formed is connected to a line. Then from these lines an area is made. So that the area formed consists of keypoints and lines. The area used for plate and line modeling is used for the modeling of the reinforcements (ivory and supports).

From the line plan drawing (station) from AutoCAD which is then converted into a line plan (ivory), the coordinates of the points that form the body plan curve can be obtained. The coordinates of these points are entered into ANSYS as a keypoint. Furthermore, the keypoints are connected into a curve to form an ivory curve (transom to ivory 84). These curves are then linked into areas. The area formed consists of keypoints and lines. Henceforth, the area formed is used for plate modeling and the curved lines forming the area are used as an enforcer (ivory).

After the hull area is formed, it is continued with the construction of the superstructure. Furthermore, from the geometric model formed, an element known as meshing is created. Before the meshing process is carried out, the element size must first be planned. In addition, it must also be determined the type of element and material properties to be used.

3.2.1 Selection and determination of elements

The elements contained in ANSYS can be categorized into 2D (2 dimensional) and or 3D (3 dimensional) element types. ANSYS elements consist of point elements, line elements, area elements, and solid elements. Several LINE elements in ANSYS can be selected according to your needs and analysis to be carried out.

For the modeling of the supports, supports, flanges, ivory, deck beams and other profiles used Beam 189_Quadratic Finite Strain Beam. Beam 189 is an element suitable for use in slender structure analysis to slightly thick structures of beams. This element is based on Timoshenko's beam theory [10]. The deformation effect of shear forces is also included. Beam 189 is a quadratic (3-Node) beam element in 3-D space. Beam 189 has six degrees of freedom, consisting of three translations and three rotations. This element is good for linear, large rotational or nonlinear strain applications.

Beam 189 is used for modeling ivory, beam, reinforcement, support, large ivory, flange and pillar because it has the ability to be a beam. In addition, the quadratic form gives more accurate results than the linear form.

3.2.2 3D Shell

The ANSYS element library contains many types of shell elements. As with line elements, these types of shell elements can be used according to needs and analysis to be carried out. For modeling composite ship plate, Shell 93_8node Structural Shell is used. Shell 93 is particularly good for modeling curved plates. This element has six degrees of freedom at each node: translation in the x, y and z directions and rotation in the x, y and z axes. The deformed form is quadratic in the plane of the element.

Shell 93 is used in ship plate modeling mainly because of its ability to model mostly curved ship plates. Also the deformed quadratic shape allows calculations in the middle of the element (mid-side node) to be more accurate. Element is formed by 8 nodes, 4 thicknesses and orthotropic material. Mid-side nodes on elements cannot be removed and thus these elements are only compatible with quadratic form elements. The orthotropic direction of the material corresponds to the direction of the element's coordinate system. All ship plates are modeled using shell 93, including flat parts such as on the superstructure or on the deck.

3.2.3 Structural mass

Mass modeling is carried out on the main engine, auxiliary engines, gear boxes, pumps, bollards, windlass, windlass foundations, hydraulic steering engines, anchors, anchor chains, and equipment with a large enough mass. These masses need to be modeled because they are part of the ship structure that must be included in the calculation. Mass modeling uses Mass 21, a point element that has six degrees of freedom and is a centered mass element.

3.2.4 Real constant and section determination

In determining the constants, the Real Constant set is used in accordance with the selection of elements used in modeling. The Real Constant set for Shell 93 is used to determine the plate thickness. Meanwhile, to determine the mass of each element, the Real Constant set for mass 21 is used. In addition to determining the constants, the beam and shell elements need to be defined sections.

In determining this section, the element size is determined in the cross section of the profile (beam) and plate (shell). For profiles, the thickness and size of the profiles are defined using the beam tool, while for plates the thickness is only defined using the real constant set for shell 93. The Real Constant set for shell 93 is used to determine the thickness of the plate. The Real Constant set for mass 21 is used to determine the mass of each element. **Figure 2** shows the beam section for the T profile.

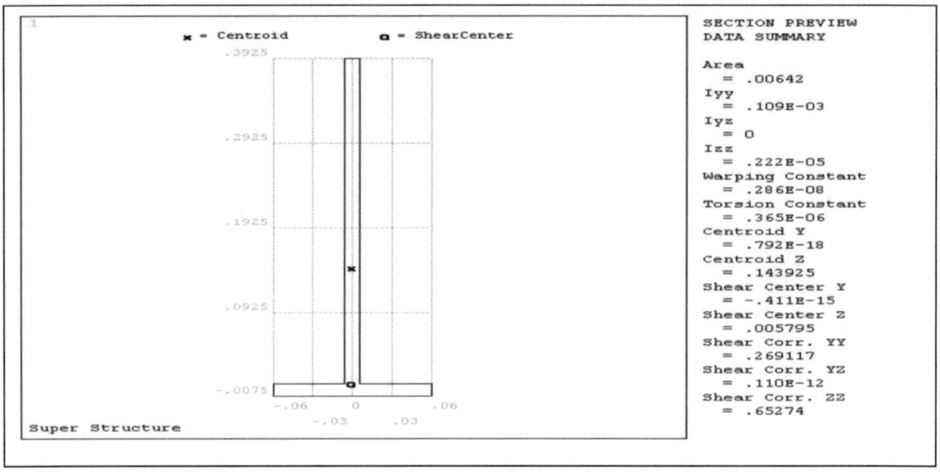

Figure 2.
Beam section for T profile.

3.3 Determination of material properties

In determining the material properties, this depends on the material used for the ship structure. The material in question is one that has a modulus of elasticity (Young's modulus), a poisson ratio and a certain density.

3.4 Finite element making (mesh density)

Determining the size of the element (mesh density) is very important. If the size of the element (mesh) is too coarse, the result may deviate considerably and may even result in an error. However, if the mesh is too fine then we will only waste computer resources, the time required to run is very long, or even the model is too large to be completed on the computer used [11].

The FPB 42 m fast boat model with a length of 42 m that has been made, has 155,988 degrees of freedom so it is hoped that the model can represent it well. The elements are tried to be the same as the example model above, namely all plate elements are expected to have a square shape, but because of the difficulties faced if all of them have to be squared, then there are elements that are made triangles or rectangles with a ratio of length to maximum width of 2.

The size of the largest element that can be created is limited by the following:

- The ivory spacing, which varies from 500 mm to 600 mm.

- Comparison of the length and width of the element for the plate in relation to the shape of the element which is good in this case the ratio of length to width is taken. 2. (Model fast boat FPB 42 m).

In the current model, it only consists of line and area elements, so only free mesh and mapped mesh are used. For the meshing area, this time, we use more meshing (free mesh) with the element length determined or the line division determined in advance. This is easier and you get the desired results. Meanwhile, for elements with identical shapes, meshing is performed using the mapped mesh, which is one of the elements that has been meshed for the first time as a reference using the free mesh. Then the next element can be meshed using a mapped mesh, with the size of the formed element the result will be the same as the element that was first meshed. However, not all areas can easily be elemented in this way. This is due to the size of the area that is too small and the various geometric shapes of the model. Thus, in making elements it is not possible to create elements with the size planned above. For that, a smaller element size is determined. If this still cannot be done, the area is redefined, that is, it is made the same area with a smaller line division but still close to the desired element shape.

To make a beam element, a line is needed, because the beam is a LINE element. The way to make it is almost the same as the meshing process on the plate elements, namely by first determining the element attributes, then meshing it using free mesh. It's just that in this meshing beam there is no need to divide the lines or determine the length of the elements, because this has been done during the meshing area. In addition, the meshing beam also has an orientation keypoint. Namely the keypoint that is used to determine the direction of the mesh section. Each line has a normal direction so that in making beam elements, the beam direction (node I and J) is meshed following the normal direction of the line. If after the mesh the beam direction is not as desired, the line must be reversed (reverse normal line). Because on a ship the entire profile faces the midship, whether the profile is on the base, deck, ivory or reinforcement, the orientation of the keypoint placement is attempted to be able to direct the section of the mesh beam (**Figure 3**) to face the midship.

For mass elements only a keypoint is needed. And for the manufacturing process, namely by selecting the keypoint closest to the location of the mass or center of gravity of the mass being modeled, then the keypoint is used as a mass element. The masses being modeled include the main motor, auxiliary motor, gear box, and other equipment which has a relatively large mass.

After the meshing process is complete, it is necessary to check the shell element whether the elements that have been made are in good condition or not. The maximum warping factor for the Shell 93 warning element is "none", the element may curve outward from the plane of the plate. From all existing tests it has been shown that all elements are in good condition, there are no warning elements or error elements. So that the model made, namely the EN AC-4310 (AlSi10Mg (b)) + SiC composite material ship, has represented the ship well, as shown in **Figure 4** is the image of the overall ship model.

The ability of a ship to float is based on Archimedes' law, the buoyancy force obtained is proportional to the weight of the water it displaces (hydrostatic support). Generally these ships are referred to as ships with hull displacement. The displacement weight is the volume weight of water displaced by the hull. So the weight of the volume of water displaced is the weight of the ship (Eq. 1) (Taggart, 1980):

$$\Delta_B(\text{Newton}) = \text{LxBxTxCBxgx}\rho \quad (1)$$

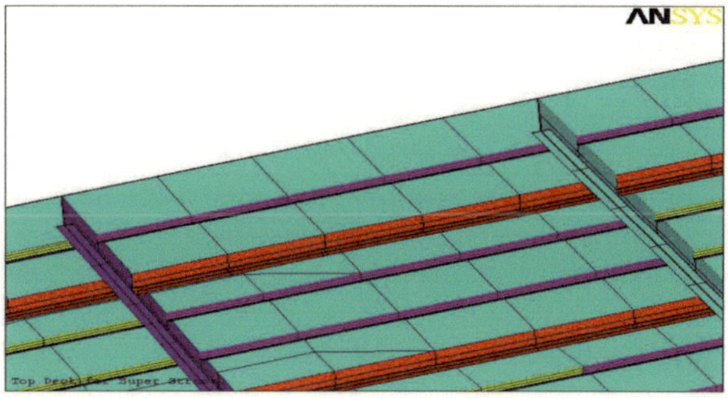

Figure 3.
Beam elements (beam and deck supports).

Figure 4.
Draw the whole ship model.

If it is used as mass displacement (ton) then it is divided by g, so that Eq. (2)

$$\Delta_m(\text{ton}) = \text{LxBxTxCBx}\rho \qquad (2)$$

Information:

Δ_B = weight of displasmen (Newton)
Δ_m = *mass displacement* (ton)
L = lenght karene
B = wide karene (is the shape of the hull that is below water level, excluding: hull thickness, keel thickness, rudder, propeller and other equipment submerged below water level).
T = loaded with ships (is the vertical distance measured from the lowest point hull to the waterline).
CB = block Coefficient (is the ratio between the content of karene and the volume of blocks with length L, width B and height T).
ρ = water density (sea water = 1025 ton/m^3, fresh water = 1 ton/m^3)

3.5 Ship's displacement weight components

The ship displacement weight component consists of the ship's dead weight (DWT) and the weight of the empty vessel (light weight). Dead weight is the carrying capacity of a ship including the weight of: cargo, fuel, lubricating oil, drinking water, foodstuffs, crew + passengers and the goods they carry. Meanwhile, the weight of an empty ship can be divided into three major parts, namely: 1) Structural weight, consisting of ship weight, superstructures, and deck houses. 2) Equipment weight, consisting of anchors, anchor chains, windlass, rigging, capstans, steering machines, winches, derrick booms, masts, vents, navigation tools, lifeboats, davits, and other. 3) The weight of the motor and its auxiliary installations consists of the main motor, auxiliary engine, boiler, pumps, compressors, separators, pressure vessels, coolers, intermediate shaft, propeller, propeller shaft, shaft bearing, reduction gear, and all equipment in the engine room. The complete component of ship displacement weight is shown in **Table 4** (composite vessel). **Table 5** shows the weight of the engine and electrical parts, the hull weight and the interior for the composite ship.

3.6 Drawing of ship models

Modeling a ship made in conditions of calm water (still wet) and wavy (waves), then modeling the behavior of water (calm and wavy water) by considering water

No.	Ship dead weight	Value (ton)
1.	Composite ship weight (6 mm thick)	61.347
2	Machinery and electric weight	
	a. Engine room equipment	37,49
	b. Pump in engine room	2.463
	c. Seat of pump	0,193
	d. Deck equipment	0,35
	e. Air conditioning room	0,35
3	Weight of Hull Outfitting	24.219
4	Weight of Interior	7.611
5	Fuel Weight	36,96
6	Freshwater Weight	16,2
7	Ship Crew Weight (ABK)	2,6
	TOTAL	**189.783**

Table 4.
Force weight on composite ship.

1. M4CHINERY & ELECTRICITY PART WEIGHT						
No.	**ITEM**	**Weight (ton)**	**AE-G (m)**	**moment (ton.m)**	**KG (m)**	**moment (ton.m)**
	ENGINE ROOM EQUIPMENT					
1	Main engine: 2 MTU 16 V4000 M90					
	with ZF7550 gear boxes (wet weight)	19.130	11.052	211.432	2.147	41.074
3	Propeller shaft	2.486	5.441	13.527	0.846	2.104
4	Propeller	0.998	1.975	1.971	0.222	0.222
5	Boss bracket	1.270	2.531	3.215	0.950	1.207
6	Stern tube	1.310	6.554	8.586	1.003	1.313
7	Genset Yanmar 6HAL2-N	1.380	15.602	21.531	1.954	2.697
8	Genset Yanmar 6HAL2-N	1.380	15.602	21.531	1.954	2.697
9	Piping and valves	6.736	16.110	108.516	2.404	16.192
10	Steering gear	2.800	1.250	3.500	2.650	7.420
	PUMP IN ENGINE ROOM					
1	Bilge pump	0.250	7.661	1.915	1.570	0.393
2	Ballast pump	0.250	7.661	1.915	1.570	0.393
3	General service/fire pump	0.620	7.661	4.750	1.570	0.973
4	Fresh water pump	0.108	15.180	1.639	1.570	0.170
5	Fresh water hydrophore	0.020	13.566	0.271	2.100	0.042
6	Sea water hydrophore	0.020	13.566	0.271	2.100	0.042
7	Oil water separator	0.120	6.602	0.792	1.760	0.211

1. M4CHINERY & ELECTRICITY PART WEIGHT						
No.	**ITEM**	**Weight (ton)**	**AE-G (m)**	**moment (ton.m)**	**KG (m)**	**moment (ton.m)**
8	FO transfer pump (PS)	0.228	15.402	3.512	1.600	0.365
	FO transfer pump (SB)	0.228	15.902	3.626	1.600	0.365
9	Main switch board	0.120	16.913	2.030	2.500	0.300
10	Lubricating oil pump	0.039	16.376	0.639	1.650	0.064
11	Air compressor	0.400	6.690	2.676	1.700	0.680
12	Compressed air tank	0.060	5.450	0.327	1.800	0.108
	SEAT OF PUMP					
1	Seat of FO transfer pump (PS)	0.015	15.402	0.237	1.400	0.022
2	Seat of FO transfer pump (SB)	0.015	15.902	0.245	1.400	0.022
3	Seat of bilge pump	0.045	7.661	0.348	1.400	0.064
4	Seat of ballast pump	0.050	7.661	0.383	1.400	0.070
5	Seat of fire GS pump	0.040	7.661	0.305	1.400	0.056
6	Seat of fresh water pump	0.028	15.180	0.431	1.400	0.040
	DECK EQUIPMENT					
1	Windlass	0.350	37.313	13.060	5.510	1.928
	AIR CONDITIONING ROOM					
1	A.C. engine	0.350	20.000	7.000	3.500	1.225
	Σ Weight =	40.848	10.776	440.182	2.019	82.456

2. HULL OUTFITTING PART WEIGHT						
No.	ITEM	Weight (ton)	AE-G (m)	moment (ton.m)	KG (m)	moment (ton.m)
1	Stairway	0.083	20.400	1.690	4.505	0.373
2	Mounting gun	0.441	32.830	14.462	5.905	2.601
3	Wooden lining in chain locker	0.992	36.999	36.703	2.701	2.680
4	Windlass foundation	0.085	37.000	3.145	4.896	0.416
5	Bollard	0.712	20.500	14.588	4.910	3.494
6	Hatch coaming	0.222	18.235	4.039	4.793	1.062
7	Under main deck	0.073	20.524	1.499	1.792	0.131
8	On main deck and navigation deck	3.824	33.481	128.037	3.787	14.480
9	Safety equipment	1.768	9.980	17.645	4.950	8.752
10	Ventilation	1.389	18.536	25.746	3.330	4.625
11	Equipment on wheelhouse	0.426	22.880	9.747	7.194	3.065
12	Radar mast	0.226	19.750	4.459	10.356	2.338
13	Floor	11.448	21.624	247.549	4.602	52.686
14	Emergency genset & battery	0.286	19.000	5.438	7.256	2.077
15	Ceiling & wall covering	2.244	23.751	53.304	4.958	11.128
	Σ Weight =	24.218	23.456	568.051	4.538	109.908

3. INTERIOR						
No.	**ITEM**	**Weight (ton)**	**AE-G (m)**	**moment (ton.m)**	**KG (m)**	**moment (ton.m)**
1	Wooden door	1.029	22.326	22.967	4.383	4.509
2	Furniture under main deck	2.352	27.562	64.830	2.635	6.199
3	Furniture on main deck	2.447	21.880	53.531	4.974	12.170
4	Furniture on navigation deck	0.577	20.996	12.112	7.371	4.253
5	Partition wall	0.636	24.987	15.899	4.598	2.925
7	Steel door	0.382	18.778	7.169	5.715	2.182
8	Steering gear construction	0.188	1.023	0.192	1.946	0.366
	Σ Weight =	7.611	23.218	176.702	4.284	32.605

Table 5.
Force/weight on composite composite.

as a series of linear-elastic springs (springs) that are not related to one another. In the depiction of this model ship, the number of springs 'fixed' is placed on the entire hull as shown in **Figure 5**. In the figure, the distance between the ivory (frame) with symbols **h** and **a** is the width of each section. So that the water surface area (wáter plan are/**Awl**) can be calculated by Eq. 4. The overall volume is the surface area of the water multiplied by the displacement / displacement of the water which is analogous to the distance (x) spring motion, shown in Eq. 4

$$Awl = h.a \tag{3}$$

$$V = Awl.a.x \tag{4}$$

So that the value of the spring constant (k) can be obtained from the spring force (Fs) shown in Eq. (5)

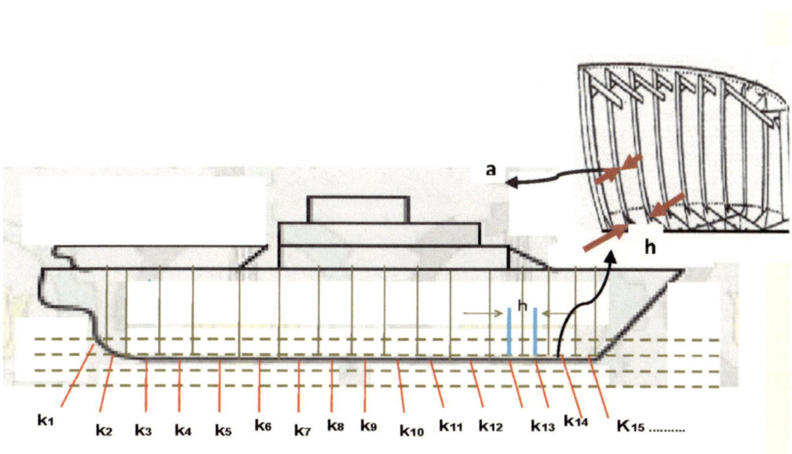

Figure 5.
Modeling of a series of springs on the hull.

$$Fs = \text{k.x} \tag{5}$$

$$m.g = k.x \tag{6}$$

Given the density equation (ρ), is:

$$\rho = \frac{m}{V} \tag{7}$$

so that with the substitution of Eqs. (4), (5) and (6), the value of the spring constant is shown in Eq. (8):

$$k = Awl.\rho.g \tag{8}$$

Information:
*Wáter plan ar*ea (Awl) = water surface area (meter^2)
k = spring constant (Newton/meter)
Fs = spring force (Newton)
x = displacement (meter)
h = distance between *frame* (meter)
a = width of each *section* (meter)
m = mass (kg)
ρ = density (kg/m^3)
g = gravity = 9806$\left(\frac{\text{m}}{\text{detik}^2}\right)$
V = volume (m^3)

3.7 Properties of material

Determining the properties of the materials to be used is taken from all the composite stress mechanical test results data (EN AC-43100 (AlSi10Mg (b)) + SiC*/15p) as numerical input modeling vessel, which are summarized in **Table 6**.

Data	**Composite Material**	
	EN AC-43100	
	AlSi10Mg(b)) + SiC*/15p	
Density	2.904 (gram/cm^3)	
Modulus Elastisity	98,902.44	(MPa)
Poisson Ratio	0.3	
Tensile Strength	225.39	(Rm) (MPa)
Permit Stress		
(sigma 0,2)	59.30	(MPa)
Ship weight		
(thick plate 6 mm)	61,347	(ton)

Source: Tensile test.

Table 6.
*Data for composite ship (EN AC-43100 (AlSi10Mg (b)) + SiC */15p).*

3.8 Loading

Generally, loads are estimated using the classification rules or direct hydrodynamic calculations. The loads that make the ANSYS version12,0 composite ship (EN AC43100 (AlSi10Mg(b)) + SiC*/15p) can be roughly divided in to two parts. Static Loads (still water) This consists of loads that do not differ with time, or even if they differ, the impact of time may be neglected; This category includes hydrostatic pressure, ship part weights, cargo and ballast loads. Besides these wave moments and forces resulting from ship components are often known as static loads. Wave Induced (Quasi Static) to consider water as a sequence of linear-elastic springs which are not connected to each other. In model ship numerics, the number of springs 'mounted' is put on the ship's entire body. Therefore it becomes important to properly understand the loads and evaluate the structure accordingly. Using ANSYS version12,0 makes the load application method very quick and manageable, also the chances of errors in combining the loads is eliminated.

Loads of wave induced *(quasi static)* what count is Coefficient Calculation, Bending Moment Wave Induced Load consists of Vertical Wave Bending Moment (MWV) or (B.M.W.V), Shear Force Wave Induced Load consists of Vertical Wave Shear Force (QWV) or (S.W.S.F), Permissible Bending Moment (S.W.B.M) and Vertical Wave Shear Force (S.F.W.V).

3.9 Component weight displacement ship

Weight component displacement vessel consists of Death Weight Tonnage / DWT and the full weight of the displacement weight component aluminum vessel and composites vessel is shown in **Table 7**. Values for the same weight, with uniform weight distribution, both for aluminum ship and composite ships.

Death Weight Tonnage	(Ton)
Weight of body aluminum ship (thick 6 mm)	54.865
Weight of body composite ship (thick 6 mm)	61.347
Weight of machinery and electricity	
a. Engine room equipment	37.49
b. Pump in engine room	2463
c. Seat of pump	0.193
d. Deck equipment	0.35
e. Air conditioning room	0.35
Weight Hull Outfitting	24,219
Weight Interior	7611
Weight fuel	36.96
Weight freshwater	16.2
Weight Crew of ship	2.6
Weight Machinery and electricity part weight consists of engine room equipment, pump in engine room, seat of pump, deck equipment, air conditioning room	40,848

Source: Fast Patrol Boats.

Table 7.
Loads on aluminum ships and composite ship.

Differential value between aluminum ship body weight and composite ship. Body weight composite ship heavier than aluminum ship, so the expected stress distribution that occurs between the aluminum ship and the composite ship is not the same.

4. Discussion

4.1 Result loads of wave induced *(Quasi Static)*

Summary results of the calculation of wave loads induced (quasi static) in **Table 8** while the chart figure of wave induce in condition hogging and sagging shown in **Figure 6** condition S.F.W.V, **Figure 7** for B.M.W.V and **Figure 8** condition for wave induced stress.

Figures 9 and **10** show the distribution of stress in the composite ship numerical model EN AC43100 (AlSi10Mg(b) + SiC*/15p) for water and wave condition induced for the entire ship body, and the plate thickness of 6 mm was used.The maximum stress occurring in composite ship numeric models is 7.24 MPa. While the maximum stress that occurs in numerical composite ship models for the conditions induced by the wave is 14.1 MPa.

Ship numerical model for base (bottom) construction, shown in **Figure 11** for still water condition and **Figure 12** for wave-induced condition. The maximum stress in numerical composite ship models, when still water conditions are 7.24 MPa. While the maximum stress that occurs in numerical composite ship models reaches 19.1 MPa for the wave induced conditions.

Distribution of stress in main deck on **Figure 13** for still water condition and **Figure 14** for wave induced condition. The maximum stress that occurs in numerical models of composite ship when conditions still water is 6.67 MPa. While for the

Condition	S.W.S.F	S.W.B.M	S.F.W.V	B.M.W.V
Max.	7593.59	0.3	969.63	622.4
Min.	2.30	0.0	−348.46	−249.5

Table 8.
The calculation of wave loads induced (quasi static).

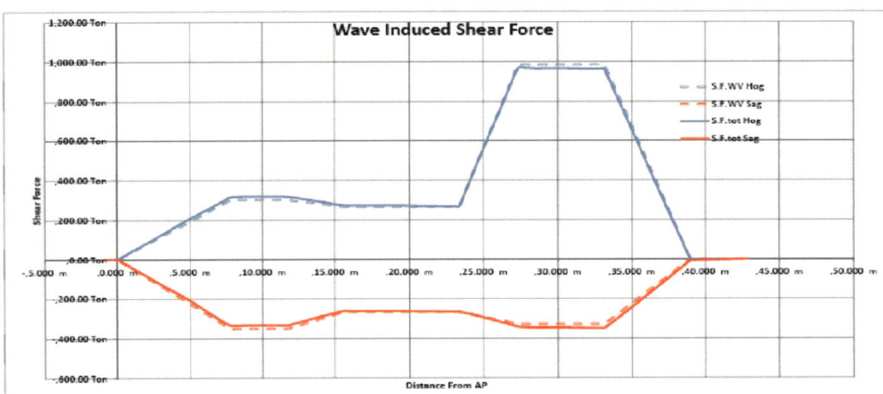

Figure 6.
Condition S.F.W.V.

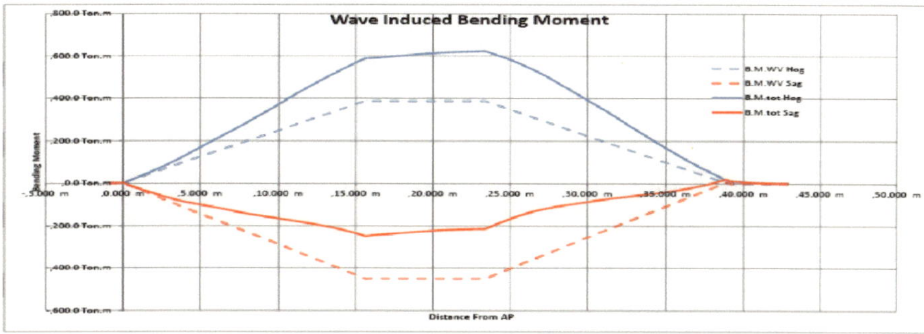

Figure 7.
Condition M.B.W.V.

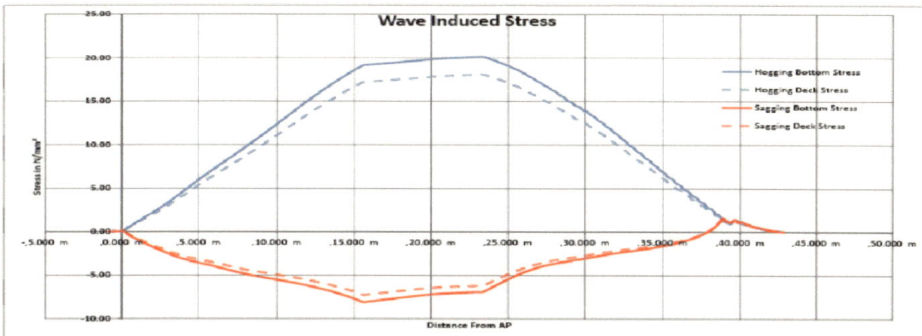

Figure 8.
Condition for wave induced stress.

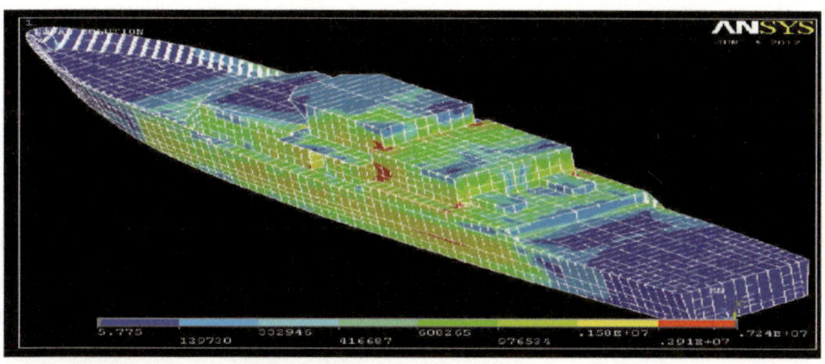

Figure 9.
Distribution of stress in model of numerical of ship composite EN AC-43100 (AlSi10Mg (b)) + SiC/15p for the full ship body for condition still water (maximum stress = 7,24 MPa).*

wave induced conditions, the maximum stress that occurs in numerical models of composite ship of 16.8 MPa.

Overall the above results are summarized in **Table 9**. From the results of the stress distribution shows that the maximum stress that occurs in induced wave conditions have higher value compared to still water conditions. This is because the load is included in the wave induced more numerous and complex than a given load on the still water. Composite ship (EN AC-43100 (AlSi10Mg (b)) + SiC*/15p), more weight than aluminum ship because in composite ship there have SiC as reinforcement, which causes the composite more heavier than the aluminum ship (for the same thickness = 6 mm). Conversely, aluminum ship lighter, so it automatically receives the maximum stress is greater than that received by a composite ship, for input the same load and the plate of the same thickness.

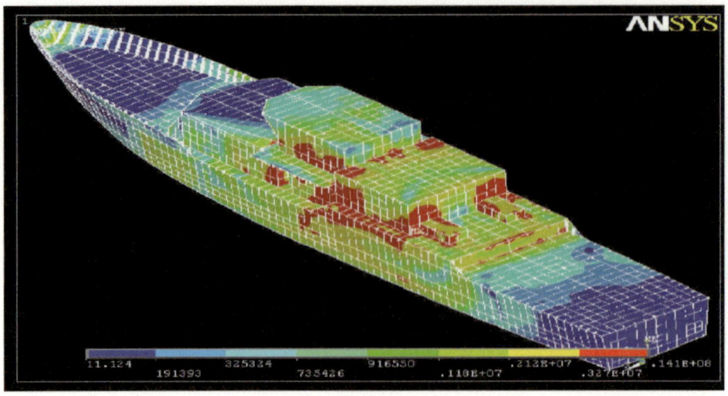

Figure 10.
Distribution of stress in model of numerical of shicomposite EN AC-43100 (AlSi10Mg (b)) + SiC/15p for full ship body for wave induced condition (maximum stress = 14,1 MPa).*

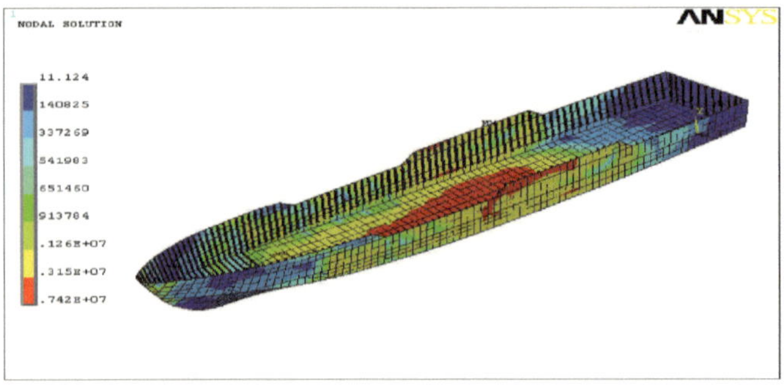

Figure 11.
Distribution of stress in model of numerical of ship composite EN AC-43100 (AlSi10Mg (b)) + SiC/15p for the hull and construction of the base (bottom) for still wave condition (maximum stress = 7,42 MPa).*

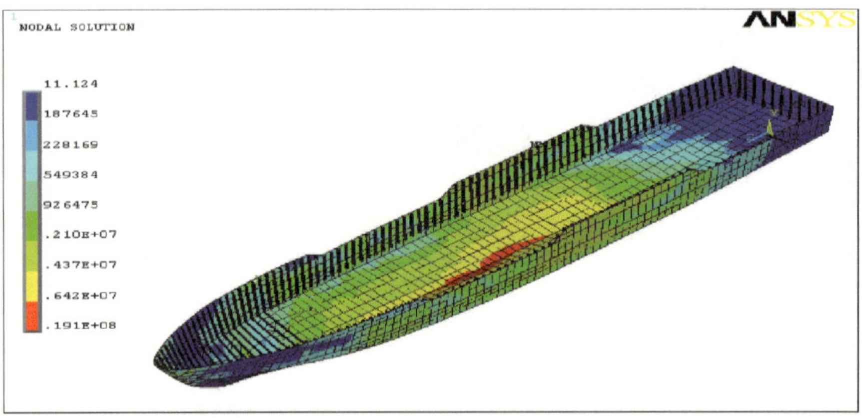

Figure 12.
Distribution of stress in model of numerical of ship composite EN AC-43100 (AlSi10Mg (b)) + SiC/15p for the hull and construction of the base (bottom) for wave induced condition (maximum stress = 19,1 MPa).*

Table 9 shows the maximum stress that occurs in composite ship more smaller than the ship of aluminum, this happens because the weight of ship of composite is heavier than ship of aluminum. So that when receiving weight distribution uniform, ship of composite are stronger hold so the impact on the value of the stress maximum is smaller. It seems that all the results obtained showed the maximum stress

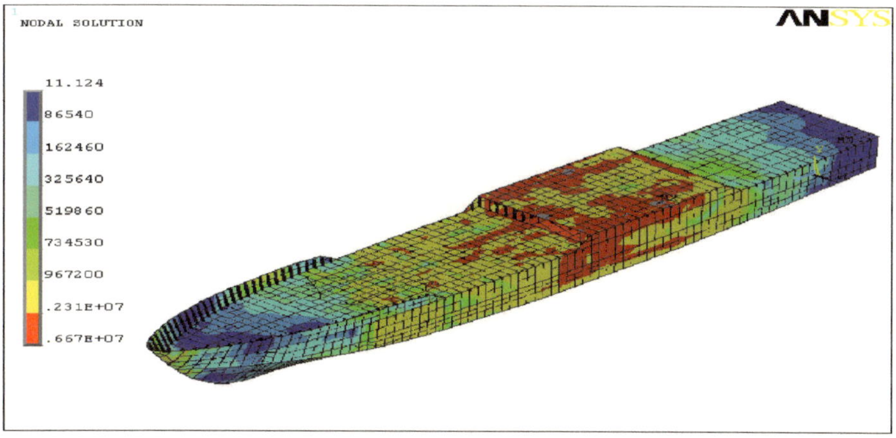

Figure 13.
Distribution of stress in model of numerical of ship composite EN AC-43100 (AlSi10Mg (b) + SiC/15p for* main deck *for still water condition (maximum stress = 6.67 MPa).*

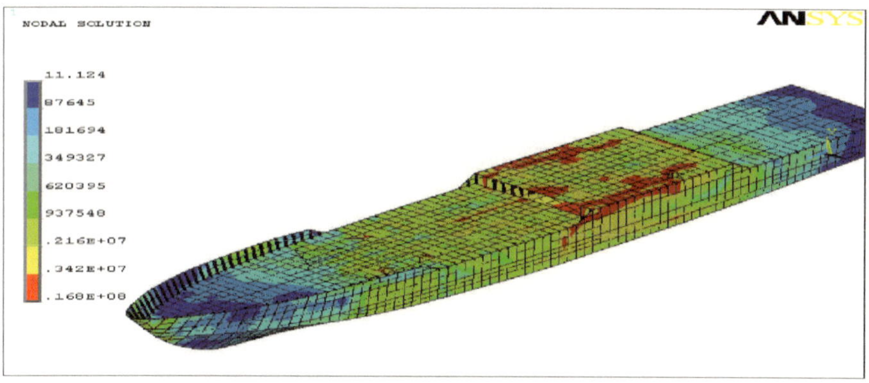

Figure 14.
Distribution of stress in model of numerical of ship composite EN AC-43100 (AlSi10Mg (b)) + SiC/15p for* main deck *for wave induced condition (maximum stress = 16.8 MPa).*

Distribusion stress	Composite material EN AC-43100 (AlSi10Mg(b)) + SiC*/15p (thickness plate 6 mm)	
	Still Water (MPa)	Wave Induce (MPa)
Full body ship area	7.24	14.1
Hull and construction		
of the base (bottom)		
area	7.24	19.1
Main deck area	6.67	16.8

Table 9.
Value of máximum stress of model of numerical of aluminum ships and the ship of composite.

does not exceed the value of the stress permits (0.2 sigma = 59.30 MPa) for composite materials. This means that composite materials EN AC-43100 (AlSi10Mg (b)) + SiC */15p) can be used throughout the full of body ship. This condition is amplified by a factor of safety which is the ratio between the material strength with strength design, where material strength is the stress permits (sigma 0.2) and the design strength is the maximum stress from the program ANSYS (FEM calculation). Factor of safety more than to 1. Factor of safety values are shown in **Table 10** indicated that the factor of safety for all conditions in still water and wave induced, has value above 1.00, so the overall material composite EN AC-43100 (AlSi10Mg (b) + SiC*/15p is safe to use.

4.2 Application analysis of modeling composite ship EN AC-AlSi10Mg (b) + SiC*/15p

Furthermore with using ANSYS software version 12.0, this research will analyze the application of composite materials EN AC-AlSi10Mg (b) + SiC*/15p that can use in ship building. Overall composite ship modeling EN AC 43100(AlSi10Mg (b)) + SiC*/15p), shows that the stress does not exceed the value limit stress 0.2 sigma. It means that this material can actually be applied to the entire body of the ship. But because it is brittle, then the selection of applications on the ship also should look at the nature of this material. Selected applications on top of the building wall (superstructure) that the wall plate height (h) = 2.2 meters and width (b) = 1.5, composite thick ship plate is 5 mm (**Figure 15**) and building applications on the deck plate (superstructure decks) on the size of 1 m x 1 m and thickness of 6 mm (**Figure 16**).

The maximum stress that occurs in ship composite EN AC 43100(AlSi10Mg (b)) + SiC*/15p) for building walls on the plate thickness 5 mm at 9.43 MPa and the deck superstructure with plate thickness 6 mm to obtain the wave-induced stress conditions maximum 10.7 MPa. Both of these results when compared with aluminum ship for the two applications (on the wall of the building) with the height and width the same, but with a thickness of 5 mm, the maximum stress value will be 9.26 MPa (**Figure 17**) and for the superstructure deck of the same size but the greater thickness of 7 mm is obtained at 10.8 MPa maximum stress (**Figure 18**). It means that the results obtained by the maximum strees between the composite ship

	Factor of Safety
	Composite material
	EN AC-43100
	(AlSi10Mg(b)) + SiC*/15p
	(thickness plate 6 mm)
Still Water:	
1. Full body ship	8.19
2. Hull and construction	
3. of the base (bottom)	8.19
Main *Deck*	8.89
Wave Induced:	
1. Full body ship	4.21
2. Hull and construction of the base (bottom)	3.11
3. Main *Deck*	3.53

Table 10.
Value factor of safety.

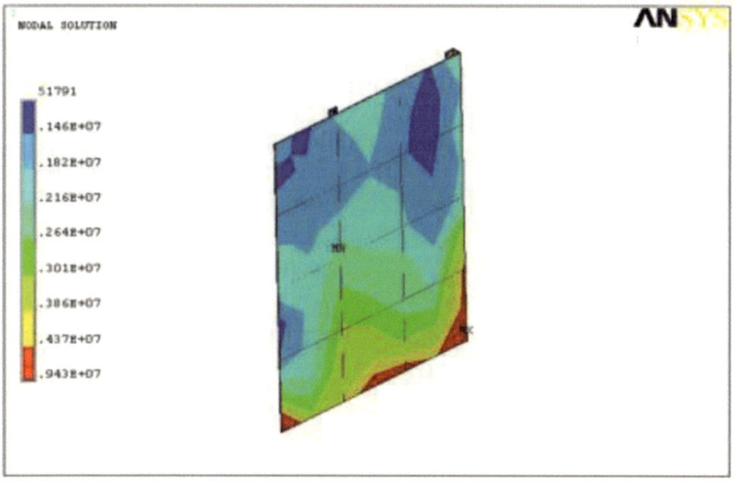

Figure 15.
Distribution stress of composite shipEN AC 43100(AlSi10Mg(b)) + SiC/15p) in superstructure wall with plate thick 5 mm (max. Stress =9.43 MPa).*

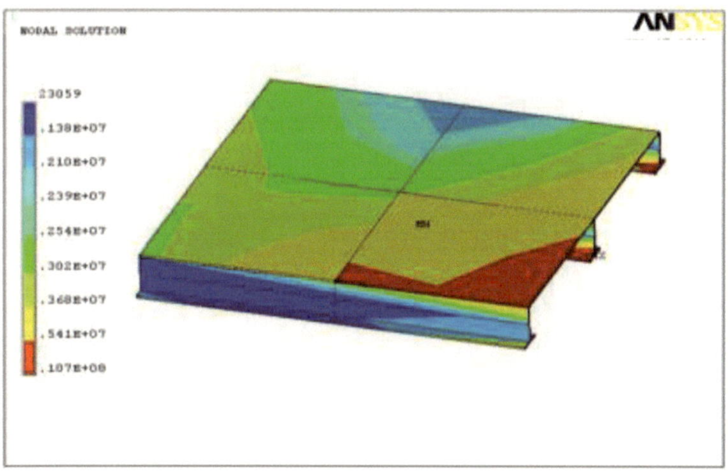

Figure 16.
Distribution stress of composite ship EN AC 43100(AlSi10Mg(b)) + SiC/15p) on superstructure deck with plate thick 6 mm for induced wave condition (max. Stress = 10.7 MPa).*

with aluminum ship are not a significant difference, in fact it can be said the maximum stress value approaching the same value.

Actually the main core of ship composite EN AC43100(AlSi10Mg(b)) + SiC*/15p) as an alternative material for building ships with reduced thickness is used in the composite material will impact on the weight loss, heavy displacement ship will be reduced, then for the length, width, and height of the vessel remains, laden vessel will be reduced. With a large reduction in the laden ship, the wetted surface area / WSA of the hull is submerged in water will also be reduced. This will reduce the size of the total water barriers experienced by vessels which in turn thrust (powering) ship engine fixed, it will increase the speed of the ship. Or conversely, if the desired speed of the ship is made permanent, this will lower the powering of the vessel and it will certainly reduce the relatively large ship main engine. So in general can decrease the volume of the cylinder marine engine. Thus the fuel consumption becomes smaller, thus making the vessel operating expenses generally become more efficient.

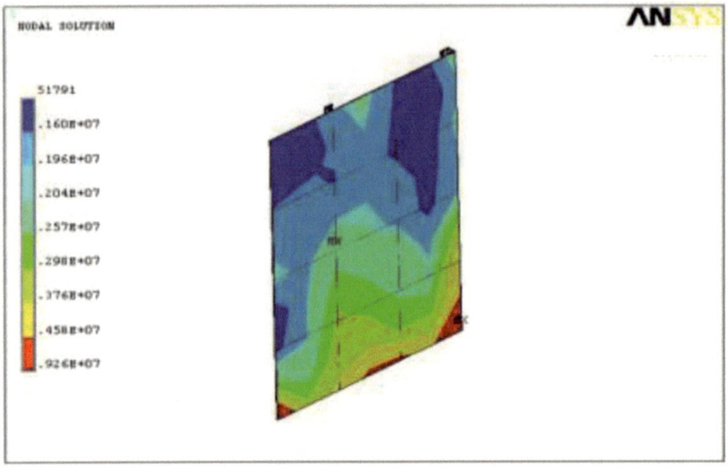

Figure 17.
Distribution stress of ship aluminum EN AC-43100(AlSi10Mg(b)) with plate thick 6 mm (max. Stress = 9.26 MPa).

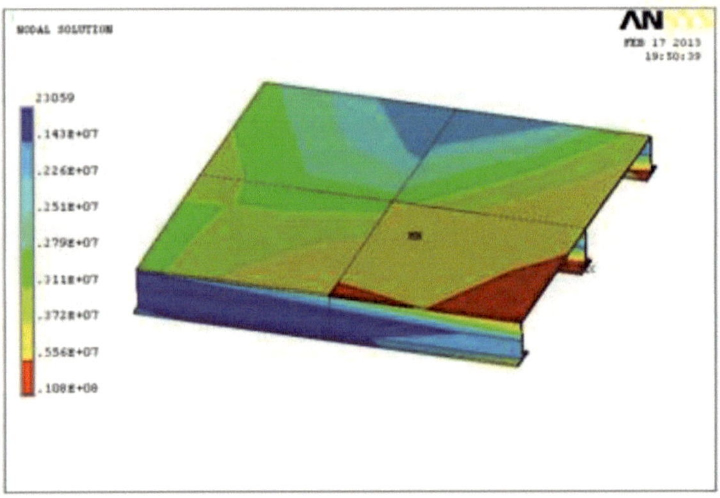

Figure 18.
Distribution stress of ship aluminum EN AC-43100(AlSi10Mg(b)) on superstructure deck with plate thick 7 mm for induced wave condition (max. Stress = 10.8 MPa).

5. Conclusion

The conclusion of this chapter is Numerical modeling of composite ship EN AC-43100 (AlSi10Mg (b)) + SiC*/15p) has successfully demonstrated the distribution stress to the full body ship, construction of the base (bottom), and main deck, for still water and wave conditions induced. The overall results of the stress distribution of model numerical of ship, its value does not exceed the stress permits (sigma 0.2) and have a factor of safety above the minimum allowable limit, so it is safe to use.In numerical modeling, the ship composite materials EN AC-43100 (AlSi10Mg (b)) + SiC*/15p) can be used as an alternative material for ship building, however is still needed comprehensive testing in the field. Reducing the thickness of the composite plate EN AC- 43100 (AlSi10Mg (b) + SiC*/15p) to be significant enough to reduce the weight of ship structure thus reducing the total water resistance experienced by the ship as a result of thrust force ship engine fixed, it will increase the speed of the ship. Conversely, if the speed of the ship is stable it will lower the thrust of force ship, so that the consumption of fuel becomes smaller, the effect on vessel operating expenses are generally becoming more efficient Generally shipbuilding from composite materials EN AC-43100 (AlSi10Mg (b) + SiC*/15p) can be made good, by using modeling ANSYS program ver.12, 0, used as an alternative material for ship building.

Author details

Prantasi Harmi Tjahjanti[1*] and Septia Hardy Sujiatanti[2]

1 Universitas Muhammadiyah Sidoarjo, Sidoarjo, Indonesia

2 Institute Technology of Sepuluh Nopember, Surabaya, Indonesia

*Address all correspondence to: prantasiharmi@umsida.ac.id

References

[1] Zhou zhao, Song Zhijian and Xu Yingkum, 1991,"*Effect of microstructure on the mechanical properties of an Al alloy 6061* ", J. material science and engineering, A132, p83.

[2] Pearce PJ, Grabovac I,1994, '*Composite Reinforcement of Ship Superstructures*', naval platform technology seminar Singapore, December.

[3] Sielski.RA, Taylor P, 2008, '*Predicting the failure of aluminum exposed to simulated fire and mechanical loading using finite element model*', Journal of Offshore and Arcric Engineering, 3(1), pp57–65.

[4] Huang Scott Xiaodi and Paxton Kip,1998, ‗*A Macrocomposite Al Brake Rotor for Reduced Weight and Improved Performance,* Journal of materials' innovations in aluminum, Part IV, August 1998.

[5] Froyen L. and Verlinden B, 1994, '*Aluminium Metal Matrix Composite Material*', Training In Application Technologies, 1994. p. 20-21.

[6] Adjiantoro Bintang, Yuswono, 1998, *Pengaruh Penambahan Unsur Magnesium Terhadap Ketahanan Aus dan Kekerasan Komposit Paduan Al-7, 14% Si dengan Penguat SiC*, Jurnal Buletin IPT No.4 VOL. IV, Oktober/November.

[7] Koczak M, Khatri SC, Allison JE., et al. 1997, *Metal Matrix Composite For Ground Vehicle, Aerospace, and Industial Aplications. "In Fundamental of Metal Matrix Composite"*, Buuterworth-Heinemann, Newmon, MA. P 297-324.

[8] Tjahjanti P.H, Darminto, Panunggal Eko,Nugroho W.H,2007,— *Perlakuan Khusus Cara Pencampuran Bahan Penguat Silikon Karbida Dan Matrik Aluminium/Aluminium Alloy Pada Bahan Aluminium Metal Matrix Cast Composite*||, Prosiding Seminar Nasional Teknoin, Jurusan Teknik Mesin F. Teknik Industri Univ.Islam Indonesia (UII) Jogjakarta, ISBN: 978–979-96964-5-8, ISSN: 0583–8697.

[9] Source: *Medjunarodna Klasifikacija Bolesti (MKB-material standards)*

[10] S.Timoshenko, 1984, *Strength of Materials: Elementary Theory and Problems*, D. Van Nostrand Company, Inc.

[11] Taggart, 1980, *Ship design and construction,* SNAME.

Chapter 12

A MATLAB-Based Symbolic Approach for the Quick Developing of Nonlinear Solid Mechanics Finite Elements

Antonio Bilotta

Abstract

A symbolic mathematical approach for the rapid early phase developing of finite elements is proposed. The algebraic manipulator adopted is MATLAB® and the applicative context is the analysis of hyperelastic solids or structures under the hypothesis of finite deformation kinematics. The work has been finalized through the production, in an object-oriented programming style, of three MATLAB® classes implementing a truss element, a tetrahedral element and plane element. The approach proposed, starting from the mathematical formulation and finishing with the code implementation, is described and its effectiveness, in terms of minimization of the gap between the theoretical formulation and its actual implementation, is highlighted.

Keywords: FEM, nonlinear solid mechanics, MATLAB®, Symbolic Math Toolbox™, object oriented

1. Introduction

The developing of finite element formulations, standard or new ones, requires a lengthy process which involves several steps.

The typical starting point is the formulation of a mathematical model where the main physical or real world phenomena to be described are established. At present time the definition of a mathematical model is at the basis of any serious attempt to obtain previsions in any engineering application [1–3], but not only in the engineering field [4, 5].

The subsequent step is the introduction of a numerical approximation technique. The most popular technique is the Finite Element Method (FEM), see [6–8], but now the number of computational approximation techniques is very large and a synthetic summary can be tried only by citing some of these less conventional methods: mixed finite element methods [9–13]; partition of unity-based discontinuous finite elements [14, 15]; meshless methods [16]; discontinuous Galerkin methods [17]. This operation leads to the identification of the needed discrete operators which define the computational model. For example, in the case of the analysis of solid mechanics problems by FEM, important discrete operators are the mechanical response vector of the finite element and its tangent stiffness matrix. This phase is characterised by the evaluation and the analysis of these operators through an algebraic manipulator such as MATLAB® [18]. MATLAB® is certainly one of the state-of-the-art mathematical softwares available for performing numeric or symbolic analyses, but it is not the only one and a quite long list, see [19], of packages offering very similar features is available.

The discrete model so defined is usually inserted into a prototype code, often by using again an algebraic manipulator but the use of compiled programming languages is also possible if not common. This prototype code allows to perform basic tests with the aim to check the effectiveness of the adopted model with respect to well known situations and to check for the presence of bugs. Often this phase can highlight also flaws in the mathematical model or in the discretization technique. In any case it is necessary to go back and to repeat the process just described.

The additional last step can be the production of an executable by using compiled programming languages such as C/C++ or fortran. This makes possible to extend the validation of the conceived numerical model by performing the analysis of larger sized problems.

As already said, the previously described work-flow is lengthy and it is often characterised by a gap between the theoretical formulation and its implementation in a numerical code. However some solutions capable to assist the developer in this process already exist ant it is worth to mention some ones. Such solutions typically refer to a specific context by keeping fixed the physical problem but letting open the specific instance of discretization technique which, however, is fixed too. This is the case of open source FEM libraries or commercial packages listed in [20]. In both cases the user must define the procedures or functions needed in order to assign the desired new finite element to be used within the analysis framework already available in the library. Other packages instead solves a generic system of Partial Differential Equations (PDEs) subjected to boundary and initial conditions. Inside this generic form the specific differential problem to be solved must to be fitted by the user, see for example [21, 22], but usually with no control over the discretization technique used by the solver.

The present work, quite far from being an alternative to the hugely developed and rich packages previously cited, proposes a basic approach for the quick early phase developing of solid mechanics finite elements formulation. Its intent is to show how to use MATLAB®, in particular by exploiting the capabilities of the Symbolic Math Toolbox™ [23], to produce numerical approximations of a given solid mechanics problem in a way that the usual gap between the theoretical formulation and its actual implementation in a code is not perceived. This result is obtained by condensing the development process going from the mathematical formulation to the prototype code implementation in a few lines of MATLAB® symbolic instructions. The applicative context is the nonlinear analysis of solids and structures, see [24, 25], by showing the formulation and the subsequent MATLAB® coding of some typical structural and solid finite elements. The mechanical formulation is based on the kinematics of finite deformation and, for the description of the material behaviour, on isotropic hyperelasticity, i.e. the stress solution is found as a derivative of some potential energy function. This allows to express the mechanical problem at hand in terms of the stationary condition of the Total Potential Energy (TPE). The stationary condition, assuming a fem discretization of the given domain, is then easily translated into a nonlinear algebraic problem whose unknowns are the position vectors of the nodes of the mesh used in the discretization.

The following finite elements are discussed: a truss element, a 3D tetrahedral element with four nodes and a four nodes quadrangular element subjected to plane strain condition. The choice allows to discuss gradually the main ingredients present in a finite element formulation and how these can be framed inside the proposed

MATLAB® approach. The latter is based on the definition of MATLAB® classes which share the same structuring and which differ only for the particular mechanical response to be implemented. In particular the generic class is structured as shown by the following instructions (Listing 1.1).

Listing 1.1. Generic class.

```
classdef Element
        properties (SetAccess = private)
          % symbolic properties

          % numeric properties
        end

        methods
        function E = Element()
        end

        function E = Initialize (E,D, i)
        end

        function E = Compute(E)
        end

        function sig = Stress(E,gx)
        end
        end
end
```

The properties section contains a group of symbolic properties devoted to handle the unknowns and the quantities depending on them used in the description of the element mechanical behaviour. The other group of numeric properties are used to handle quantities that are known and then they can have a numeric value. Beyond the constructor, that must to be present in any class, we have the function *Initialize,* belonging to the *pre-processing phase* of a FEM code, whose main task is the inizialization of the i-th element on the basis of the assigned data structure D. This is the moment also for evaluating the element operators, in a symbolic format, needed to the analysis. In the subsequent *analysis phase,* the function Compute evaluates the numeric instances of the symbolic operators previously prepared. The function Stress is typical of the *post-processing phase* of any FEM code and its task is to compute the stress solution inside the generic element starting from the kinematic global solution represented by vector gx.

Before proceeding with the description of the proposed work, it is noteworthy to observe that the use of MATLAB® to perform mathematical and numeric analyses is not certainly new and several books are dedicated to this subject, see [26–28] just to cite a few. Moreover the already cited book [24] employs MATLAB® for the implementation of a FEM software. However the present less comprehensive work is different because it carry out the formulation of the FEM operators by exploiting the potentiality of the symbolic manipulator and advising an object-oriented programming style.

A last further annotation regards the use of the symbolic approach which, with respect to the expected performance of final codes, represents a weakness. This aspect however is to be considered less important in a work regarding the early phase developing of a FEM formulation. Nevertheless techniques, [29–31], for the automatic generation of efficient and highly compressed code is a research theme which is attracting increasingly interest, making viable the up-scaling of the proposed approach.

The chapter is organised as follows. Section 2 presents the FEM formulation of the Total Potential Energy for a generic structure or solid, showing also the evaluation of the gradient needed to define the discrete equilibrium equations and the evaluation of the Jacobian necessary for their solution. Sections 3, 4 and 5 describes, respectively, the truss element, the tetrahedral element and plane quadrangular element. The closing section furnishes some additional final comments.

2. Total Potential Energy

An effective description of a generic mechanical problem can be obtained through the stationary condition of its Total Potential Energy (TPE) which, see for example [24], can be expressed as follows

$$\prod(\boldsymbol{x}) \equiv \prod_{\text{int}}(\boldsymbol{x}) + \prod_{\text{ext}}(\boldsymbol{x}) = \text{stat.} \tag{1}$$

$\boldsymbol{x}$ is the global vector of the current positions of the nodal points defining the mesh used to describe the geometry of the solid. $\prod_{\text{int}}(\boldsymbol{x})$, excluding dynamic and dissipative effects, is given only by the strain energy obtained by summing all the contribution from all the finite elements, i. e.

$$\prod_{\text{int}}(\boldsymbol{x}) = \sum_{\text{e}} \Psi_{\text{e}}(\boldsymbol{x}), \tag{2}$$

being $\Psi_{\text{e}}(\boldsymbol{x})$ the hyperelastic strain energy relative to the generic finite element. $\prod_{\text{ext}}(\boldsymbol{x})$ is the potential energy of external forces. For simplicity the case of a solid body subjected only to external punctual forces will be considered, in this case the potential energy can be written as

$$\prod_{\text{ext}}(\boldsymbol{x}) = -\boldsymbol{f} \cdot \boldsymbol{x}, \tag{3}$$

where $\boldsymbol{f}$ is the global vector of the applied forces in each node of the mesh. $\boldsymbol{f}$ has the same length of $\boldsymbol{x}$, however it is mainly composed by null entries.

On this basis the equilibrium equations can be easily formulated with respect the degrees of freedom involved in the FEM description of the body. In particular, by imposing the stationary condition (1), the equilibrium equations can be derived, obtaining

$$\underset{\text{e}}{\text{A}} \boldsymbol{g}_{\text{e}}(\boldsymbol{x}) - \boldsymbol{f} = \boldsymbol{0}. \tag{4}$$

where the assembly operator A is used to build up the global response vector by using the gradient vector of each finite element strain energy contribution, i.e.

$$\boldsymbol{g}_{\text{e}}(\boldsymbol{x}) = \frac{\partial \Psi_{\text{e}}(\boldsymbol{x})}{\partial \boldsymbol{x}}. \tag{5}$$

The solution of Eq. (4), a typically nonlinear algebraic system whose unknowns are the components of vector $\boldsymbol{x}$, is based on a Newton–Raphson iteration which can be formulated as follows

$$\underset{e}{\mathrm{A}} \boldsymbol{J}_{\mathrm{e}}(\boldsymbol{x}_j)(\boldsymbol{x}_{j+1} - \boldsymbol{x}_j) = -\left(\underset{e}{\mathrm{A}} \boldsymbol{g}_{\mathrm{e}}(\boldsymbol{x}_j) - \boldsymbol{f}\right), \tag{6}$$

where $\boldsymbol{x}_j$ and $\boldsymbol{x}_{j+1}$ are the estimated solutions at j-th and $(j+1)$-th iterations and $\boldsymbol{J}_{\mathrm{e}}(\boldsymbol{x})$ is the Jacobian matrix of the finite element given by

$$\boldsymbol{J}_{\mathrm{e}}(\boldsymbol{x}) = \frac{\partial \boldsymbol{g}_{\mathrm{e}}(\boldsymbol{x})}{\partial \boldsymbol{x}}. \tag{7}$$

The gradient vector $\boldsymbol{g}_{\mathrm{e}}(\boldsymbol{x})$ and the Jacobian matrix $\boldsymbol{J}_{\mathrm{e}}(\boldsymbol{x})$ can be used as basic building blocks for the finite element formulation. This is the approach at the basis of the MATLAB® implementations to be described in the following sections.

3. Truss element

The strain energy of the truss element is defined, see [24], as follows

$$\Psi_{\mathrm{e}}(\boldsymbol{x}) = \frac{1}{2} E \varepsilon^2 V, \varepsilon = \varepsilon(\boldsymbol{x}) = \ln\left(\frac{l}{L}\right), \tag{8}$$

where E is the Young modulus, L and V are the length and the volume of the bar in the reference configuration, l is the length of the bar in the current configuration. The geometric quantities just described are depicted in **Figure 1** where the coordinate vectors of the nodal points are also shown.

The implementation of the MATLAB® class Truss can stem from the properties reported in Listing 1.2. Some of them, those describing the reference configuration, can be numeric because are fixed. The other properties, which describe the current configuration, are expressed in symbolic form in order to be used as quantities whose the strain energy of truss element depends on.

Listing 1.2. Truss class: properties.

```
properties (SetAccess = private)
        % symbolic properties
        xa % current coordinates of node a
        xb % current coordinates of node b
        xe % current element coordinates
        ge % gradient g (xe)
        Je % Jacobian J (xe)
        eps % strain eps (xe)
        N % axial force N (xe)

        % numeric properties
        a % node a global index
        b % node b global index
        Xe % reference element coordinates
        % ...
end
```

During the pre-processing phase the numeric properties of the class are appropriately assigned and the symbolic properties are evaluated as shown in the following listing (Listing 1.3). This happens inside the function Initialize belonging to the methods of the class.

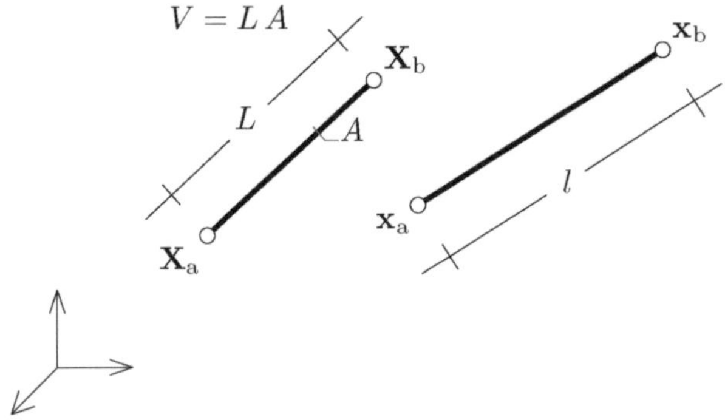

Figure 1.
Truss element: definition of the geometric quantities relative to the reference configuration (upper-case letters) and the current configuration (lower-case letters).

Listing 1.3. Truss class: function Initialize.

```
function T = Initialize (T,D, i)
        % D brings all the problem data and its use
        % is not shown here
        T. xa = sym ('xa', [3 1], 'real');
        T. xb = sym ('xb', [3 1], 'rea1');
        T. xe = [T. xa; T. xb];

        % reference configuration
        L = dot(Xb–Xa, Xb–Xa);
        L = sqrt(L);

        % current configuration
        1 = dot (T. xb–T. xa, T. xb–T. xa);
        1 = sqrt(1);

        T. eps = log(1/L);
        Psi = 1/2 * E * T.eps^2 * (L*A);
        T. ge = gradient(Psi, T.xe);
        T. Je = jacobian (T.ge, T.xe);

        % preliminary symbolic evaluation of N
        T .N = E*A*T. eps;
end
```

Listing 1.3 shows that, after the computation of the strain energy using Eq. (8), symbolic properties $\boldsymbol{g}_e$ and $\boldsymbol{J}_e$ are evaluated on the basis of Eqs. (5) and (7), respectively, by simply calling the function gradient and the function jacobian both belonging to the Symbolic Math Toolbox™. This highlights the *short distance* between the formulation and its code implementation.

After having prepared each Truss object in the way described above, it is possible to evaluate, whenever it is needed during the solution of the nonlinear equilibrium equations, the gradient and the Jacobian of the generic element with respect to estimated solution $\boldsymbol{x}_j$, see Eq. (6), by calling the following class method (Listing 1.4).

Listing 1.4. Truss class: function Compute.

```
function T = Compute(T)
```

```
        T.se = subs (T.ge, T.xe, T.xxe);
        T.Ke = subs (T.Je, T.xe, T.xxe);
end
```

The function Compute uses the numeric property T.xxe previously filled with the current nodal coordinates values and it stores the resulting numeric expressions of the gradient and Jacobian in the class properties se and Ke. The desired result is obtained by calling MATLAB® function subs which substitutes the symbolic variable T.xe with its numeric value T.xxe.

The post-processing phase of any FEM codes certainly comprehends the evaluation of the stress solution. In the case of the truss element, the axial force must to be computed with respect to each vector $\boldsymbol{x}$ calculated by the solution of the equilibrium Eqs. (4). Listing 1.5 shows the very simple function implementing the required computation.

Listing 1.5. Truss class: function Stress.

```
function N = Stress (T,gx)
        % extraction of local vector lx from global gx
        N = subs(T.N, T.xe, lx);
end
```

The complete listing of the class can be found in [32].

4. Tetrahedral element

The discussion of the implementation of a tetrahedral element, in particular a 4 nodes tetrahedron, allows to introduce an important ingredient of all finite element formulations: the interpolation chosen for the kinematic description. Standard approaches are hinged on the interpolation of the displacement field, in the present approach the focus is on the interpolation of the element coordinates in the reference configuration and in the current one. In the previous section regarding the truss element, this aspect remained hidden because the element elongation is easily formulated with respect to the element nodal coordinates.

Another important aspect which the tetrahedral element bring into play is the use of the continuum mechanics instruments, see [24, 25], and how these can be smoothly framed inside the proposed MATLAB® implementation.

Let us consider the geometry of the 4 node tetrahedron as illustrated in **Figure 2**. The description of the reference and current configurations of the tetrahedron are as follows.

$$\begin{aligned} \boldsymbol{X}(\zeta_1, \zeta_2, \zeta_3, \zeta_4) &= N_1\boldsymbol{X}_1 + N_2\boldsymbol{X}_2 + N_3\boldsymbol{X}_3 + N_4\boldsymbol{X}_4 \\ &= \zeta_1\boldsymbol{X}_1 + \zeta_2\boldsymbol{X}_2 + \zeta_3\boldsymbol{X}_3 + \zeta_4\boldsymbol{X}_4, \end{aligned} \tag{9}$$

$$\begin{aligned} \boldsymbol{x}(\zeta_1, \zeta_2, \zeta_3, \zeta_4) &= N_1\boldsymbol{x}_1 + N_2\boldsymbol{x}_2 + N_3\boldsymbol{x}_3 + N_4\boldsymbol{x}_4 \\ &= \zeta_1\boldsymbol{x}_1 + \zeta_2\boldsymbol{x}_2 + \zeta_3\boldsymbol{x}_3 + \zeta_4\boldsymbol{x}_4. \end{aligned} \tag{10}$$

The element local coordinates $\boldsymbol{\zeta} = [\zeta_1\ \zeta_2\ \zeta_3\ \zeta_4]^{\mathrm{T}}$ are the standard tetrahedral coordinates whose definition can be found in several resources, for example [6, 8]. On this basis the description of the deformation gradient over the tetrahedron can be formulated as follows

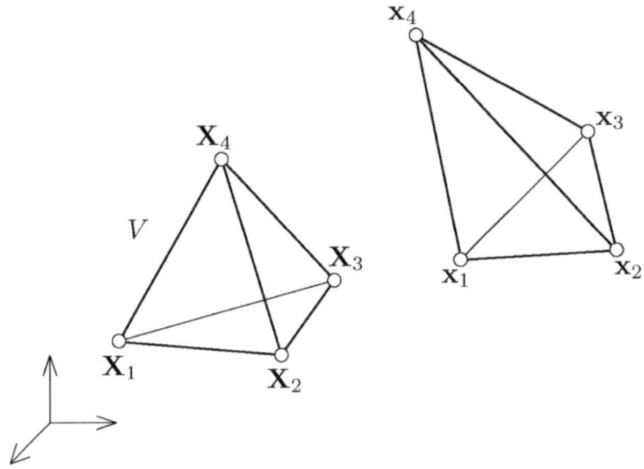

Figure 2.
Tetrahedral element: definition of the geometric quantities relative to the reference configuration (upper-case letters) and the current configuration (lower-case letters).

$$F = \frac{\partial \boldsymbol{x}}{\partial \boldsymbol{X}} = \frac{\partial \boldsymbol{x}}{\partial \zeta}\frac{\partial \zeta}{\partial \boldsymbol{X}} = \frac{\partial \boldsymbol{x}}{\partial \zeta}\left(\frac{\partial \boldsymbol{X}}{\partial \zeta}\right)^{-1} = \boldsymbol{F}(\boldsymbol{x}), \tag{11}$$

with the operator

$$\frac{\partial \boldsymbol{x}}{\partial \zeta} = [\,\boldsymbol{x}_1\boldsymbol{x}_2\boldsymbol{x}_3\boldsymbol{x}_4\,] \tag{12}$$

containing in its columns the four coordinate vectors relative to the current configuration, unknown vectors to be expressed in MATLAB® symbolic format, and the operator

$$\frac{\partial \boldsymbol{X}}{\partial \zeta} = [\,\boldsymbol{X}_1\boldsymbol{X}_2\boldsymbol{X}_3\boldsymbol{X}_4] \tag{13}$$

containing the four coordinate vectors relative to the reference configuration to be evaluated numerically for each tetrahedron of the mesh. The apparent problem represented by the evaluation of inverse $\left(\frac{\partial \boldsymbol{X}}{\partial \zeta}\right)^{-1}$ starting from a 3×4 matrix is a standard matter in FEM procedures, see for example [6], and it can be easily calculated as shown in Appendix A.

The geometric formulation described above is directly inserted inside the Tetra4 class which can be implemented by following the same scheme already adopted for the class Truss. In particular the geometrical properties of the class are listed below (Listing 1.6).

Listing 1.6. Tetra4 class: properties.

```
properties (SetAccess = private)
        % symbolic propetries
        xl % current coordinates of node 1
        x2 % current coordinates of node 2
        x3 % current coordinates of node 3
        x4 % current coordinates of node 4
        xe % current element coordinates
        Fe % deformation gradient
```

```
        % numeric properties
        Xe % reference element coordinates
end
```

On this basis the evaluation of the deformation gradient can be performed during the initialisation of the generic element by carrying out the following instructions (Listing 1.7).

Listing 1.7. Tetra4 class:function Initialize (evaluation of the deformation gradient).

```
function T = Initialize (T,D, i)
        % D brings all the problem data and its use
        % is not shown here ...

        % reference configuration
        dXdzeta = [X1 X2 X3 X4];
        A = [1 1 1 1; dXdzeta];
        V = det(A)/6;
        iA = inv(A);
        dzetadX = iA(1:4 ,2:4);

        % current configuration (symbolic)
        dxdzeta = [T.xl T.x2 T.x3 T.x4];

        % deformation gradient
        T.Fe = dxdzeta*dzetadX;

        % ...
end
```

It is now possibile to discuss the strain energy of the tetrahedral element. The choice is for a compressible neo-Hookean material, see [25], which allows to express the strain energy of the generic tetrahedron as follows

$$\Psi_e(\boldsymbol{x}) = \int_{\Omega_e} \Psi(\boldsymbol{C})dV = \left(\frac{\mu}{2}(I_1 - 3) - \mu \ln J + \frac{\lambda}{2}(\ln J)^2\right) V. \tag{14}$$

$\boldsymbol{C} = \boldsymbol{C}(\boldsymbol{x}) = \boldsymbol{F}^{\mathrm{T}}\boldsymbol{F}$ is the right Cauchy strain tensor and I_1 its first invariant, J = det $\boldsymbol{F}$, λ and μ are the Lamè parameters of the material. Thanks to the constant pattern of **F** over the element domain Ω_e, the strain energy of the element is simply given by the product between the strain energy density and the reference volume of the element. Such a evaluation, together with the derivation of the gradient vector and Jacobian matrix is implemented inside the function Initialize as shown in Listing 1.8.

Listing 1.8. Tetra4 class: function Initialize (strain energy).

```
function T = Initialize (T,D, i )
        % ...
        C = T. Fe.'*T.Fe;
        Il = trace (C);
        J = det (T.Fe);

        Psi = (mi/2*(Il-3)-mi*log(J) + lam/2*log (J)^2) *V;
        T.ge = gradient(Psi, T.xe);
```

```
        T.Je = jacobian(T.ge, T.xe);
        % ...
end
```

The symbolic gradient vector and Jacobian matrix evaluated in the initialization phase are then numerically computed during the analysis using a function identical to the function already presented in Listing 1.4 for the Truss class.

The basic operation of the post-processing is the computation of the Cauchy stress solution which, as **F**, is constant over the element domain. This step requires the evaluation of the second Piola-Kirchhoff stress tensor

$$\boldsymbol{S} = 2\frac{\partial \Psi}{\partial \boldsymbol{C}} = \boldsymbol{S}(\boldsymbol{C}) \tag{15}$$

and, by applying a push-forward operation to **S** [24, 25], the computation of the Cauchy stress tensor is

$$\sigma = J^{-1}\boldsymbol{F}\boldsymbol{S}(\boldsymbol{C})\boldsymbol{F}^{\mathrm{T}}. \tag{16}$$

The MATLAB® implementation of Eqs. (15) requires the introduction of a symbolic matrix for **C** to be used to perform another evaluation of the strain energy depending, this time, from the components of **C**. The obtained expression, $\Psi(\boldsymbol{C})$, can be derived in order obtain **S**. This step can be performed only one time during the initialisation of the Tetra4 class. Listing 1.9 shows these instructions together with the declaration of the necessary symbolic properties.

Listing 1.9. Tetra4 class: function Initialize (second Piola-Kirchhoff stress tensor).

```
        properties (SetAccess = private)
        % ...

        % symbolic properties
        Se % second Piola-Kirchhof stress tensor S(C)
        Ce % symbolic tensor C from which Se depends
end

function T = Initialize (T,D, i )
        % ...
        T.Ce = sym ('C', [3, 3], 'real');
        I1 = trace (T. Ce);
        I3 = det (T.Ce);
        Psi = mi/2*(Il – 3)–mi*log ( sqrt (I3))+ ...
              lam/2*log (sqrt( I3))^2;
        T.Se = reshape (2* gradient (Psi ,T.Ce (:)) ,3 ,3 );
end
```

Eq. (16) is used to compute the stress solution for each tetrahedron with respect to all the solutions $\boldsymbol{x}$ found by means of equilibrium equations (6). The class function implementing the required operations is reported in Listing 1.10.

Listing 1.10. Tetra4 class: function Stress.

```
function sig = Stress (T,gx)
        % extraction of local vector lx from global gx
        F = subs (T.Fe, T.xe, lx);
```

```
        C = F.'*F;
        sig = F*subs (T.Se, T.Ce, C)*F.'/det(F);
    end
```

The complete listing of the class can be found in [33].

5. Plane strain 4 nodes element

The use of finite elements specifically formulated for the analysis of problems which admit a 2D reduction is very common and quadrangular elements play an important role in the case of simple geometries. In this section a 4 nodes quadrangular element subjected to plain strain condition is discussed. The element is very basic but it allows to discuss also the use of the Gauss integration points in the calculation of the required FEM operators. The use of the Gauss integration point is an important cornerstone for all finite element formulations.

Plane strain condition stems from the following assumption on the transformation defining the new configuration of each point of the body

$$x_1 = x_1(X_1, X_2), \tag{17}$$

$$x_2 = x_2(X_1, X_2), \tag{18}$$

$$x_3 = X_3. \tag{19}$$

Consequently, the associated deformation gradient takes the following form

$$\boldsymbol{F} = \begin{bmatrix} F_{11} & F_{12} & 0 \\ F_{21} & F_{22} & 0 \\ 0 & 0 & 1 \end{bmatrix}, \boldsymbol{F}_{2\times2} \begin{bmatrix} F_{11} & F_{12} \\ F_{21} & F_{22} \end{bmatrix}. \tag{20}$$

Eqs. (18)–(20) allow the dealing with a 2D kinematic description. The stress solution, however, is not strictly plane because Eq. (19) constitutes an internal constraint determining also the presence of the component σ_{33}. This component anyway depends only from the 2D kinematic solution as it will be shown in the following.

The standard shape function of the four nodes plane element are

$$\begin{aligned} N_1 &= \frac{1}{4}(1-\zeta_1)(1-\zeta_2), N_2 = \frac{1}{4}(1+\zeta_1)(1-\zeta_2) \\ N_3 &= \frac{1}{4}(1+\zeta_1)(1+\zeta_2), N_4 = \frac{1}{4}(1-\zeta_1)(1+\zeta_2) \end{aligned} \tag{21}$$

being $\boldsymbol{\zeta} = [\zeta_1 \zeta_2]^T$ the element local coordinates used for quadrangular elements, see for example [6]. Shape functions (21) can be properly used to describe, see **Figure 3**, the reference configuration and the current configuration of the element giving

$$\boldsymbol{X}(\zeta_1, \zeta_2) = N_1\boldsymbol{X}_1 + N_2\boldsymbol{X}_2 + N_3\boldsymbol{X}_3 + N_4\boldsymbol{X}_4. \tag{22}$$

$$\boldsymbol{x}(\zeta_1, \zeta_2) = N_1\boldsymbol{x}_1 + N_2\boldsymbol{x}_2 + N_3\boldsymbol{x}_3 + N_4\boldsymbol{x}_4. \tag{23}$$

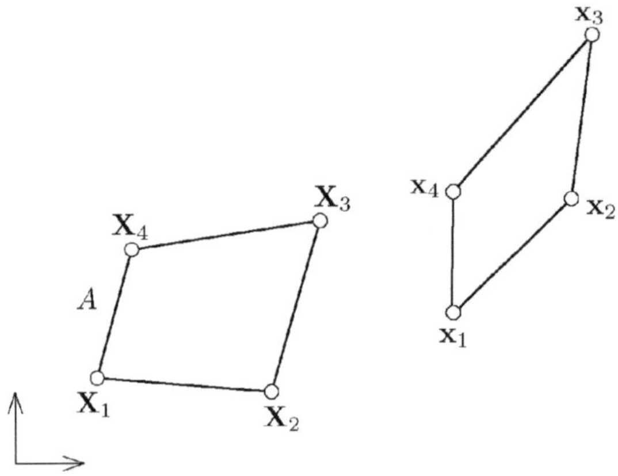

Figure 3.
Four nodes plane element: definition of the geometric quantities relative to the reference configuration (uppercase letters) and the current configuration (lower-case letters).

We have exactly the same pattern of the tetrahedron element, see Eqs. (9) and (10), except for the meaning of the shape function and the 2D dimension of the symbolic vectors $\boldsymbol{x}_i (i = 1 \ldots 4)$ and numeric vectors $\boldsymbol{X}_i (i = 1 \ldots 4)$. The deformation gradient can be evaluated using always Eq. (11) where now the operators are

$$\frac{\partial \boldsymbol{x}}{\partial \zeta} = \left[\left(-\frac{(1-\zeta_2)}{4}\boldsymbol{x}_1 + \frac{(1-\zeta_2)}{4}\boldsymbol{x}_2 + \frac{(1+\zeta_2)}{4}\boldsymbol{x}_3 - \frac{(1+\zeta_2)}{4}\boldsymbol{x}_4 \right) \left(-\frac{(1-\zeta_1)}{4}\boldsymbol{x}_1 - \frac{(1+\zeta_1)}{4}\boldsymbol{x}_2 + \frac{(1+\zeta_1)}{4}\boldsymbol{x}_3 + \frac{(1-\zeta_1)}{4}\boldsymbol{x}_4 \right) \right] \tag{24}$$

and

$$\frac{\partial \boldsymbol{X}}{\partial \zeta} = \left[\left(-\frac{(1-\zeta_2)}{4}\boldsymbol{X}_1 + \frac{(1-\zeta_2)}{4}\boldsymbol{X}_2 + \frac{(1+\zeta_2)}{4}\boldsymbol{X}_3 - \frac{(1+\zeta_2)}{4}\boldsymbol{X}_4 \right) \left(-\frac{(1-\zeta_1)}{4}\boldsymbol{X}_1 - \frac{(1+\zeta_1)}{4}\boldsymbol{X}_2 + \frac{(1+\zeta_1)}{4}\boldsymbol{X}_3 + \frac{(1-\zeta_1)}{4}\boldsymbol{X}_4 \right) \right] \tag{25}$$

are 2×2 matrices depending on the local coordinates of the element. Then the necessity to use the Gauss integration points in the evaluation of the strain energy of the element and, as a consequence, of the gradient and Jacobian of the element, see Eqs. (5) and (7). In particular four Gauss points are used, their coordinates and weights can be found in any FEM text book and are also shown in the complete listing of the class available in [34].

Previous discussion introduces the implementation details of the MATLAB® class PF4, PF stays for Plane **F**, whose kinematic properties, see Listing 1.11, are similar to those used for the class Tetra4 plus other properties required for the Gauss integration points. These properties are used to implement Eqs. (11), (24) and (25) which must to be evaluated in each Gauss point (bulky details are not shown but they can be found in the complete listing of the class, see [34]).

Listing 1.11. PF4 class: kinematic properties and evaluation of F.

properties (Constant)

```
            nG = 4;
            xiG = % Gauss coordinates , values not shown here
            wG = [1 1 1 1];
      end

      properties (SetAccess = private)
            % symbolic properties
            xl % current coordinates of node 1
            x2 % current coordinates of node 2
            x3 % current coordinates of node 3
            x4 % current coordinates of node 4
            xe % current element coordinates
            Fe % deformation gradient F(xe) (nGP times)

            % numeric properties
            Xe % reference element coordinates
      end

      function PF = Initialize (PF,D, i)
            % D brings all the problem data and its use
            % is not shown here

            PF . Fe = sym( zeros (2 ,2 ,PF .nG));
            for g = l:PF.nG
              % dzetadX evaluation in g
              % ...

              % dxdzeta evaluation in g
              % ...

              % F in g
              F = dxdzeta * dzetadX;
              PF.Fe(:, :, g) = F;

              % ...
            end
      end
```

In each Gauss integration point the strain energy, the compressible neo-Hookean form is used again, must to be evaluated by taking into account the simplification determined by the plane form assumed by tensor $\boldsymbol{F}$, then

$$J = \det\boldsymbol{F} = \det\boldsymbol{F}_{2\times2}, \tag{26}$$

and by tensor $\boldsymbol{C}$

$$\boldsymbol{C} = \begin{bmatrix} C_{11} & C_{12} & 0 \\ C_{21} & C_{22} & 0 \\ 0 & 0 & 1 \end{bmatrix}, \quad \boldsymbol{C}_{2\times2} = \begin{bmatrix} C_{11} & C_{12} \\ C_{21} & C_{22} \end{bmatrix}, \tag{27}$$

from which

$$I_1 = \mathrm{tr}\mathbf{C} = \mathrm{tr}\mathbf{C}_{22} + 1. \tag{28}$$

Then the expression of the strain energy density valid for the plane strain condition is

$$\Psi_{PF} = \frac{\mu}{2}(I_1 - 2) - \mu \ln J + \frac{\lambda}{2}(\ln J)^2, \tag{29}$$

where I_1 and J are calculated on the basis of the plane form of kinematic tensors. The resulting strain energy of the generic element can be then evaluated by using the following formula

$$\Psi_e(\boldsymbol{x}) = \int_{\Omega_e} \Psi_{PF}\, dV = \sum_{g=1}^{4} [\Psi_{PF}] A_g\, th\, w_g = \sum_{g=1}^{4} \Psi_g, \tag{30}$$

where $A_g = \det\left[\frac{\partial \boldsymbol{X}}{\partial \zeta}\right]_g$ is the part of the reference domain pertaining to the Gauss point, w_g is the Gauss point weight and th is the domain thickness usually assumed unitary under plane strain condition. Using Eq. (30), (5), and (7) the following results are valid for the generic element

$$\boldsymbol{g}_e = \sum_{g=1}^{4} \frac{\partial \Psi_g}{\partial \boldsymbol{x}} = \sum_{g=1}^{4} \boldsymbol{g}_g, \quad \boldsymbol{J}_e = \sum_{g=1}^{4} \frac{\partial \boldsymbol{g}_g}{\partial \boldsymbol{x}} = \sum_{g=1}^{4} \boldsymbol{J}_g, \tag{31}$$

where $\boldsymbol{g}_g$ and $\boldsymbol{J}_g$ are the gradient and Jacobian, respectively, pertaining to the generic Gauss point.

The following MATLAB® instructions, Listing 1.12, implements, inside the function Initialize of PF4 class, the operations required by Eq. (31).

Listing 1.12. PF4 class: evaluation of ge and Je.

```
function PF = Initialize (PF,D, i)
        % ...

        PF. Je = sym (zeros (8 ,8 , PF .nG));
        PF. Fe = sym ( zeros (2 ,2 ,PF .nG));
        for g = l:PF.nG
        % ...
        C = F.'*F;
        I1 = trace (C);
        J = det (F);
        Psi = (mi/2*(Il–2)–mi*log(J) + lam/2*log (J)^2) ...
        A*PF.wG(g)*th;
        PF.ge(: , 1, g) = gradient (Psi, PF. xe);
        PF.Je(:, :, g) = jacobian (PF. ge(: , 1, g), PF. xe);
        % ...
        end
        % ...
end
```

During the analysis the main task to be performed by the element is the numerical evaluation of $\boldsymbol{g}_e$ and $\boldsymbol{J}_e$ that now must to be performed, see Eq. (31), on the basis of the following implementation of the class function Compute.

Listing 1.13. PF4 class:function Compute.

```
function PF = Compute(PF)
        PF. se = zeros (8, 1);
        PF. Ke = zeros (8, 8);
        for g = l:PF.nG
          PF.se = PF.se + subs (PF.ge(:, 1, g), PF.xe, PF.xxe);
          PF.Ke = PF.Ke + subs (PF.Je(:, :, g), PF.xe, PF.xxe);
        end
end
```

The last part of the class to be discussed regards the evaluation of stress solution. As already observed in the beginning of this section, Eq. (19) constitutes an internal constraint determining the presence of also the stress component σ_{33} to be evaluated together with the plane part of the stress tensor. The plane part can be calculated using Eqs. (15) and (16) where the plane version of C and F must be used starting from the strain energy expression given by Eq. (29). The σ_{33} component, stems from the plane solution, and is given by

$$\sigma_{33} = J^{-1}S_{33} = J^{-1}\frac{\lambda}{2}\ln(\det\mathbf{C}) \tag{32}$$

A simple derivation of this expression through MATLAB® is reported in Appendix A. The implementation of the operations required for the evaluation of the stress solution are reported below, Listing 1.14.

Listing 1.14. PF4 class: function Stress.

```
function sig = Stress (T, gx)
        % retrieve local vector lx from global
        % solution gx
        sig = zeros(3, 3, PF.nG);
        for g = l:PF.nG
          F = subs (PF.Fe(:, :, g), PF.xe, lx);
          C = F.'*F;
           sig(l:2, l:2, g) = F*subs (PF. Se, PF.Ce,C)*F.'/det(F);
          sig (3, 3, g) = subs (PF.Se33, PF.Ce, C)/det (F);
        end
end
```

Listing 1.14 shows the use of the symbolic properties PF.Se which is initialised in way similar to the property T.Se shown in Listing 1.9 for the tetrahedral element. Anyway the complete listing of the class can be found in [34].

6. Conclusions

The early phase developing of finite elements can be a lengthy and error prone processes involving the use of different tools. The MATLAB® symbolic approach here presented can be effectively used to test a produce new finite element formulation reducing a lot the distance between the formulation and its actual implementation. In order to be more illustrative the presentation regarded basic solid mechanics finite elements, a truss, tetrahedral and plane quadrangular element, but the developing of finite elements for more specific engineering applications is an objective worth to be pursued and it is the subject of the author's current work.

The weakness of the proposed approach is the low performance of the final codes making difficult the analysis of real sized problems by using common hardware resources which, however, are adequate if small but significative test cases are chosen. A workaround, already tested by the author but not presented here, is the generation and storing on files of MATLAB® functions for the evaluation of the element operators. This must happens before, and one time for all, the execution of the analysis. The MATLAB® functions so obtained can be called during the analysis for evaluating the required finite element operators avoiding the calls to time-consuming function subs. Anyway the tuning of this operation is less automatic because the generation of the required MATLAB® functions can be, depending on the size of the operator to be translated into a MATLAB® function, time consuming, specially if the optimization flag is active. Then techniques quite common in the field of the symbolic and /or algorithmic differentiation should be exploited for the most intricate cases.

A. Appendix

A.1 Tetrahedron reference configuration operator inversion

The problem of the evaluation of the inverse of matrix $\frac{\partial \boldsymbol{X}}{\partial \zeta}$ present in Eq. (11) is circumvented by evaluating the Jacobian of the following system of equations

$$\begin{aligned} 1 &= \zeta_1 + \zeta_2 + \zeta_3 + \zeta_4 \\ \boldsymbol{X}(\zeta_1, \zeta_2, \zeta_3, \zeta_4) &= \zeta_1 \boldsymbol{X}_1 + \zeta_2 \boldsymbol{X}_2 + \zeta_3 \boldsymbol{X}_3 + \zeta_4 \boldsymbol{X}_4 \end{aligned} \tag{33}$$

whose linearisation gives

$$\begin{bmatrix} 0 \\ d\boldsymbol{X} \end{bmatrix} = \begin{bmatrix} 1 & 1 & 1 & 1 \\ \boldsymbol{X}_1 & \boldsymbol{X}_2 & \boldsymbol{X}_3 & \boldsymbol{X}_4 \end{bmatrix} [d\zeta]. \tag{34}$$

By inverting this relationship, i. e.

$$[d\zeta] = \begin{bmatrix} 1 & 1 & 1 & 1 \\ \boldsymbol{X}_1 & \boldsymbol{X}_2 & \boldsymbol{X}_3 & \boldsymbol{X}_4 \end{bmatrix}^{-1} \begin{bmatrix} 0 \\ d\boldsymbol{X} \end{bmatrix} = \begin{bmatrix} - & \frac{\partial \zeta_1}{\partial X_1} & \frac{\partial \zeta_1}{\partial X_2} & \frac{\partial \zeta_1}{\partial X_3} \\ - & \frac{\partial \zeta_2}{\partial X_1} & \frac{\partial \zeta_2}{\partial X_2} & \frac{\partial \zeta_2}{\partial X_3} \\ - & \frac{\partial \zeta_3}{\partial X_1} & \frac{\partial \zeta_3}{\partial X_2} & \frac{\partial \zeta_3}{\partial X_3} \\ - & \frac{\partial \zeta_4}{\partial X_1} & \frac{\partial \zeta_4}{\partial X_2} & \frac{\partial \zeta_4}{\partial X_3} \end{bmatrix} \begin{bmatrix} 0 \\ d\boldsymbol{X} \end{bmatrix}. \tag{35}$$

the evaluation of the desired 4×3 matrix, $\left(\frac{\partial \boldsymbol{X}}{\partial \zeta}\right)^{-1}$, is obtained. Moreover the volume of the tetrahedron in its reference configuration is an additional result thanks to relationship

$$6V = \det \begin{bmatrix} 1 & 1 & 1 & 1 \\ \boldsymbol{X}_1 & \boldsymbol{X}_2 & \boldsymbol{X}_3 & \boldsymbol{X}_4 \end{bmatrix}. \tag{36}$$

A.2 Out-of-plane normal component for the plane strain condition

The following MATLAB® instructions allow to find an explicit expression of the S_{33} component, from which $\sigma_{33} = J^{-1}S_{33}$.

```
syms lam mi 'real'
syms C [3 3] 'real'
C(l, 3) = 0; C(3, l) = 0; C(3, 2) = 0; C(2, 3) = 0;
I1 = trace(C); I3 = det(C);
Psi = mi/2*(I1–3)–mi*log(sqrt (I3))+ ...
                    lam/2*log (sqrt (13))^2;
S33 = simplify (2*diff (Psi ,C(3, 3)));
S33 = subs(S33, C(3, 3), 1);
```

Author details

Antonio Bilotta
Department of Informatics, Modeling, Electronics and System Engineering (DIMES), University of Calabria, Rende 87036, CS, Italy

*Address all correspondence to: antonio.bilotta@unical.it

References

[1] Cassel K W. Variational Methods with Applications in Science and Engineering. Cambridge University Press; 2013.

[2] Kreyszig E. Advanced Engineering Mathematics, 9th edition. John Wiley & Sons; 2006.

[3] Gu E Y L. A Journey from Robot to Digital Human (Mathematical Principles and Applications with MATLAB Programming). Springer-Verlag Berlin Heidelberg; 2013.

[4] Dym C L. Principles of Mathematical Modeling, 2nd edition. Elsevir Academic Press; 2004.

[5] Samarskii A A, Mikhailov A P. Principles of Mathematical Modeling (Ideas, Methods, Examples). CRC Press: Taylor & Francis Group; 2018.

[6] Zienkiewicz O C, Taylor R L, Zhu J Z. The Finite Element Method: Its Basis and Fundamentals (7th edition). Butterworth-Heinemann; 2013.

[7] Bathe K-J. Finite Element Procedures. Prentice-Hall; 1996.

[8] Hughes J R. The Finite Element Method. Dover Publications, Inc.; 2000.

[9] Brezzi F, Fortin M. Mixed and Hybrid Finite Element Methods. Springer-Verlag New York; 1991.

[10] Bilotta A, Casciaro R. Assumed stress formulation of high order quadrilateral elements with an improved in-plane bending behaviour. Comput. Methods Appl. Mech. Engrg. 2002; 191/15–16:1523–1540.

[11] Bilotta A, Casciaro R. A high performance element for the analysis of 2D elastoplastic continua. Computer Methods in Applied Mechanics and Engineering 2007; 196:818–828.

[12] Bilotta A, Leonetti L, Garcea G. Three field finite elements for the elastoplastic analysis of 2D continua. Finite Elements in Analysis & Design. 2011;47:1119–1130.

[13] Bilotta A, Turco E. Elastoplastic analysis of pressure-sensitive materials by an effective three-dimensional mixed finite element. ZAMM Zeitschrift fur Angewandte Mathematik und Mechanik 2017; 97(4): 382–396, doi 10.1002/zamm.201600051

[14] Simone A. Partition of unity-based discontinuous finite elements: GFEM, PUFEM, XFEM. Revue Europeenne de Genie Civil. 2007; 11/7–8:1045–1068. DOI: 10.1080/17747120.2007.9692976

[15] Fortino S, Bilotta A. Evaluation of the amount of crack growth in 2D LEFM problems. Engineering Fracture Mechanics. 2004; 71/9–10:1403–1419. DOI: 10.1016/S0013-7944(03)00161-9

[16] Belytschko T, Krongauz Y, Organ D, Fleming M, Krysl P. Meshless methods: An overview and recent developments. Computer Methods in Applied Mechanics and Engineering. 1996; 139/1: 3–47. DOI: 10.1016/S0045-7825(96)01078-X

[17] Cockburn B, Karniadakis G E, Shu C-W, editors. Discontinuous Galerkin Methods. Springer-Verlag Berlin Heidelberg; 2000.

[18] MATLAB® main documentation resource. Available from: https:// www.mathworks.com/help/matlab/ [Accessed: 2020-10-01]

[19] List of computer algebra systems. Available from: https://en.wikipedia.org/Eist_of_computer_algebra_systems [Accessed: 2020-10-01]

[20] List of finite element software packages. Available from: https://en.

wikipedia.org/wiki/List_of_finite_ element_software_packages [Accessed: 2020-10-01]

[21] Partial Differential Toolbox. Available from: https://www. mathworks.com/products/pde.html [Accessed: 2020-10-01]

[22] Rackauckas C, A Comparison Between Differential Equation Solver Suites In MATLAB, R, Julia, Python, C, Mathematica, Maple, and Fortran. The Winnower 6:e153459.98975 (2018). DOI: 10.15200/winn.153459.98975

[23] Symbolic Math Toolbox™main documentation resource. Available from: https://www.mathworks.com/products/symbolic.html [Accessed: 2020-10-01]

[24] Bonet J, Gil A J, Wood R D. Mechanics for finite element analysis: Statics. Cambridge University Press; 2016.

[25] Holzapfel G A. Nonlinear solid mechanics. John Wiley and Sons; 2000.

[26] Cooper J. A MATLAB® Companion for Multivariable Calculus. Harcourt/Academic Press; 2001.

[27] Yang W Y, Choi Y K, Kim J, Kim M C, Kim H J, Im T. Engineering Mathematics with MATLAB®. CRC Press Taylor & Francis Group; 2018.

[28] Wilson H B, Turcotte L H, Halpern D. Advanced Mathematics and Mechanics Applications Using MATLAB®.

[29] Korelc J, Wrigger P. Automation of Finite Element Methods. Springer International Publishing Switzerland; 2016.

[30] Naumann U. The Art of Differentiating Computer Programs. Society for Industrial and Applied Mathematics; 2012.

[31] Griewank A, Walther A. Evaluating Derivatives. Society for Industrial and Applied Mathematics; 2008.

[32] Complete implementation of class Truss. Available from: https:// antoniobilotta-structuralengineer.github.io/MyWebSite/MATLAB/Truss.m [Uploaded: 2020-10-30]

[33] Complete implementation of class Tetra4. Available from: https:// antoniobilotta-structuralengineer.github.io/MyWebSite/MATLAB/Tetra4.m [Uploaded: 2020-10-30]

[34] Complete implementation of class PF4. Available from: https:// antoniobilotta-structuralengineer.github.io/MyWebSite/MATLAB/PF4. m [Uploaded: 2020-10-30]

Chapter 13

A Posteriori Error Analysis in Finite Element Approximation for Fully Discrete Semilinear Parabolic Problems

Younis Abid Sabawi

Abstract

This Chapter aims to investigate the error estimation of numerical approximation to a class of semilinear parabolic problems. More specifically, the time discretization uses the backward Euler Galerkin method and the space discretization uses the finite element method for which the meshes are allowed to change in time. The key idea in our analysis is to adapt the elliptic reconstruction technique, introduced by Makridakis and Nochetto 2003, enabling us to use the a posteriori error estimators derived for elliptic models and to obtain optimal order in $L_\infty(H^1)$ for Lipschitz and non-Lipschitz nonlinearities. In this Chapter, some challenges will be addressed to deal with nonlinear term by employing a continuation argument.

Keywords: A posteriori error estimates, semilinear parabolic problems, finite element approximation, L_∞ (H^1) bounds in finite element approximation, fully discrete semilinear parabolic approximation

1. Introduction

The finite element method (FEM) consider is the most of flexibility common technique used for dealing with various kinds of application in many fields, for instance, in engineering, in chemistry and in biology. The derivation of a posteriori error estimates for linear and nonlinear parabolic problems are gaining increasing interest and there is a significant implementation of the method now are understandable and available in the literature [1–9]. However, There is less progress has been made comparatively in the proving of a posteriori error bounds for semilinear parabolic problems [10–13]. These estimations play a crucial rule in designing adaptive mesh refinement algorithms and consequently leading to a good accuracy while reducing the computational cost of the scheme.

The key technique used in the proofs is the elliptic reconstruction idea, introduced by Makridakis and Nochetto for spatially discrete conforming FEM [2] and extended to fully discrete conforming FEM by Lakkis and Makridakis [3] These ideas have been carried forward also to fully discrete schemes involving spatially non-conforming/dG methods in [14]. The choice of this technique for deriving a posteriori error for parabolic problem is motivated by the following factors.

First, elliptic reconstruction allows us to utilise the readily available elliptic a posteriori estimates [2] to bound the main part of the spatial error. Second, this technique combines the energy approach and appropriate pointwise representation of the error in order to arrive to optimal order a posteriori estimators in the $L_\infty(L^2)$-norm. As a result, this approach will lead to optimal order in both $L^2(H^1)$ and $L_\infty(L^2)$-type norms, while the results obtained by the standard energy methods are only optimal order in $L^2(H^1)$-type norms.

The aim of this Chapter is to derive a posteriori error bounds for the fully discrete in two cases Lipschitz and non Lipschitz. Continuation Argument will be used to deal with nonlinear forcing terms.

2. Preliminaries

Before we proceed with the error analysis, we require some auxiliary results that will be used in our analysis.

2.1 Functional spaces

Let $z(t, \ x)$ is a function of time t and space χ, we introduce the Bochner space $L_P(0, \ T, \cdot \ X)$ where (X is some real Banach space equipped with the norm $\|\cdot\|_X$) which is the collection of all measurable functions v: $(0, \ T) \to X$, more precisely, for any number $r \geq 1$

$$L_P(0, \ T; \ X) = \left\{ z \ : \ (0, \ T) \to X \ : \ \int_0^T \|z\|^2 dt \leq \infty \right\}, \tag{1}$$

such that

$$\begin{aligned} \|z\|_{L_P(0,T;X)} &:= \left(\int_0^T \|z\|^2 dt \right)^{1/2} < \infty \quad \text{for } 1 \leq p < \infty, \\ \|z\|_{L_P(0,T;X)} &:= \max_{t \in [0, T]} \|z(t)\|_X < \infty \quad \text{for} p = \infty. \end{aligned} \tag{2}$$

Lemma 1.1 (Continuous Gronwall inequality). Let $C_0, \ C_1 \in L^1(0, \ T)$ for all $T > 0$ and $z \in W^{1,1}$, then for almost every $t \in (0, \ T]$, reads

$$z'(t) \leq C_0(t) + C_1(t) z(t) \ , \tag{3}$$

then

$$z(t) \leq F(0, \ T) z(0) + \int_0^T F(0, \ T) z(s) ds, \tag{4}$$

where $F(0, \ T) = \exp\left(\int_0^T C_1(\xi(t) d\xi\right)$. Furthermore, if C_0 and C_1 are non-negatives, gives

$$z(T) \leq F(0, \ T) \left(z(0) + \int_0^T C_0(s) ds \right). \tag{5}$$

Proof: See [15].

DOI: http://dx.doi.org/10.5772/intechopen.94369

Theorem 1.2 Given some $p \geq 2$, we have

$$\|v\|^{p}_{L^p(\Omega)} \leq C\|\nabla v\|^{\frac{pd-2d}{2}}\|_v\|^{\frac{2p+2d-pd}{2}}$$
$$\|v\|^{p}_{L^p(\Omega)}p \leq C\|\nabla v\|^{p-2}\|v\|^2, d=2$$
$$\|v\|^{p}_{L^p(\Omega)}p \leq C\|\nabla v\|^{\frac{3p-6}{2}}\|v\|^{\frac{6-p}{2}},\ d=3, p\leq 6.$$

Proof: See [16].

3. Model problem

Consider the semilinear parabolic problem as

$$\begin{aligned}
&\frac{\partial u}{\partial t} - \Delta u = f(u), && \text{in } \Omega \cup [0,\ T], \\
&u = 0, && \text{on } \partial\Omega, \\
&u(0,x) = u_0(x), && \text{on } \{0\} \times \Omega,
\end{aligned} \tag{6}$$

where Ω is a plane convex domain subset of $\mathbb{R}^k$, $\Omega \subset \mathbb{R}^k$ with smooth boundary condition $\partial\Omega$, where $u_t = \partial u/\partial t$, $T > 0$ and $f \in C^1(\mathbb{R})$. Let $L_p(\omega)$, $1 \leq p \leq \infty$ and $H^r(\omega)$, $r \in \mathbb{R}$, denote the standard Lebesgue and Hilbertian Sobolev spaces on a domain $\omega \subset \Omega$. For brevity, the norm of $L_2(\omega) \equiv H^0(\omega)$, $\omega \subset \Omega$, will be denoted by $\|\cdot\|_\omega$, and is induced by the standard $L_2(\omega)$-inner product, denoted by $(\cdot,\cdot)_\omega$; when $\omega = \Omega$, we shall use the abbreviations $\|\cdot\| \equiv \|\cdot\|_\Omega$ and $(\cdot,\cdot) \equiv (\cdot,\cdot)_\Omega$.

Returning to the (6), multiplying by a test function $v \in H^1_0(\Omega)$ and then integrate by parts, we arrive to (7) in weak form, which reads: find $u \in L_2(0,\ T, H^1_0)(\Omega) \cap H^1_0(0,\ T,\ L_2(\Omega)$ for almost every $t \in (0,\ T]$, this becomes

$$\int_\Omega \frac{\partial z}{\partial t} v dx + D(t; z, v) = \int_\Omega f(z) v dx, \tag{7}$$

for all $v \in H^1_0(\Omega)$. Here,

$$D(t; z,\ v) = \int_\Omega \nabla z \cdot \nabla v dx. \tag{8}$$

By using Cauchy-Schwarz inequality, the convercitivity and continuity of the bilnear form D, viz.

$$\begin{aligned}
&D(v,\ v) \geq C_{\text{coer}}\|\nabla v\|^2 \quad \text{for all } v \in H^1_0(\Omega), \\
&|D(v,\ w)| \leq C_{\text{cont}}\|\nabla v\|\|\nabla w\| \quad \text{for all } v,\ w \in H^1_0(\Omega),
\end{aligned} \tag{9}$$

with $C_{\text{cont}}, C_{\text{coer}}$ positive constants independent of $w,\ v$.

4. Fully discrete backward Euler formulation

To introduce a backward Euler approximation of the time derivative paired with the standard conforming finite element method of the spatial operator. To this end, we will discretize the time interval $[0, \ T]$ into subintervals $(t_{n-1}, \ t_n], \ n = 1, \ \dots, \ N$ with $t^0 = 0$ and $t_N = T$, and we denote by $\kappa_n = t_n - t_{n-1}$ the local time step. We associate to each time-step t_N a spatial mesh $\mathcal{T}^n$ and the respective finite element space $V^n; \ = V_h^p(\mathcal{T}^n)$. The fully discrete scheme is defined as follows. Set $Z(0)$ to be a projection of z_0 onto some space V^0 subordinate to a mesh $\mathcal{T}^0$ employed for the discretization of the initial condition. For $k = 1, \ \dots, \ n$, find $Z \in S^n$ such that the fully discrete, then reads as follows

$$\left(\frac{Z^n - Z^{n-1}}{K_n}, \ \phi^n\right) + D(Z^n, \ \phi^n) = (\ f^n(Z^n), \ \phi^n), \ \forall \phi^n \in V^n \tag{10}$$

where $D^n(\cdot, \cdot) = D(t_n, \ \cdot, \ \cdot)$ denotes the cG bilinear form defined on the mesh $\mathcal{T}^n$. Since $Z^n \in V^n$, there exist $\alpha_i(t) \in \mathbb{R}, j = 0, 1, 2, \ \dots, \ N_h$, so that

$$Z^n(x, \ t) = \sum_{j=0}^{N_{loc}N_{el}} \alpha_j^n(t)\Phi_j(x), \ \Phi_j, \ \ j = 0, 1, 2 \dots N_h \tag{11}$$

is the basis functions. After plugging (11) into (10), yields a nonlinear system of ordinary differential equations

$$\begin{aligned} &(M \ + \kappa_n A)\alpha_j^n(t) = M\alpha_j^{n-1}(t) + \kappa_n F \\ &\alpha(0) = \delta, \end{aligned} \tag{12}$$

where $M_{i,j} = (\Phi_j, \ \Phi_j)$ and $A_{i,j} = D(\Phi_j, \ \Phi_j)$ are called the mass and stiffness matrices with element $F_{\,j,k} = (\ f(\Phi_j), \ \Phi_k)$. We define the piecewise linear interpolant Z and time-dependent elliptic reconstruction $w(t)$ as by the linear interpolant with respect to t of the values Z^{n-1} and Z^n, viz.,

$$Z(t) := \ell_{n-1}(t)Z^{n-1} + \ell_n(t)Z^n, \qquad w(t) := \ell_{n-1}\mathrm{R}_{be}^{n-1}Z^{n-1} + \ell_n\mathrm{R}_{be}^n Z^n, \tag{13}$$

where $\{\ell_{n-1}, \ell_n\}$ denotes the linear Lagrange interpolation basis on the interval I_n are defined as

$$\ell_n := \frac{t_n - t}{K_n}, \quad \ell_{n-1} := \frac{t - t_{n-1}}{K_n}. \tag{14}$$

We give here some essential definitions in the error analysis of the discrete parabolic equations.

i. L^2 projection operator Π_0^n; The operator defined Π_0^n: $L^2 \to V^n, \ 1 \le n \le N$ such that

$$(\Pi_0^n v, \ \phi^n) = (v, \ \phi^n) \ \ \forall \phi^n \in V^n, \tag{15}$$

for all $v \in L^2(\Omega)$.

ii. Discrete elliptic operator: The elliptic operator defined A_h^n: $H_0^1(\Omega) \to V^n$ such that for $v \in H_0^1(\Omega)$, reads

$$(\mathrm{A}_h^n v, \ \phi^n) = D(v, \ \phi^n) \ \forall \phi^n \in V^n. \tag{16}$$

Using the above projections, (10) can be expressed in distributional form as

$$\frac{Z^n - \Pi_0^n Z^{n-1}}{K_n} + \mathrm{A}_h^n Z^n = \Pi_0^n f^n(Z^n). \tag{17}$$

5. Elliptic reconstruction

The aim of this section will be introduced the elliptic reconstruction operator and then discuss the related aposteriori error analysis for the backward Euler approximation. To do this, we define the elliptic reconstruction $\mathrm{R}_{be}^n \in H_0^1(\Omega)$ of Z^n as the solution of elliptic problem

$$D\left(\mathrm{R}_{be}^n v,\ \phi\right) = (g^n,\ \phi), \tag{18}$$

for a given $v \in V^n$ and $g^n = \Pi_0^n f^n(Z^n) - \frac{Z^n - \Pi_0^n Z^{n-1}}{k_n}$. The crucial property, this operator R_{be}^n is orthogonal with respect to D such that

$$D\left(u - \mathrm{R}_{be}^n u,\ v\right) = 0 \quad u,\ v \in V^n. \tag{19}$$

The following lemma is the elliptic reconstruction error bound in the H^1 and L_2-norms To see the proof, we refer the reader to [3] for details.

Lemma 1.3 (Posteriori error estimates). For any $Z^n \in V^n$, the following elliptic a posteriori bounds hold:

$$\begin{aligned} &\|\langle R_{be}^n Z^n - Z^n\| \leq C\Phi_{n,L_2}^2 \\ &\|\nabla\left(\langle R_{be}^n Z^n - Z^n\right)\| \leq C\Phi_{n,\mathrm{H}^1}^2 \end{aligned} \tag{20}$$

where

$$\begin{aligned} \Phi_{n,L_2}^2 &:= \|h_n^2(g^n + \Delta^n Z^n)\| + \|h_n^{3/2}[\![Z^n]\!]\|_{\Sigma_n}, \\ \Phi_{n,\mathrm{H}^1}^2 &:= \|h_n(g^n + \Delta^n Z^n)\| + \|h_n^{1/2}[\![Z^n]\!]\|_{\Sigma_n}, \end{aligned} \tag{21}$$

and g^n defined in (18).

Lemma 1.4 (Main semilinear parabolic error equation). The following error bounds hold

$$\begin{aligned} \left(\frac{\partial \rho}{\partial t},\ \psi\right) + D(\rho,\ \phi) &= (\,f(z) - f^n(Z^n),\ \phi) + \left(\frac{\partial \varepsilon}{\partial t},\ \phi\right) + D(w - w^n,\ \phi) \\ &+ \left(\Pi_0^n f^n(Z^n) - f^n(Z^n) + \frac{\Pi_0^n Z^{n-1} - Z^{n-1}}{K_n},\ \phi\right). \end{aligned} \tag{22}$$

Proof: To begin with, we first decompose the error as

$$e := \rho - \varepsilon, \quad \rho := z - w, \quad \varepsilon := w - Z. \tag{23}$$

By recalling (17), this becomes

$$\left(\frac{\partial Z}{\partial t},\ \phi\right) + D(w^n,\ \phi) = \left(\frac{\Pi_0^n Z^{n-1} - Z^{n-1}}{k_n},\ \phi\right) + (\Pi_0^n f^n(Z^n),\ \phi) \ \ \forall \phi \in H_0^1(\Omega), \tag{24}$$

where $\frac{\partial Z}{\partial t} = \frac{Z^{n-1} - Z^n}{\kappa_n}$. Subtracting (24) from (7), gives

$$\left(\frac{\partial}{\partial t}[Z - z],\ \phi\right) + D(w^n - z,\ \phi) = (\Pi_0^n f^n(Z^n) - f(z),\ \phi) + \left(\frac{\Pi_0^n Z^{n-1} - Z^{n-1}}{\kappa_n},\ \phi\right). \tag{25}$$

Using elliptic reconstruction to split the error, gives

$$\begin{aligned}\left(\frac{\partial}{\partial t}[-z - w + w + Z^n],\ \phi\right) + D(w^n - w + w - z,\ \phi) = (\Pi_0^n f^n(Z^n) - f^n(Z^n),\ \phi)\\ +(f^n(Z^n) - f(z),\ \phi) + \left(\frac{\Pi_0^n Z^{n-1} - Z^{n-1}}{K_n},\ \phi\right).\end{aligned} \tag{26}$$

After using triangle inequality, the proof will be concluded.
The proof of the following Lemmas 1.5, 1.6, 1.7 in details, we refer to [3].
Lemma 1.5 (Temarol error estimate). Let $T_{n,1}, 1 \le n \le N$ be given by

$$T_{n,1} := \int_{t_{n-1}}^{t_n} \left| D\left(w - w^n,\ \frac{\partial \rho}{\partial t}\right)\right| dt, \tag{27}$$

then

$$T_{n,1} \le \left(\int_{t_{n-1}}^{t_n} \left\|\frac{\partial \rho}{\partial t}\right\|^2 dt\right)^{1/2} (\kappa_n)^{1/2} \Phi_{n,2}, \tag{28}$$

where

$$\Phi_{n,2} := \begin{cases} \frac{\sqrt{3}}{3}\partial\left(\left\|\prod_0^n f^n(Z^n) - \frac{Z^n - \prod_0^n Z^{n-1}}{k_n}\right\|\right) & \text{for} n \in [2:N], \\ \frac{\sqrt{3}}{3}\left(\left\|\prod_0^1 f^1(Z^1) - \frac{Z^1 - \prod_0^1 Z^0}{k1}\right\|\right) & \text{for} n = 1. \end{cases} \tag{29}$$

Lemma 1.6 (Space-mesh error estimate). Let $T_{n,2}, 1 \le n \le N$ is defined by

$$T_{n,2} := \int_{t_{n-1}}^{t_n} \left| \left(\frac{\partial \varepsilon}{\partial t},\ \frac{\partial \rho}{\partial t}\right) \right| dt, \tag{30}$$

we have

$$T_{n,2} \le \left(\int_{t_{n-1}}^{t_n} \left\|\frac{\partial \rho}{\partial t}\right\|^2 dt\right)^{1/2} (\kappa_n)^{1/2} \Upsilon_{n,2}, \tag{31}$$

where

$$\Upsilon_{n,2} := C\left(\frac{d}{dt}\|h_n^2(g^n + \Delta^n Z^n)\|\right) + C\|\tilde{h}_n^{3/2}[\![Z^n - Z^{n-1}]\!]\|_{\tilde{\Sigma}_n} + C\|\tilde{h}_n^{3/2}[\![Z^n - Z^{n-1}]\!]\|_{\tilde{\Sigma}_n\ \hat{\Sigma}_n}. \tag{32}$$

Lemma 1.7 (Mesh change estimates). Let $T_{n,3}, 1 \le n \le N$ is given by

$$T_{n,3} := \int_{t_{n-1}}^{t_n} \left| \left(\Pi_0^n f^n(Z^n) - f^n(Z^n) + \frac{\Pi_0^n Z^{n-1} - Z^{n-1}}{\kappa_n},\ \frac{\partial \rho}{\partial t}\right) \right| dt, \tag{33}$$

such that

$$T_{n,3} \le \kappa_n \max_{t\in[0,t_m]} \|\nabla\rho\| \left(\delta_{n,\infty} + \sum_{n=2}^{m} \kappa_n \delta_{n,1} + \delta_{\infty,1} \right), \tag{34}$$

where

$$\begin{aligned} \delta_{n,1} &:= \|h_n^{\wedge}\partial\left(\Pi_0^n - I\right)\left(f^n(Z^n) - \kappa_n Z^{n-1}\right)\|, \\ \delta_{n,\infty} &:= \|h_n\left(\Pi_0^n - I\right)\left(f^n(Z^n) - \kappa_n Z^{n-1}\right)\|. \end{aligned} \tag{35}$$

6. A posteriori error bound for fully discrete semilinear parabolic problems

The aim of this section is to study a posteriori error bound in $L_\infty(H^1)$-norm for nonlinear forcing terms. Both globally and locally Lipschitz continuous nonlinearities are considered.

6.1 A posteriori error analysis for the globally Lipschitz continuity case

Let us suppose that f is defined on the whole of and satisfies globally Lipschitz continuous

$$|f(z_1) - f(z_2)| \le C_g |z_1 - z_2|, \tag{36}$$

where $|\cdot|$ denotes the standard Euclidean norm on $(\mathrm{R} \ge 1)$.

Lemma 1.8 (Data approximation error estimate). Suppose that the nonlinear reaction f satisfying the globally Lipschitz continuous defined in (36), then, the following error bounds hold:

$$\begin{aligned} T_{n,4} = \int_{t_{n-1}}^{t_n} \left| \left(f(z) - f^n(Z^n), \frac{\partial\rho}{\partial t} \right) \right| dt &\le \frac{\sqrt{C_g}}{2\beta} \kappa_n \|\nabla\rho\|^2 + \frac{\beta\sqrt{C_g}}{2} \int_{t_{n-1}}^{t_n} \left\| \frac{\partial\rho}{\partial t} \right\|^2 dt \\ + \kappa_n \Psi_{n,1} \left(\int_{t_{n-1}}^{t_n} \left\| \frac{\partial\rho}{\partial t} \right\|^2 dt \right)^{1/2} &+ \kappa_n \Psi_{n,2} \left(\int_{t_{n-1}}^{t_n} \left\| \frac{\partial\rho}{\partial t} \right\|^2 dt \right)^{1/2}, \end{aligned} \tag{37}$$

where

$$\begin{cases} \Psi_{n,1} := \sqrt{C_g}\left\{ \|\varepsilon^{n-1}\|, \ \|\varepsilon^n\| \right\}, \\ \Psi_{n,2} := \dfrac{1}{\kappa_n} \displaystyle\int_{t_{n-1}}^{t_n} \|f(Z) - f^n(Z^n)\|. \end{cases} \tag{38}$$

Proof: Using triangle inequality, $T_{n,4}$ written as

$$\begin{aligned} T_{n,4} = \int_{t_{n-1}}^{t_n} \left| \left(f(z) - f^n(Z^n), \frac{\partial\rho}{\partial t} \right) \right| dt &\le \int_{t_{n-1}}^{t_n} \left| \left(f(z) - f(w), \frac{\partial\rho}{\partial t} \right) \right| dt \\ + \int_{t_{n-1}}^{t_n} \left| \left(f(w) - f(Z), \frac{\partial\rho}{\partial t} \right) \right| dt &+ \int_{t_{n-1}}^{t_n} \left| \left(f(Z) - f^n(Z^n), \frac{\partial\rho}{\partial t} \right) \right| dt \\ := L_{n,1} + L_{n,2} + L_{n,3}. \end{aligned} \tag{39}$$

Applying Cauchy–Schwarz inequality and (36) along with Young's inequality and Poincar'e-Friedrichs inequality, $L_{n,1}$ gives

$$\begin{aligned} L_{n,1} &= \int_{t_{n-1}}^{t_n} \left| \left(f(z) - f(w), \frac{\partial \rho}{\partial t} \right) \right| dt \leq \int_{t_{n-1}}^{t_n} \| f(z) - f(w) \| \left\| \frac{\partial \rho}{\partial t} \right\| dt \\ &\leq \frac{\sqrt{C_g}}{2\beta} \kappa_n \| \nabla \rho \|^2 + \frac{\beta \sqrt{C_g}}{2} \int_{t_{n-1}}^{t_n} \left\| \frac{\partial \rho}{\partial t} \right\|^2 dt. \end{aligned} \tag{40}$$

The second term $L_{n,2}$, reads

$$\begin{aligned} L_{n,2} &= \int_{t_{n-1}}^{t_n} \left| \left(f(w) - f(Z), \frac{\partial \rho}{\partial t} \right) \right| dt \leq \int_{t_{n-1}}^{t_n} \| w - Z \| \left\| \frac{\partial \rho}{\partial t} \right\| dt \\ &\leq \sqrt{C_g} \int_{t_{n-1}}^{t_n} \left(\left| \frac{t_n - t}{\kappa_n} \right| \| \varepsilon^{n-1} \| + \left| \frac{t - t_{n-1}}{\kappa_n} \right| \| \varepsilon^n \| \right) \left\| \frac{\partial \rho}{\partial t} \right\| dt \\ &\leq \frac{\sqrt{C_g}}{2} \kappa_n \left(\| \varepsilon^{n-1} \| + \| \varepsilon^n \| \right) \left(\int_{t_{n-1}}^{t_n} \left\| \frac{\partial \rho}{\partial t} \right\|^2 dt) \right)^{1/2}. \end{aligned} \tag{41}$$

Finally, $L_{n,3}$ can be bounded by using Cauchy–Schwarz inequality, to obtain

$$L_{n,3} = \int_{t_{n-1}}^{t_n} \left| \left(f(Z) - f^n(Z^n), \frac{\partial \rho}{\partial t} \right) \right| dt \leq \| f(Z) - f^n(Z^n) \| \left(\int_{t_{n-1}}^{t_n} \left\| \frac{\partial \rho}{\partial t} \right\|^2 dt) \right)^{1/2}. \tag{42}$$

Collecting all the results together, the proof will be finished.

Lemma 1.9 Let z be the exact solution of (7) and let Z^n be its finite element approximation obtained by the backward Euler approximation (10). Then, for $1 \leq n \leq N$, the following a posteriori error bounds hold:

$$\left(\max_{t \in [0, t_m]} \| \nabla \rho(t) \|^2 + \int_0^{t_m} \left\| \frac{\partial \rho}{\partial t} \right\|^2 dt \right)^{1/2} \leq \{ 2 \mathcal{E}_G(m) \| \nabla \rho \|^2 \} \}^{1/2} + 2 \mathcal{E}_G(m) \left(\mathcal{F}_{1,m}^2 + \mathcal{F}_{2,m}^2 \right) \tag{43}$$

where

$$\begin{aligned} \mathcal{F}_{1,m} &:= 2 \max_{t \in [0, t_m]} \delta_{m,\infty} + 2 \sum_{n=2}^{m} \kappa_n \delta_{n,1}, \\ \mathcal{F}_{2,m}^2 &:= \sum_{n=1}^{m} \kappa_n \left(\Phi_{n,2}^2 + \Upsilon_{n,2}^2 + \Psi_{n,1}^2 + \Psi_{n,2}^2 \right). \end{aligned} \tag{44}$$

Proof: Now, setting $\phi = \frac{\partial \rho}{\partial t}$ in 22, gives

$$\begin{aligned} \frac{1}{2} \frac{d}{dt} \| \nabla \rho(t) \|^2 + \frac{C_{coer}}{2} \left\| \frac{\partial \rho}{\partial t} \right\|^2 &\leq \left| \left(\frac{\partial \varepsilon}{\partial t}, \frac{\partial \rho}{\partial t} \right) \right| + \left| \left(f(z) - f^n(Z^n), \frac{\partial \rho}{\partial t} \right) \right| + \left| D \left(w - w^n, \frac{\partial \rho}{\partial t} \right) \right| \\ &+ \left| \left(\Pi_0^n f(Z^n) - f^n(Z^n) + \frac{P_0^n Z^{n-1} - Z^{n-1}}{k_n}, \frac{\partial \rho}{\partial t} \right) \right|. \end{aligned} \tag{45}$$

Integrate the above from t_{n-1} to t_n then, we have

$$\frac{1}{2}\|\nabla\rho(t_n)\|^2 - \frac{1}{2}\|\nabla\rho(t_{n-1})\|^2 + \frac{C_{coer}}{2}\int_{t_{n-1}}^{t_n}\left\|\frac{\partial\rho}{\partial t}\right\|^2 dt \le T_{n,1} + T_{n,2} + T_{n,3} + T_{n,4}, \quad (46)$$

where $T_{n,i}, i = 1, 2, 3, 4$ defined in Lemmas 1.5, 1.6, 1.7 and 1.8, respectively. Summing up over $n = 1{:}\, m$ so that

$$\|\nabla\rho(t_m)\|^2 + C_{coer}\int_0^{t_m}\left\|\frac{\partial\rho}{\partial t}\right\|^2 dt \le \|\nabla\rho(0)\|^2 + 2\sum_{n=1}^{m}(T_{n,1} + T_{n,2} + T_{n,3} + T_{n,4}). \quad (47)$$

By introducing

$$\|\nabla(\rho_m^*)\|; = \|\nabla\rho(t_m^*)\| = \max_{t\in[0,t_m]}\|\nabla\rho(t)\|, \quad (48)$$

therefore

$$\max_{t\in[0,t_m]}\|\nabla\rho(t)\| + C_{coer}\int_0^{t_m}\left\|\frac{\partial\rho}{\partial t}\right\|^2 dt \le 2\|\nabla\rho(0)\|^2 + 4\sum_{n=1}^{m}(T_{n,1} + T_{n,2} + T_{n,3} + T_{n,4}). \quad (49)$$

Now, using Lemmas 1.5, 1.6, 1.7 and 1.8, reads

$$\max_{t\in[0,t_m]}\|\nabla\rho(t)\|^2 \le 2\|\nabla\rho(0)\|^2 + \left(2\beta\sqrt{C_g} - C_{coer}\right)\int_0^{t_m}\left\|\frac{\partial\rho}{\partial t}\right\|^2 dt + 2\max_{t\in[0,t_m]}\|\nabla\rho(t)\|\mathcal{F}_{1,m}$$
$$+\frac{2\sqrt{C_g}}{\beta}\sum_{n=1}^{m}K_n\max_{t\in[0,t_m]}\|\nabla\rho(t)\|^2 + 4\left(\int_{t_{n-1}}^{t_n}\left\|\frac{\partial\rho}{\partial t}\right\|^2 dt\right)^{1/2}(\kappa_n)^{1/2}(\Phi_{n,2} + \Upsilon_{n,2} + \Psi_{n,1} + \Psi_{n,2}). \quad (50)$$

Selecting now $\beta > 0$ be such that $\left(2\beta\sqrt{C_g} - C_{coer}\right) > 0$ and using Gronwall's inequality, imply

$$\max_{t\in[0,t_m]}\|\nabla\rho(t)\|^2 + \mathcal{E}_G(m)\int_0^{t_m}\left\|\frac{\partial\rho}{\partial t}\right\|^2 dt \le 2\mathcal{E}_G(m)\|\nabla\rho(0)\|^2 + 2\mathcal{E}_G(m)\max_{t\in[0,t_m]}\|\nabla\xi(t)\|\mathcal{F}_{1,m}$$
$$+4\mathcal{E}_G(m)\sum_{n=1}^{m}\left(\int_{t_{n-1}}^{t_n}\left\|\frac{\partial\rho}{\partial t}\right\|^2 dt\right)^{1/2}(\kappa_n)^{1/2}(\Phi_{n,2} + \Upsilon_{n,2} + \Psi_{n,1} + \Psi_{n,2}), \quad (51)$$

with $\mathcal{E}_G(m) := \left\{1, \sum_{n=1}^{m}\frac{2\sqrt{C_g}}{\beta}\kappa_n \exp\left(\frac{2\sqrt{C_g}}{\beta}\left(\Sigma_{n<j<m}k_j\right)\right)\right\}$. To finish the proof of lemma, we use a standard inequlty. For $(a_0, a_1, \ldots, a_n), (b_0, b_1, \ldots, b_n) \in \mathbb{R}^{m+1}$.

$$|a|^2 \le c^2 + ab, \quad (52)$$

then

$$|a| \le |c| + |b|, \quad (53)$$

and by taking

$$a_0 := \max_{t\in[0,t_m]} \|\nabla\rho(t)\|, \quad a_n := \left\{ \mathcal{E}_G(m)\int_0^{t_m} \left\|\frac{\partial\rho}{\partial t}\right\|^2 dt \right\}^{1/2}, \quad c := \left\{2\mathcal{E}_G(m)\|\nabla\rho(0)\|^2\right\}^{1/2}$$
$$b_0 := \sqrt{2}\mathcal{E}_G(m)\mathcal{F}_{1,m}, \quad b_n := 4\mathcal{E}_G(m)\sum_{n=1}^{m}(\kappa_n)^{1/2}(\Phi_{n,2}+\Upsilon_{n,2}+\Psi_{n,1}+\Psi_{n,2}). \tag{54}$$

The proof already will be finished.

Theorem 1.10 Let z be the exact solution of (7) and let Z^n be its finite element approximation obtained by the backward Euler approximation (10). Then, for $1 \le n \le N$, the following a posteriori error bounds hold:

$$\max_{t\in[0,t_m]} \|\nabla(z(t)-Z(t))\|^2 \le 2\mathcal{E}_G(m)\left(\Phi^2_{n,\mathrm{H}^1}(0)+\|\nabla(z(0)-Z(0))\|^2\right)$$
$$+2\mathcal{E}_G(m)\left(\mathcal{F}^2_{1,m}+\mathcal{F}^2_{2,m}\right)) + 2\max_{t\in[0,t_m]}\Phi^2_{n,\mathrm{H}^1}, \tag{55}$$

where Φ^2_{n,H^1} defined in (20).

Proof: By decomposing $Z(t)-z(t)$ into ρ and ε, so that

$$\|\nabla(Z(t)-z(t))\|^2 \le 2\|\nabla\varepsilon\|^2 + 2\|\nabla\rho\|^2. \tag{56}$$

To be able to bound the first term on the right hand side of (56), using (13), this becomes

$$\|\nabla\varepsilon(t)\|^2 = \|\nabla(w(t)-Z(t))\|^2 = \|\nabla\left(\ell_n \mathrm{R}^n_{be}Z^n + \ell_{n-1}\mathrm{R}^{n-1}_{be}Z^{n-1} - \ell_{n-1}(t)Z^{n-1} - \ell_n(t)Z^n\right)\|^2$$
$$\le \ell_n\|\nabla(\mathrm{R}^n_{be}Z^n - Z^n)\|^2 + \ell_{n-1}\|\nabla(\mathrm{R}^{n-1}_{be}Z^n - Z^{n-1})\|^2$$
$$\le \max_{t\in[0,t_m]}\left\{\left\|\nabla(\mathrm{R}^{n-1}_{be}Z^{n-1} - Z^{n-1})\right\|^2, \left\|\nabla(\mathrm{R}^n_{be}Z^n - Z^n)\right\|^2\right\} \tag{57}$$
$$\le \max_{t\in[0,t_m]}\left\{\|\nabla(\mathrm{R}^n_{be}Z^n - Z^n)\|^2\right\}$$
$$\le \max_{t\in[0,t_m]}\Phi^2_{n,\mathrm{H}^1}.$$

and $\|\nabla\rho(0)\|^2 = \|\nabla(w(0)-z(0))\|^2 \le 2\|\nabla\varepsilon(0)\|^2 + 2\|\nabla(z(0)-Z(0))\|^2$. Finally, the second term on the right hand side of (56) will be estimated via Lemma 1.9.

6.2 A posteriori error analysis for the locally Lipschitz continuity case

Let $f: R \to R$ is locally Lipschitz continuous for a.e. $(x, t) \in \Omega \cup [0, T]$, in the sense that there exist real numbers $C_L > 0$ and $\gamma \ge 0$ such that

$$|f(u) - f(v)| = C_L(t)\left(1+|u|^\gamma + |v|^\gamma\right)|u-v|. \tag{58}$$

Lemma 1.11 (Estimation of the nonlinear term). If the nonlinear reaction f is satisfying the growth condition (58) with $0 \le r < 2$ for $d = 2$, and with $0 \le r \le 4/3$ for $d = 3$, we have the bound

$$\|f(z) - f^n(Z^n)\| \le \mathcal{N}_1(t)\left\{\mathcal{N}_2(Z)(\|\rho\| + \|\varepsilon\|) + \sqrt{3}\|\rho\|\|\nabla\rho\|^\gamma + \sqrt{5}\|\varepsilon\|\|\nabla\varepsilon\|^\gamma\right\}$$
$$+\Theta_{n,3}\left(\int_{t_{n-1}}^{t_n}\left\|\frac{\partial\rho}{\partial t}\right\|^2 dt\right)^{1/2}, \tag{59}$$

where $\mathcal{N}_1(t) := \frac{1}{\sqrt{2}} C_L(t) \max\{1, 4^\gamma\}$, $N(Z) := \frac{1}{\sqrt{2}} \sqrt{1 + 4\gamma |Z|_\infty^{2\gamma}}$ and

$$\Theta_{n,3} := \frac{1}{\kappa_n} \int_{t_{n-1}}^{t_n} \|(f(Z) - f^n(Z^n)\|.$$

Proof: Applying triangle inequality, reads

$$\begin{aligned} T_{L,4} = \int_{t_{n-1}}^{t_n} \left| \left(f(z) - f^n(Z^n), \frac{\partial \rho}{\partial t} \right) \right| dt &\leq \int_{t_{n-1}}^{t_n} \left| \left(f(z) - f(Z), \frac{\partial \rho}{\partial t} \right) \right| dt \\ + \int_{t_{n-1}}^{t_n} \left| \left(f(Z) - f^n(Z^n), \frac{\partial \rho}{\partial t} \right) \right| dt &:= \mathcal{J}_{n,1} + \mathcal{J}_{n,2}. \end{aligned} \tag{60}$$

$\mathcal{J}_{n,1}$ can be bounded as follows

$$\begin{aligned} \mathcal{J}_{n,1} &= \int_{t_{n-1}}^{t_n} \left(f(z) - f(Z), \frac{\partial \rho}{\partial t} \right) \leq \int_{t_{n-1}}^{t_n} \|(f(z) - f(Z)\| \left\| \frac{\partial \rho}{\partial t} \right\| dt \\ &\leq \frac{1}{2} \left\| (f(z) - f(Z)\|^2 + \frac{1}{2} \int_{t_{n-1}}^{t_n} \left\| \frac{\partial \rho}{\partial t} \right\|^2 dt. \end{aligned} \tag{61}$$

Now, we have

$$\begin{aligned} \|f(z) - f(Z)\|^2 = \int_{t_{n-1}}^{t_n} \|f(z) - f(Z)\|^2 dt &\leq \int_{t_{n-1}}^{t_n} \|f(z) - f(w)\|^2 dt + \int_{t_{n-1}}^{t_n} \|f(w) - f(Z)\|^2 dt \\ &:= Z_{1,n} + Z_{2,n}. \end{aligned} \tag{62}$$

To estimate $Z_{1,n}$ on the first term in the right hand side of (62), we use the Cauchy–Schwarz inequality and (58) to obtain

$$\begin{aligned} Z_{1,n} &= \int_{t_{n-1}}^{t_n} \|f(z) - f(w)\|^2 dt = C_L^2(t) \int_{t_{n-1}}^{t_n} \left(1 + |z|^{2\gamma} + |w|^{2\gamma}\right) |z - w|^2 \\ &\leq \int_{t_{n-1}}^{t_n} \left(1 + |z|^{2\gamma}\right) |z - w|^2 dt + \int_{t_{n-1}}^{t_n} |w|^{2\gamma} |z - w|^2 dt. \end{aligned} \tag{63}$$

Applying the elementary inequality $|C_a + C_b|^{2\alpha} \leq C\left(|C_a|^{2\alpha} + |C_b|^{2\alpha}\right)$ with $C_a = z - w$ and $C_b = w$, so that $|z|^{2\alpha} \leq C|z - w|^{2\alpha} + C|w|^{2\alpha}$, this becomes

$$\begin{aligned} Z_{1,n} &\leq C_L^2(t) C \int_{t_{n-1}}^{t_n} \left(1 + |z - w|^{2\gamma}\right) |z - w|^2 dt + C_L^2(t) C \int_{t_{n-1}}^{t_n} \left(2|w - Z|^{2\gamma} + 2|Z|^{2\gamma}\right) |z - w|^2 dt \\ &\leq C_L^2(t) C \max\{1, 16^\gamma\} \left(\left(1 + 4^\gamma |Z|^{2r}\right) \|\rho\|^2 + \|\rho\|_{2+2\gamma}^{2+2\gamma} + 2 \int_{t_n 1}^{t_n} \|\varepsilon\|^{2\gamma} \|\rho\|^2 \right). \end{aligned} \tag{64}$$

Similarly, $Z_{2,n}$ follows as

$$\begin{aligned} Z_{2,n} &= \int_{t_{n-1}}^{t_n} \|f(w) - f(Z)\|^2 dt = C_L^2(t) C \int_{t_{n-1}}^{t_n} \left(1 + |w|^{2\gamma} + |Z|^{2\gamma}\right) |w - Z|^2 \\ &\leq C_L^2(t) C \max\{1, 16^\gamma\} \left(\left(1 + 4^r |Z|_\infty^{2\gamma}\right) \|\varepsilon\|^2 + \|\varepsilon\|_{2+2\gamma}^{2+2\gamma} \right). \end{aligned} \tag{65}$$

Collecting all these terms, we obtain

$$\|f(z)-f(Z)\|^2 \leq C_L^2(t)C\max\{1,16^\gamma\}\left(1+4^\gamma|Z|_\infty^{2\gamma}\right)(\|\rho\|^2+\|\varepsilon\|^2)$$
$$+C_L^2(t)C\max\{1,16^\gamma\}\left(\|\rho\|_{2+2\gamma}^{2+2\gamma}+3\|\varepsilon\|_{2+2\gamma}^{2+2\gamma}+2\int_{t_{n-1}}^{t_n}\|\varepsilon\|^2\|\rho\|^{2\gamma}dt\right). \tag{66}$$

Using Holder's inequality and Young's inequality, we deduce that

$$\int_{t_{n-1}}^{t_n}\|\alpha\|^{2r}\|\beta\|^2dx \leq \frac{\|\alpha\|_{2+2r}^{2+2r}}{r+1}+\frac{r\|\beta\|_{2+2r}^{2+2r}}{r+1}. \tag{67}$$

Therefore,

$$\int_{t_{n-1}}^{t_n}\|\varepsilon\|^{2r}\|\rho\|^2 \leq \frac{\|\varepsilon\|_{2+2\gamma}^{2+2\gamma}}{\gamma+1}+\frac{\gamma\|\rho\|_{2+2\gamma}^{2+2\gamma}}{\gamma+1}$$
$$\leq \|\varepsilon\|_{2+2\gamma}^{2+2\gamma}+\|\rho\|_{2+2\gamma}^{2+2\gamma}. \tag{68}$$

Substituting this into our grand inequality yields

$$\|f(z)-f(Z)\|^2 \leq \mathcal{N}_1^2(t)\left(\mathcal{N}_2^2(Z)(\|\rho\|^2+\|\varepsilon\|^2)+3\|\rho\|_{2+2\gamma}^{2+2\gamma}+5\|\varepsilon\|_{2+2\gamma}^{2+2\gamma}\right), \tag{69}$$

where $\mathcal{N}_1^2(t)=\frac{1}{2}C_L^2(t)C\max\{1,16^\gamma\}$ and $\mathcal{N}_2^2(Z)=\frac{1}{2}\left(1+4^r|Z|_\infty^{2r}\right)$. From Gagliardo-Nirenberg inequality in Theorem 1.2, implies that

$$\|\rho\|_{2+2\gamma} \leq C\|\nabla\rho\|^{\frac{(2+2\gamma)d-2d}{2}}\|\rho\|^{\frac{4+4\gamma+2d-2d-2d\gamma}{2}}, \tag{70}$$

valid for all $\gamma\geq 0$ for $d=2$ and $0\leq\gamma\leq 2$ for $d=3$. Combining this with the Poincar'e-Friedrichs inequality $\|\rho\|\leq C\|\nabla\rho\|$, yields

$$\|\rho\|_{2+2\gamma}\leq C\|\nabla\rho\|. \tag{71}$$

Finally,

$$\mathcal{J}_{n,2}=\int_{t_{n-1}}^{t_n}\left|\left(f(Z)-f^n(Z^n),\frac{\partial\rho}{\partial t}\right)\right|dt \leq \left\|(f(Z)-f^n(Z^n)\|\left(\int_{t_{n-1}}^{t_n}\left\|\frac{\partial\rho}{\partial t}\right\|^2dt\right)^{1/2}\right.. \tag{72}$$

Putting all of the results together the proof will be finished.

Theorem 1.12 Let z be the exact solution of (7) and let Z^n be its finite element approximation obtained by the backward Euler approximation (10). Then, for $1\leq n\leq N$, the following a posteriori error bounds hold

$$\max_{t\in[0,t_m]}\|\nabla(z(t)-Z(t))\|^2 \leq 4\mathcal{E}(t_n,\ Z)\left(\|\nabla(z(0)-Z(0))\|^2+\Phi_{n,\mathrm{H}^1}^2(0)\right)$$
$$+4\mathcal{E}(t_n,Z)\sum_{n=1}^{m}\mathcal{F}_{1,m}^2+4\mathcal{E}(t_n,\ Z)\sum_{n=1}^{m}\kappa_n^2\{\Phi_{n,2}^2+\Upsilon_{n,2}^2+\Psi_{n,1}^2+\Psi_{n,2}^2\}$$
$$+4\mathcal{N}_1^2(t)\mathcal{E}(t_n,\ Z)\sum_{n=1}^{m}\left(\mathcal{N}_2^2(Z)\Phi_{n,L_2}^2+\Phi_{n,L_2}^2\Phi_{n,\mathrm{H}^1}^{2\gamma}\right)+2\max_{t\in[0,t_m]}\Phi_{n,\mathrm{H}^1}^2, \tag{73}$$

where Φ_{n,L_2}^2 and Φ_{n,H^1}^2 are given in (20).

Proof: Now, setting $v = \frac{\partial \rho}{\partial t}$ in 22, and integrate from t_{n-1} to t_n along with summing up over $n = 1{:}\, m$ we have

$$\max_{t\in[0,t_m]} \|\nabla\rho(t)\|^2 + C_{coer}\int_0^{t_m} \|\frac{\partial\rho}{\partial t}\|^2 dt \le \|\nabla\rho(0)\|^2 + 2\sum_{n=1}^{m}\int_{t_{n-1}}^{t_n} \|f(z) - f^n(Z^n)\|^2$$
$$+2\sum_{n=1}^{m}(T_{n,1} + T_{n,2} + T_{n,3}). \tag{74}$$

Using Lemma 1.11, along with lemmas 1.3, 1.5, 1.6 and 1.7, imply

$$\max_{t\in[0,t_m]} \|\nabla\rho(t)\|^2 + \int_0^{t_m} \left\|\frac{\partial\rho}{\partial t}\right\|^2 dt \le \|\nabla\rho(0)\|^2 + \sum_{n=1}^{m}\mathcal{F}_{1,m}^2$$
$$+\sum_{n=1}^{m}\kappa_n^2\left(\Phi_{n,2}^2 + \Upsilon_{n,2}^2 + \Psi_{n,1}^2 + \Psi_{n,2}^2\right) + \mathcal{N}_1^2(t)\sum_{n=1}^{m}\left(\mathcal{N}_2^2(Z)\Phi_{n,L_2}^2 + 5\Phi_{n,L_2}^2\Phi_{n,\mathrm{H}^1}^{2\gamma}\right)$$
$$+\sum_{n=1}^{m}\int_{t_{n-1}}^{t_n}\left(\mathcal{N}_1^2(t)\mathcal{N}_2^2(Z)\|\nabla\rho\|^2 + 3\mathcal{N}_1^2(t)\|\rho\|^2\|\nabla\rho\|^{2\gamma}\right). \tag{75}$$

Setting

$$\mathcal{F}(t_n,\ Z,\ \varepsilon)^2 := \|\nabla\rho(0)\|^2 + \sum_{n=1}^{m}\mathcal{F}_{1,m}^2 + \sum_{n=1}^{m}\kappa_n^2\{\Phi_{n,2}^2 + \Upsilon_{n,2}^2 + \Psi_{n,1}^2 + \Psi_{n,2}^2\}$$
$$+\mathcal{N}_1^2(t)\sum_{n=1}^{m}\left(\mathcal{N}_2^2(Z)\Phi_{n,L_2}^2 + 5\Phi_{n,L_2}^2\Phi_{n,\mathrm{H}^1}^{2\gamma}\right). \tag{76}$$

Upon observing that

$$\int_{t_{n-1}}^{t_n} \|\nabla\rho\|^{2r}\|\rho\|^2 \le \max_{t\in[0,t_m]} \|\nabla\rho\|^{2\gamma}\int_{t_{n-1}}^{t_n} \|\rho\|^2)ds$$
$$\le \left(\max_{t\in[0,t_m]} \|\nabla\rho\|^2 + \int_{t_{n-1}}^{t_n} \|\rho\|^2\right)dt)^{\gamma+1}. \tag{77}$$

Now combining two equations, we obtain

$$\max_{t\in[0,t_m]} \|\nabla\rho(t)\|^2 + \int_0^{t_m} \left\|\frac{\partial\rho}{\partial t}\right\|^2 dt \le \mathcal{F}(t_m,\ Z,\ \varepsilon)^2 + \sum_{n=1}^{m}\int_{t_{n-1}}^{t_n} \mathcal{N}_1^2(t)\mathcal{N}_2^2(Z)\|\nabla\rho\|^2$$
$$+3\mathcal{N}_1^2(t)\sum_{n=1}^{m}\left(\max_{t\in[0,t_m]} \|\nabla\rho(t)\|^2 + \int_{t_{n-1}}^{t_n} \|\rho\|^2 dt)\right)^{\gamma+1}. \tag{78}$$

To bound of the nonlinear term of above equation, we shall employ a continuation argument in the spirit of [17, 18]. To do that, we consider the set

$$\mathcal{M}_n = \left\{\lim_{t\in[0,t_m]} \|\nabla\rho(t)\|^2 + C_{coer}\int_0^{t_m} \left\|\frac{\partial\rho}{\partial t}\right\|^2 dt \le 4\mathcal{F}(t_m,\ Z,\ \varepsilon)^2\mathcal{E}(t_m,Z)\right\}, \tag{79}$$

where $\mathcal{E}(t_m,\ Z) = exp\left(\int_0^{t_m}\mathcal{N}_1^2(t)\mathcal{N}_2^2(Z)dt\right)$. Since the left hand side of (78) depends continuously on t, and our aim is to show that $\mathcal{M}_n = [0,\ T]$. To do this, assuming $t_m^* = \max \mathcal{M}_n > 0$ and $t_m^* < T$, imply

$$\max_{t\in[0,t_m^*]} \|\nabla\rho(t)\|^2 + \int_0^{t_m^*}\left\|\frac{\partial\rho}{\partial t}\right\|^2 dt \leq \mathcal{F}(t_n,\ Z,\ \varepsilon)^2 + \{4\mathcal{F}(t_m,\ Z,\ \varepsilon)\mathcal{E}(t_m,\ Z)\}^{\gamma+1}$$
$$+\mathcal{N}_1^2(t)\mathcal{N}_2^2(Z)\int_0^{t_m^*}\|\nabla\rho\|^2 dt, \tag{80}$$

and Grönwall inequality, thus, implies

$$\max_{t\in[0,t_m^*]} \|\nabla\rho(t)\|^2 + \int_0^{t_m^*}\left\|\frac{\partial\rho}{\partial t}\right\|^2 dt \leq$$
$$\mathcal{E}\ (t_m,\ Z)\left\{\left(4\mathcal{N}_1^2(t)\mathcal{F}\left(t_m, Z,\ \varepsilon\right)^2\mathcal{E}(t_m,Z)\right)^{\gamma+1} + \mathcal{F}^2(t_m,\ Z,\ \varepsilon)^2\right\}. \tag{81}$$

Since $\mathcal{E}(t_m^*,\ Z) \leq \mathcal{E}(t_m,\ Z)$ and, suppose that the maximum size h_{max} of the mesh is small enough that, for $h < h_{\max}$, satisfy

$$\mathcal{F}(t_m,Z,\ \varepsilon) \leq \left(\frac{1}{\mathcal{N}_1^2(t)}\right)^{\gamma}\left(\frac{1}{4\mathcal{F}(t_m,Z,\varepsilon)^2\mathcal{E}(t_m,Z)}\right)^{\gamma+1}. \tag{82}$$

This leads to

$$\mathcal{N}_1^2(t)\ \left(4\mathcal{F}(t_m,\ Z,\ \varepsilon)^2\mathcal{E}(t_m,Z)\right)^{\gamma+1} \leq \mathcal{F}(t_m,\ Z,\ \varepsilon)^2. \tag{83}$$

Then, (81), becomes

$$\max_{t\in[0,t_m^*]} \|\nabla\rho(t)\|^2 + \int_0^{t_m^*}\left\|\frac{\partial\rho}{\partial t}\right\|^2 dt \leq 2\mathcal{E}(t_m,\ Z)\mathcal{F}(t_m,\ Z,\ \varepsilon)^2. \tag{84}$$

This leads to contradictions, because of t_m^* suppose to be $t_m^* = \max \mathcal{M}_n$. The triangle inequality along with Lemma 1.3, imply that

$$\max_{t\in[0,t_m]} \|\nabla e\|^2 \leq 2\max_{t\in[0,t_m]} \|\nabla\rho\|^2 + 2\max_{t\in[0,t_m]} \|\nabla\varepsilon\|^2$$
$$\leq 4\mathcal{F}(t_m,\ Z,\ \varepsilon)^2\mathcal{E}(t_m,Z) + 2\max_{t\in[0,t_m]} \Phi_{n,H^1}^2. \tag{85}$$

By recalling (76), the proof already finished.

7. Adaptive algorithms

This section aims to explain an adaptive algorithm aiming to investigate the performance of the presented a posteriori bound from Theorems 1.10 and 1.12 for the backward-Euler cG method for the semilinear parabolic problem (6). To this end, the implementation of the adaptive algorithm will be based on the deal. II finite element library [19] to the present setting of semilinear problems. We shall write algorithm for Theorem 1.10. For the Theorem 1.12 will follow the same with some modifcations. To begin with, we have

$$\Psi^j_{ini} := \|\nabla(z(0) - Z(0))\| + \|\nabla\varepsilon(0)\|$$

$$\Psi^j_{time} := \sum_{j=1}^{m}\left(\kappa_j\frac{\sqrt{3}}{3}\partial\left\|\Pi^j_0 f^j(Z^j) - \frac{Z^j - \Pi^j_0 Z^{j-1}}{\kappa_j}\right\| + \int_{t_{j-1}}^{t_j}\|f(Z) - f^j(Z^j)\|\right) \tag{86}$$

$$\Psi^j_{space} := \|h_j(g^j + \Delta^j Z^j)\| + \|h_j^{1/2}[\![Z^j]\!]\|_{\Sigma j}.$$

The adaptive algorithm from [15], starts with an initial uniform mesh in space and with a given initial time step. Starting from a uniform square mesh of 16×16 elements, the algorithm adapts the mesh to improve approximation to the initial condition using the initial condition estimator Ψ_{ini} until some tolerance is satisfied. To adapt the timestep κ_j, the algorithm bisects a time interval not satisfying a user-defined temporal tolerance $\Psi^j_{time} \leq$ **ttol**, and leaves a time-interval unchanged if $\Upsilon^j_{time} \leq$ **ttol**.

Once the time-step is adapted, the algorithm performs spatial mesh refinement and coarsening, determined by the space indicator Ψ^j_{space} using the user-defined tolerances $\mathbf{stol}^+$ and $\mathbf{stol}^-$, corresponding to refinement and coarsening, respectively. More specifically, we select the elements with the largest local contributions which result to $\Psi^j_{space} > \mathbf{stol}^+$ for refinement. The spatial coarsening threshold is set to $\mathbf{stol}^- = 0.001 * \mathbf{stol}^+$; we select the elements with the smallest local contributions which result to $\Psi^j_{space} < \mathbf{stol}^-$ for coarsening. The algorithm iterates for each time-step. We refer to [15] for the algorithm's workflow and all implementation details. The following two algorithms give the backward Euler method to the ODE system (12) and space-time adaptivity for Theorem 1.10.

Algorithm 1. The backward Euler method for solving the semilinear parabolic equation

---

1: Create a mesh with n elements on the interval I_n.
2: We disctize I_n as $0 = t_1 < t_2 < t_{3,\dots,<t_n} = T$, where n is time step defined as $\kappa_n = t_n - t_{n-1}$.
3: Setting $\alpha^0 = \alpha(0)$.
4: **for** $k = 1, 2, \quad \dots, \quad n$ **do**
5: Calculate the mass and stiffness matrices M and A, and the load vector F with entries

$$M_{i,j} = \int_{I_n}\phi_j\phi_i dx, \quad A_{i,j} = \int_{I_n}\phi'_j\phi'_i dx, \quad F_{i,j} = \int_{I_n} f\left(\phi_j\right)\phi_i dx. \tag{87}$$

6: Solve

$$(M + \kappa_n A)\alpha^n_i(t) = M\alpha^{n-1}_i(t) + \kappa_n F. \tag{88}$$

7: **end for**

Algorithm 2. Space-time adaptivity.

Input $a,\ b, f, z^0,\ T,\ \Omega, n,\ \mathcal{T},\ ttol, stol^{+}, stol^{-}$
Pick $\kappa_1,\ \dots,\ \kappa_n = \frac{T}{n}$.
Compute Z^0.
Compute Z^1 from Z^0.
while $\left(\Psi^1_{time}\right)^2 > ttol^{+}$ **or** $\max\left(\Psi^1_{space}\right)^2 > stol^{+}$ do bisction $\mathcal{T}^0$ by refining all elements such that $\left(\Psi^1_{space}\right)^2 > stol^{+}$ and coarsening all elements such that $\left(\Psi^1_{space}\right)^2 < stol^{-}$
if $\left(\Psi^1_{time}\right)^2 > ttol$, **then.**
$n-1 \leftarrow n$.
$K_n = K_{n-1},\ \dots,\ \kappa_2 = \kappa_1$.
$\kappa_2 = \frac{\kappa_1}{2}$.
$\kappa_1 \leftarrow \frac{\kappa_1}{2}$.
end if.
Compute Z^0.
Compute Z^1 from Z^0.
end while
put $j = 1,\ \mathcal{T}^1 = \mathcal{T}^0,\ time = \kappa_1$.
while $time < T$ do
Calculute Z^j from Z^{j-1}.
while $\left(\Psi^i_{time}\right)^2 > ttol$ **do**
if $\left(\Psi^1_{time}\right)^2 > ttol$ **then**
$n-1 \leftarrow n$.
$\kappa_n = \kappa_{n-1},\ \dots, \kappa_{j+2} = \kappa_{j+1}$.
$\kappa_{j+1} = \frac{\kappa_j}{2}$.
$\kappa_j \leftarrow \frac{\kappa_j}{2}$.
end if
Compute Z^j from Z^{j-1}.
end while
Create $\mathcal{T}^j$ from $\mathcal{T}^{j-1}$ by refining all elements such that $\left(\Psi^i_{space}\right)^2 > stol^{+}$ and coarsening all elements such that $\left(\Psi^i_{space}\right)^2 < stol^{-}$.
Compute Z^j from Z^{j-1}.
$time \leftarrow time + \kappa_j$.
$j-1 \leftarrow j$.
end while

8. Conclusion

The aim of this Chapter is to derive an optimal order a posteriori error estimates in term of the $L_\infty(H^1)$ for the fully semilinear parabolic problems in two cases when $f(u)$ Lipschitz and non Lipschitz are proved. The crucial tools in proving this error is the elliptic reconstruction techniques introduced by Makridakis and Nochetto 2003. This is consequently enabling us to use a posteriori error estimators derived for elliptic equation to obtain optimal order in terms of $L_\infty(H^1)$ norm for Lipschitz and non-Lipschitz nonlinearities. Some challenges have to be overcome due to nonlinearity on the forcing term depending on Gronwall's Lemma and Sobolev embedding through continuation argument. Furthermore, this will give insight about designing adaptive algorithm, which allow use to control the cost of computations. In the future, this Chapter can be extended to the fully discrete case for semilinear parabolic interface problems in $L_\infty(L_2) + L_2(H^1)$ and $L_\infty(L_2)$ norms [18, 20–22].

Notes/thanks/other declarations

It is pleasure to thank Prof. E. Greogoulis (Department of Mathematics, University of Leicester, UK), and Assistant Prof. A. Cangiani (Department of Mathematics, University of Nottingham, UK) for their help and encouragement.

Author details

Younis Abid Sabawi[1,2]

1 Department of Mathematics, Faculty of Science and Health, Koya University Koya KOY45, Kurdistan Region – F.R. Iraq

2 Department of Mathematics Education, Faculty of Education, Tishk International University, Kurdistan – Iraq

*Address all correspondence to: younis.abid@koyauniversity.org; younis.sabawi@tiu.edu.iq

References

[1] Slimane Adjerid, Joseph E. Flaherty, and Ivo Babuska, A posteriori error estimation for the finite element method- of-lines solution of parabolic problems, Math. Models Methods Appl. Sci. 9 (1999), no. 2, 261–286.

[2] Charalambos Makridakis and Ricardo H. Nochetto, Elliptic reconstruction and a posteriori error estimates for parabolic problems, SIAM J. Numer. Anal. 41 (2003), no. 4, 1585–1594.

[3] Omar Lakkis and Charalambos Makridakis, Elliptic reconstruction and a posteriori error estimates for fully discrete linear parabolic problems, Math. Comp. 75 (2006), no. 256, 1627–1658.

[4] Charalambos Makridakis, Space and time reconstructions in a posteriori analysis of evolution problems, ESAIMPro- ceedings. [Journées d'Analyse Fonctionnelle et Numérique en l'honneur de Michel Crouzeix], ESAIMProc., vol. 21, EDP Sci., Les Ulis, 2007, pp. 31–44.

[5] E. Buä nsch, F. Karakatsani, and Ch. Makridakis, A posteriori error control for fully discrete Crank-Nicolson schemes, SIAM J. Numer. Anal. 50 (2012), no. 6, 2845–2872.

[6] Alan Demlow, Omar Lakkis, and Charalambos Makridakis, A posteriori error estimates in the maximum norm for parabolic problems, SIAM J. Numer. Anal. 47 (2009), no. 3, 2157–2176. MR 2519598 (2010e:65142).

[7] INatalia Kopteva and Torsten Linss, Maximum norm a posteriori error estimation for parabolic problems using elliptic reconstructions, SIAM J. Numer. Anal. 51 (2013), no. 3, 1494–1524.

[8] Sutton, O.J., Long-time $L_t(L^2)$ a posteriori error estimates for fully discrete parabolic problems. IMA Journal of Numerical Analysis 2018.

[9] Schulz J. Machine grading and moral Cangiani, A., Georgoulis, E.H., Kyza, I. and Metcalfe, S., 2016. Adaptivity and blow-up detection for nonlinear evolution problems. SIAM Journal on Scientific Computing, (2016) 38(6), 3833–3856.

[10] Kyza Irene and Stephen Metcalfe, Pointwise a posteriori error bounds for blow-up in the semilinear heat equation, arXiv preprint arXiv:1802.07757(2018).

[11] Cangiani A, Georgoulis EH, Morozov AY, Sutton OJ. Revealing new dynamical patterns in a reaction–diffusion model with cyclic competition via a novel computational framework. Proceedings of the Royal Society A: Mathematical, Physical and Engineering Sciences. 2018 May 31;474(2213): 20170608.

[12] Younis A. Sabawi, "A Posteriori $L_\infty(H^1)$ -Error Bound in Finite Element Approximation of Semilinear Parabolic Problems, "2019 First International Conference of Computer and Applied Science (CAS), Baghdad,Iraq, 2019, pp. 102–106, doi: 10.1109/ CAS47993.2019.9075699.

[13] Younis A. Sabawi, A Posteriori $L_\infty(L_2) + L_2(H^1)$ Error Bounds In Discontinuous Galerkin Methods For Semidiscrete Semilinear Parabolic Interface Problems Baghdad science journal 2019. Accepted for Publication.

[14] Emmanuil H. Georgoulis, Omar Lakkis, and Juha M. Virtanen, A posteriori error control for discontinuous Galerkin methods for parabolic problems, SIAM J. Numer. Anal. 49 (2011), no. 2, 427–458.

[15] Stephen Metcalfe. Adaptive discontinuous Galerkin methods for nonlinear parabolic problems. PHD Thesis, University of Leicester, 2015.

[16] Slimane Adjerid, Joseph E. Flaherty, and Ivo Babuška. A posteriori error estimation for the finite element method-of-lines solution of parabolic problems. Math. Models Methods Appl. Sci., 9(2):261–286, 1999.

[17] Bartels S. A posteriori error analysis for time-dependent Ginzburg-Landau type equations. Numerische Mathematik. 2005 Feb 1;99(4):557–83.

[18] Cangiani A, Georgoulis EH, Jensen M. Discontinuous Galerkin methods for mass transfer through semipermeable membranes. SIAM Journal on Numerical Analysis. 2019;51 (5):2911–34.

[19] Wolfgang Bangerth, Ralf Hartmann, and Guido Kanschat. deal. iiâĂŤa general-purpose object-oriented finite element library. ACM Transactions on Mathematical Software (TOMS), 33(4): 24, 2007.

[20] Younis A. Sabawi, Adaptive discontinuous Galerkin methods for interface problems, PhD Thesis, University of Leicester, Leicester, UK (2017).

[21] Cangiani, Andrea, Emmanuil H. Georgoulis, and Younis A. Sabawi Adaptive discontinuous Galerkin methods for elliptic interface problems, Math. Comp. 87 (2018), no. 314, 2675–2707.

[22] Andrea Cangiani, Emmanuil H. Georgouils, and Younis A. Sabawi, Convergence of an adaptive discontinuous Galerkin method for elliptic interface problems, Journal of Computational and Applied Mathematics 367.

Chapter 14

Finite Element Magnetic Method for Magnetorheological Based Actuators

Ubaidillah and Bhre Wangsa Lenggana

Abstract

Magnetorheological materials based actuators have been currently exciting research topic for more than half-decades. Some actuators have been developed based on magnetorheological fluids and elastomers such as dampers, brakes, haptic devices, clutches, mountings, etc. These devices have their exciting properties which are capable of changing characteristic based on the amount of magnetic flux applied to them. Due to this capability, they are usually called semi-active devices. These devices employ an electromagnetic coil for magnetic flux production. Therefore, during the design process, magnetostatic simulation using the finite element method magnetic is carried out to make a better magnetic circuit. This chapter will consider several discussions such as necessary magnetostatic using free software finite element method magnetic (FEMM); design consideration for the magnetic circuit of the device and case studies of several type simulation in magnetorheological materials based devices.

Keywords: finite element, magnetostatic, magnetorheology, actuator, magnetic flux

1. Introduction

The finite element method (F.E.M.) is a numerical procedure that can be applied to solve various problems in engineering and science. In general, this method is used to solve steady, transient, linear, and nonlinear problems in electromagnetics, structural analysis, and fluid dynamics [1]. The finite element method has the main advantage of being able to handle all kinds of geometries and non-homogeneous materials without the need to change computer code formulations. The idea of this method is to break the problem into a large number of areas, each with simple geometry to facilitate problem-solving. As a result, the domain breaks down into a number of small elements, and the problem goes from small but challenging to solve into large and relatively easy to solve. Through the process of discretization, linear algebra problems are formed with many unknowns. In the case of electromagnetics, a discretization scheme, as implied by F.E.M., which implicitly combines most of the theoretical features of the problem analyzed is the best solution for obtaining accurate results in problems with complex, nonlinear geometries, etc. [2, 3]. This method can also be used for complex differential equations that are very difficult to solve. In the case of electromagnetic or magnetic fields, the finite element method is also known as FEMM.

FEMM is a suite of programs for solving low-frequency electromagnetic problems on two-dimensional planar and axisymmetric domains. The program currently addresses linear/nonlinear magnetostatic problems, linear/nonlinear time-harmonic magnetic problems, linear electrostatic problems, and steady-state heat flow problems. In the problem of this method, there are generally three parts to the problem [4].

Interactive shell program is a Multiple Document Interface pre-processor and a post-processor for various types of problems that are solved by FEMM. This case is a CAD-like interface for laying out the geometry of the problem to be solved and for defining material properties and boundary conditions [5]. The program also allows the user to inspect a field at specific points, as well as evaluate several different integrals and plot varying amounts of interest along with a user-defined contour [6]. Triangle breaks down the solution region into a large number of triangles, a vital part of the finite element process. Furthermore, Solvers, each solver takes a set of data files that describe problems and solves the relevant partial differential equations to obtain values for the desired field throughout the solution domain [7].

Finite Element Method Magnetics (FEMM) software has been developed for reasons of dealing with some of the limiting cases of Maxwell equations. The magnetic problem that is handled can be considered as a low frequency (L.F.) problem. In some cases, this problem can ignore displacement currents. This program discusses 2D planar and 3D axisymmetric linear and nonlinear harmonic magnetic, magnetostatic, and linear electrostatic problems [8].

Computer-assisted field distribution analysis for electromagnetic devices or component performance has become a simple, profitable, and fast method with good accuracy [9]. The magnetic field calculation problem aims to determine the value of one or more unknown functions, such as magnetic field intensity, magnetic flux density, scalar magnetic potential, and magnetic vector potential.

From a mathematical point of view, Maxwell equations can generally explain physical electromagnetic phenomena. Specifically, this point of view is a differential equation with specific boundary conditions. With this method, the correct solution to the problem is obtained. It is an analytical method that can be used to solve problems [10]. Analytical methods (method of appropriate representation, method of variable separation) are often applied to solve relatively simple problems. However, the problems that occur in practice are sometimes more complex regarding loading conditions, boundary conditions, geometric construction, and material heterogeneity, so that the integration of differential equations is challenging to solve by analytical methods. Therefore, the analytic solution can only be done by making a simplified model that allows the integration of the differential Equations [11]. Sometimes it is better to come up with a more realistic estimate of the value, rather than a precise solution from a simplified model. The approximate solution by the finite element method obtained by the numerical method reflects reality better than the exact solution of the simplified model [12].

Specific forms of electromagnetic field law for static magnetic fields are overcome by considering solving the magnetic problem through FEMM. Some of them are considering the model of the relationship between magnetic induction and the intensity of its magnetic field, the enunciation of static magnetic fields, passing conditions through discontinuity surfaces, enunciation of scalar magnetic potential - magnetostatic field problems and enunciation using magnetic vector potentials [13]. Some geometric configurations conform to the general formula for the unique conditions of a particular shape. The solution to this problem also depends on the relationship between magnetic intensity and magnetic field induction, the choice of material types such as linear and non-isotropic materials, linear and isotropic materials, nonlinear and non-isotropic materials with hysteresis, and nonlinear and isotropic materials, without permanent magnetization [14].

In other cases, such as in the development of magnetorheological devices (dampers, brakes, mounting, etc.), FEMM is used to solve the problem of magnetic flux density. Using the help of FEMM, solving magnetic problems can be solved quickly. The magnetic flux density, which is complex and challenging to be solved by numerical methods, can be determined by simulating the FEMM by using the material properties data to obtain the magnitude of the magnetic flux density. The simulation results are then used to determine the predicted values such as the pressure difference (in the case of the damper valve) using numerical or calculation methods [15].

2. General description of M.R. devices

The magnetorheology (M.R.) device is a device that implements intelligent materials as a working medium such as magnetorheological fluids (MRFs) and magnetorheological elastomers (M.R.E.s). M.R. devices are types of the controllable (semi-active) category. During its development, this device has been developed into a working medium such as M.R. damper, brake, and mounting for various applications. On the commercialization side, this device is not popular enough because of several things such as higher costs, more difficult production levels, and still under development. However, compared to other types of devices in its application (active and passive), M.R. devices have more advantages. MRFs and M.R.E.s are materials that are often used for research and development of M.R. devices.

As new technologies are developed, these materials have been discovered and developed in several applications. This material is unique because external stimuli can alter it. In this case, magnetorheological fluids are materials with properties that can be controlled by magnetic fields [16]. The MR fluids condition can be altered by using a varying magnitude of the magnetic field. This fluid is composed of magnetic particles that are pressed into a viscosity fluid. The absence of a magnetic field in this fluid causes its lower viscosity. These particles have a tiny size, ranging from 3 to 10 microns [17]. The magnetic particles of M.R. fluids are equipped with a special coating to weaken their magnetism and reduce the tendency to bond with each other between the particles. One of the weaknesses in M.R. fluid is the deposition, which occurs due to differences in density and gravitational force so that the fluid only focuses on the point where it is treated. Another disadvantage is the possibility of leakage into unwanted areas in the mechanism and thickening after long-term use, so component replacement is required. However, the application of M.R. fluid is extensive due to its precise control capabilities and dynamic response [17, 18]. The resulting output is relatively faster and more accurate because it uses an electric current as a conductor when compared to conventional mechanical mechanisms [19].

The structure and properties of the M.R. fluid outside or under the influence of the magnetic field are shown in **Figure 1**. The changes that occur when the M.R. fluid is under the influence of a magnetic field occurs in less than ten milliseconds. M.R. fluids regain their properties in the temperature range − 40 to 150 C, while the yield points of M.R. fluids range from 50 to 100 kPa [20].

The particle chain blocks the flow and converts the liquid to a semi-solid state in milliseconds. This phenomenon develops yield stress which increases with the magnitude of the applied magnetic field [21]. M.R. devices typically consist of hydraulic cylinders containing micron-sized magnetically polarized particles suspended in the fluid [17, 18].

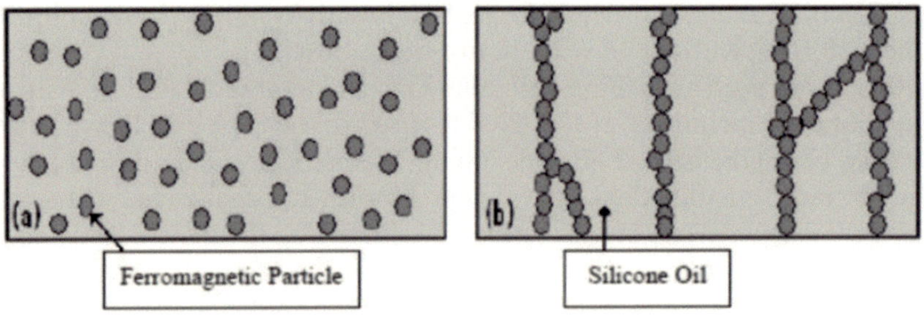

Figure 1.
Structures of M.R. fluid, ferromagnetic particles in silicon oil suspension: (a) without magnetic field effect, and (b) with magnetic field effect [17].

M.R. fluids work in several modes, including shear mode, valve mode, and squeeze mode [22]. MRF has been widely applied through shear mode and valve mode. Meanwhile, the application of MRFs which work with the new squeeze mode, has recently been developed. Also, MRFs can be operated in a combination of common MRFs working modes.

The shear mode is an operating mode in which the MRFs are influenced by a magnetic field between two parallel surfaces. One of the surfaces will move, and the other will be in a fixed condition. The shear mode is mostly applied to brakes and clutches. However, some dampers use a shear mode. The second is flow mode or valve mode; this mode is an operating mode in which the MRFs flow between two parallel surfaces that are at rest and simultaneously subjected to a magnetic field perpendicular to the direction of flow. Many applications of valve mode are found in dampers. Squeeze mode is an operational mode in which the MRFs flow-through two parallel surfaces and are subjected to a magnetic field that is perpendicular to the direction of flow. Squeeze mode is different from shear mode, the force exerted by one of the surfaces is the compression force, while in the shear mode it provides the shear force. **Figure 2** shows an illustration of the working principle of each MRFs working mode.

The commercialization of the use of MRFs technology was first used in 1995 for braking on stationary bicycles. MRFs technology tends to be cheaper and easier to use when compared to previous eddy-current-based braking technologies [24]. The world is full of potential applications for MRFs. Systems that require fluid motion control by changing viscosity, solutions based on MRFs technology may be applied to save functionality as well as costs. Simple and smart technology that can produce better products is the crucial factor of MRFs technology. Superior features such as fast response, simple application of electrical power input and mechanical power output, and controllability make MRFs technology the choice of many engineering

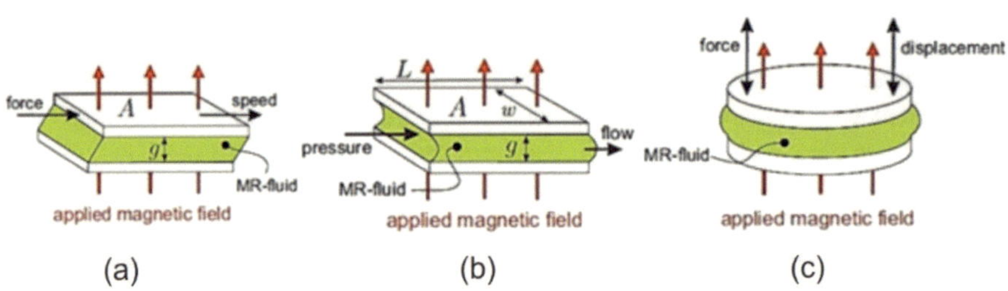

Figure 2.
MRFs working mode; (a) shear mode; (b) flow mode; (c) squeeze mode [23].

technologies. The sliding mode (used in brake and clutch) and valve mode (used in shock breakers) have been thoroughly studied, and several products are already on the market [25].

Besides MRFs, magnetorheological elastomers (M.R.E.s) are also intelligent materials that are currently a topic of development. In the last 20 years, the number of publications related to the creation, characterization, and application of M.R.E. has increased significantly. This significant increase occurred after 1995 regarding the viscoelasticity properties of M.R.E. initiated by Rigby and Jilken in their 1983 publication [26] when it is compared with the number of publications in the field of MRF and MRF applications.

The development of intelligent components based on M.R.E.s must pay attention to the composition of M.R.E.s because it can be formed with a variety of fill materials. The characteristics of the pre-blended matrix greatly influence the physical properties of M.R.E.s, which can make M.R.E.s solid or hollow. However, in general, M.R.E.s use a non-hollow matrix. To obtain a non-hollow matrix, the degassing method can be used to remove air bubbles or voids in the matrix. The magnetizable particles have an essential role in the magnetic induction properties of M.R.E.s. Much research has focused on these magnetized particles to achieve better rheological properties. Particles that are generally used are iron particles because they have a high permeability value and can be magnetized well [27]. M.R. effect is greatest due to the relationship between iron particles, this property can be achieved with high permeability and particle saturation. However, high saturation is also followed by an increase in the residual magnetic field that appears [28]. Therefore, the use of alloy particles in M.R.E.s, such as iron and cobalt or nickel alloys, is not as widely used as the use of C.I.P. The residual magnetic field in the particles will remain after the magnetic field has been lost so that the M.R. properties cannot return to their original state [29]. The size of the particles must be considered because it affects the properties of M.R.E. in receiving several magnetic domains.

3. Reluctance circuit for M.R. devices

Magnetic reluctance, or magnetic resistance, is a concept used in the analysis of magnetic circuits. It is defined as the ratio of magnetomotive force (mmf) to magnetic flux. It represents the opposition to magnetic flux and depends on the geometry and composition of an object.

Magnetic reluctance in a magnetic circuit is analogous to electrical resistance in an electrical circuit in that resistance is a measure of the opposition to the electric current. The definition of magnetic reluctance is analogous to Ohm law in this respect. However, the magnetic flux passing through a reluctance does not give rise to the dissipation of heat as it does for current through a resistance. Thus, the analogy cannot be used for modeling energy flow in systems where energy crosses between the magnetic and electrical domains. An alternative analogy to the reluctance model, which correctly represents energy flows is the gyrator–capacitor model. The magnetic circuit is derived using Kirchoff law, as illustrated in **Figure 3** [30, 31].

The symbols 1 dan Mrfluid are used to illustrate the reluctance of the design. So that it can be obtained as in the Eq. (1):

$$\Re = \frac{L}{\mu A} \tag{1}$$

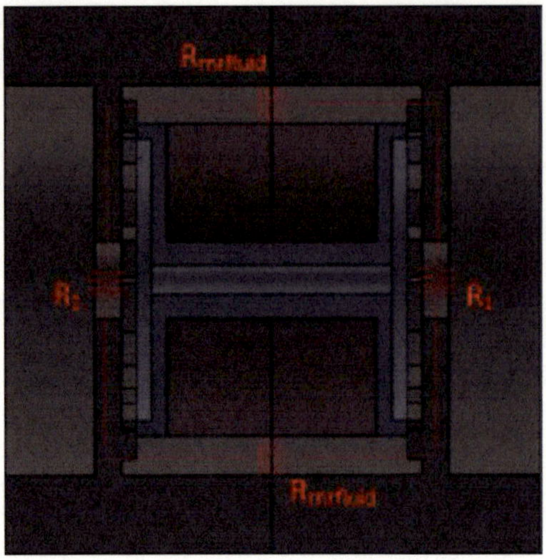

Figure 3.
Illustration of reluctance circuit on M.R. device.

where L is the effective distance that magnetic flux passes in each slice, μ is the magnetization property, and A is the effective area of the magnetic flux. Eq. 2 shows the total magnetomotive force generated from the sum of the magnetomotive force on all parts contained in one loop. So that we get the direct magnetomotive force for magnetic flux and reluctance as illustrated below,

$$\Phi_1 + \Phi_2 - (2\Re_1) - \left(2\Re_{mrfluid}\right) = 0 \tag{2}$$

Magnetic flux depends on a large number of copper coils and the current flowing in the coil so that Eq. (2) can be rewritten as Eq. (3),

$$NI - (2\Re_1) - \left(2\Re_{mrfluid}\right) = 0 \tag{3}$$

where N and I are the numbers of copper turns on the coil and the current flowing in the coil.

4. FEMM simulation procedure

4.1 Two-dimensional device sketch

The dimensions of the M.R. device depend on the target performance required, the function, and the space to be used. Dimensions determine the level of difficulty or ease in the M.R. device manufacturing process. Besides, according to the control classification of M.R. devices, dimensions will affect the value of pressure drop and damping force as well as the appearance of the device. One example is the configuration of a geometric arrangement that relies on the length of the fluid flow path in the equation to determine the predicted value for pressure drop.

4.2 Considering the target to achieve the predicted value

In this case, suppose that the target to be achieved is the pressure drop and damping force. The target pressure drop and damping force should be considered according to the needs and functions of the device. The milestones are related to the dimensions of the devices that have been designed. It is the determination of the number of devices that must be used with the available space and the targets that must be achieved.

4.3 FEMM simulation

FEMM can be simulated with some software. In general, all simulation procedures in some software are almost the same, such as FEMM and Ansoft Maxwell. The simulation process starts by making a design that will be simulated in a two-dimensional sketch. However, to perform a FEMM simulation, in general, the design to be simulated is made in two dimensions. Next is the material selection stage, coil configuration, meshing, and simulating as described below:

4.3.1 Set the initial simulation settings in the FEMM software

The initial settings made in the FEMM software are problem settings to be simulated. In this case, the problem to be simulated is a magnetic problem with axisymmetric.

4.3.2 Material selection

After making the initial setup and exporting the 2D design, then select the material according to the design that has been made. Material selection can be done

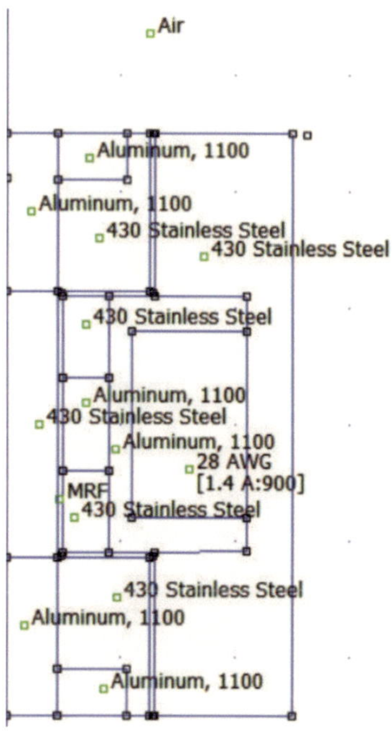

Figure 4.
Material selection.

by taking the existing software library or creating materials that have not been provided by FEMM by inputting all material property data to be used. Material selection can be seen in **Figure 4** above.

4.3.3 Coil configuration

The coil that will be used is inserted when performing the magnetic simulation with the FEMM software. In determining the coil, it is needed to input the type of wire, the number of turns, and the current that will be used.

4.3.4 Meshing

This process is an essential part of the simulation. The meshing process is a process of dividing an area which is divided into several areas to simplify the simulation process. **Figure 5** shows the results of the FEMM simulation meshing.

4.3.5 Running and plotting the result

The simulation software used is the Finite Element Method Magnetics (FEMM). The software is used to simulate the magnetic valve design that has been made. 2D designs that have been created are then exported to FEMM. Next, the problem setting to be simulated is determined, namely the magnetic problem with the symmetrical type. After the basic settings for the simulation are carried out, then adjust the material selection and coil selection according to the design that has been made.

The materials selection in the valve circuit is considered to get optimal results. Material selection is based on a predetermined valve design. Thus the direction of magnetic flux can be bent by nonmagnetic materials and produce a magnetic flux direction that is perpendicular to the direction of fluid flow. This is under the coil

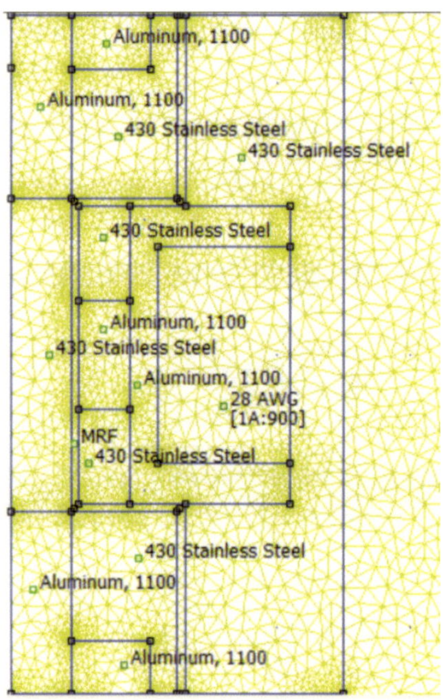

Figure 5.
Meshing result.

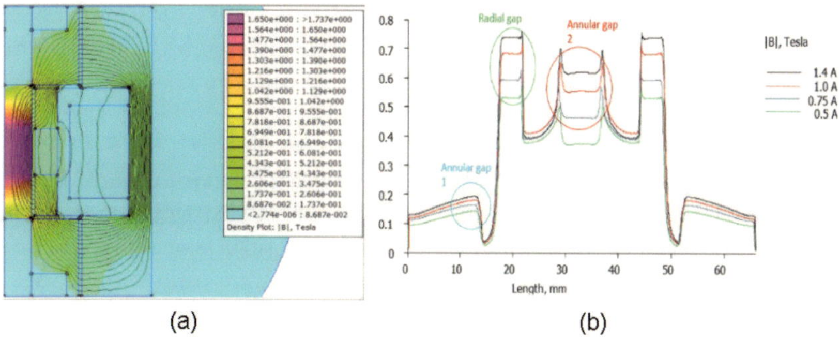

Figure 6.
(a) Magnetic flux density results from the FEMM simulation; (b) magnetic flux density plot.

configuration, and the fluid flow path geometry arrangement used to obtain the magnetic flux direction perpendicular to the fluid direction.

After all the parameters have been adjusted, proceed with the meshing process and simulate to get the magnetic flux density (B) results, as shown in **Figure 6**.

5. M.R. devices simulation

5.1 Magnetorheological multi-coil brake

This study describes a 3D magnetic simulation design of a magnetorheological multi-coil brake (M.R.B.). The design used in this study is an axial M.R.B. design with a configuration of more than one coil that is placed outside the casing. The placement of the device aims to simplify the brake maintenance process. **Figure 7** shows the multi-coil M.R. brake design in vertical and horizontal views. The simulation process is only carried out on a pair of coils that represent the entire coil and can distribute the magnetic flux to the entire electromagnetic part. The purpose of this simulation is to determine the results of the magnetic flux on the surface of the disc brake rotor. This simulation uses the FEMM modeling approach assisted by Ansoft Maxwell software.

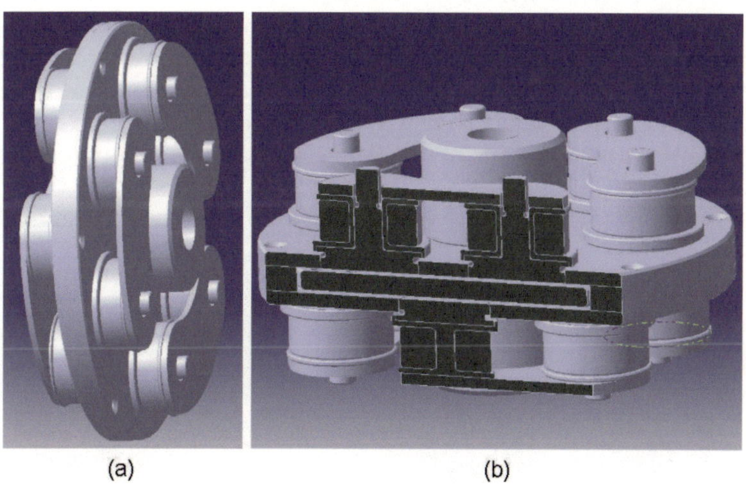

Figure 7.
Multi-coil MR brake design; (a) vertical view; (b) horizontal view.

The result is that the magnetic flux value of M.R.B. with a multi-coil configuration is higher than the magnetic flux value in conventional M.R.B. which only uses one coil with a larger size. Furthermore, the simulation results that have been obtained are used to determine the effect of different fluids on each variation. This study used several types of magnetorheological fluids (MRFs), MRF-122EG, MRF-132DG, and MRF-140CG, which were injected into each device design. Variations in the electric current input of 0.25 amperes, 0.50 amperes, 0.75 amperes, and 1.00 amperes are given in the simulation process. The results of magnetic flux distribution for MRF-132DG with the difference in current input can be seen in **Figure 8** below.

The resulting magnetic flux values were obtained from the FEMM simulation. The simulation is carried out by taking several variations of the electric current input and the difference in the fluid flow gap given to the device. The results show an increase in magnetic flux with each increase in electric current input and an increase with each narrower gap. As an example is the MRF-132DG design simulation for the MRF-132DG type, as shown in **Figure 9** below.

5.2 Fabrication and morphological characterization of anisotropic magnetorheological elastomer (M.R.E.)

In this study, silicone R.T.V. based anisotropic magnetorheological elastomer with 70% weight fraction of iron particle were fabricated using a validated mold and capable of aligning the particle in several angles (0°, 45°, dan 90°). This study begins with the fabrication of anisotropic M.R.E. curing mold, which covers the stage of design, simulation, prototype fabrication, and validation. Anisotropic M.R.E. mold was designed using Autodesk Fusion 360. To determine the value of magnetic flux density and distribution throughout the print, it was examined using simulations on Ansoft Maxwell. The simulation results show that the best magnetic flux density value on the mold is 0.3 T to form a good particle alignment in the matrix. At the same time, the magnetic flux density value of 0.3 T can be achieved by providing an electric current input of 0.2, 0.1, and 1 ampere respectively for the mold angles of 0°, 45°, and 90° during the curing chamber. This curing process is carried out for three hours under a magnetic field and left for one day before the sample is taken.

Magnetostatic simulation has a vital role in this research. The simulation process is carried out using Ansoft Maxwell software. This simulation is useful in estimating the magnetic flux density value in the curing chamber and knowing the direction of the magnetic field vector formed. The mold design that has been made will be simulated with various current values so that it can be seen as the current value needed to

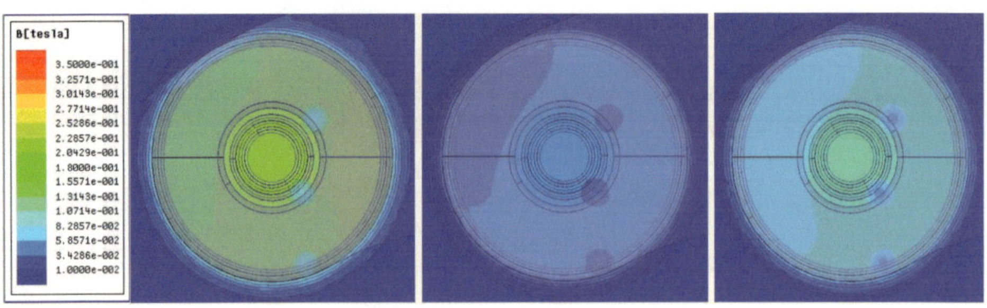

Figure 8.
Comparison of magnetic flux distribution to variation of MRFs; (a) 0.5 amperes; (b) 0.75 amperes; (c) 1 ampere.

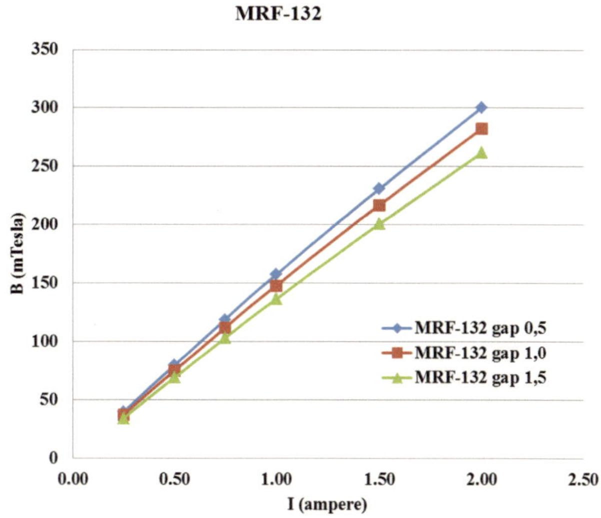

Figure 9.
FEMM simulation results for magnetic flux.

generate a magnetic flux density value of 0.3 T in the curing chamber. The magnetic properties data from V.S.M. are used to create new materials in the simulation. Thus, the material formed in the simulation is the same as the material used as the mold material. After the material in the simulation is the same as the actual condition, it is expected that the results of the simulation will not differ much from the measurement using a gauss-meter.

The simulation was carried out by providing variations in the angle of formation of M.R.E. with 0°, 45°, and 90°. One of the simulation results using an angle of 45° is shown in **Figure 10** below.

Figure 10 shows the distribution of the magnetic flux over a 45° curing space. The distribution of magnetic flux density in the curing chamber is marked in green color, which means that the value of the magnetic flux density in the area is medium. By changing the angle of the curing space by 45° relative to the direction of the magnetic field vector, anisotropic M.R.E. with a particle arrangement of 45° can be produced. After simulating several current values, the current required to produce 0.3 T in the curing chamber is 0.1 A. The graph in **Figure 11** shows the low magnetic flux density values on the left and right of the graph. This is because the measuring line of the magnetic flux density value touches the wall of the curing chamber, which is made of nonmagnetic aluminum.

5.3 Characterization torque of T-shaped magnetorheological brake

In recent research, M.R.B. T-shaped usually used more than one wire coil electromagnetic to maximize magnetic flux reaching all Magnetorheological Fluids (MRFs) gap. This research was focused on the reduction of wire coil on Magnetorheological Brake (M.R.B.). Serpentine flux was used to maximize all MRFs gaps that only use a single coil. The research was begun by designing M.R.B. design, followed by magnetostatic simulation using Finite Element Method Magnetics, calculate braking torque based on simulation, prototyping M.R.B. to get real braking torque measurement, and the last was measure braking torque using a torque sensor with constant angular velocity. The result of magnetostatic simulation shows the magnetic flux that reaches all MRFs gap. The most excellent magnetic flux density was 0,45 T at 1 A current on the outer annular. This result was used to

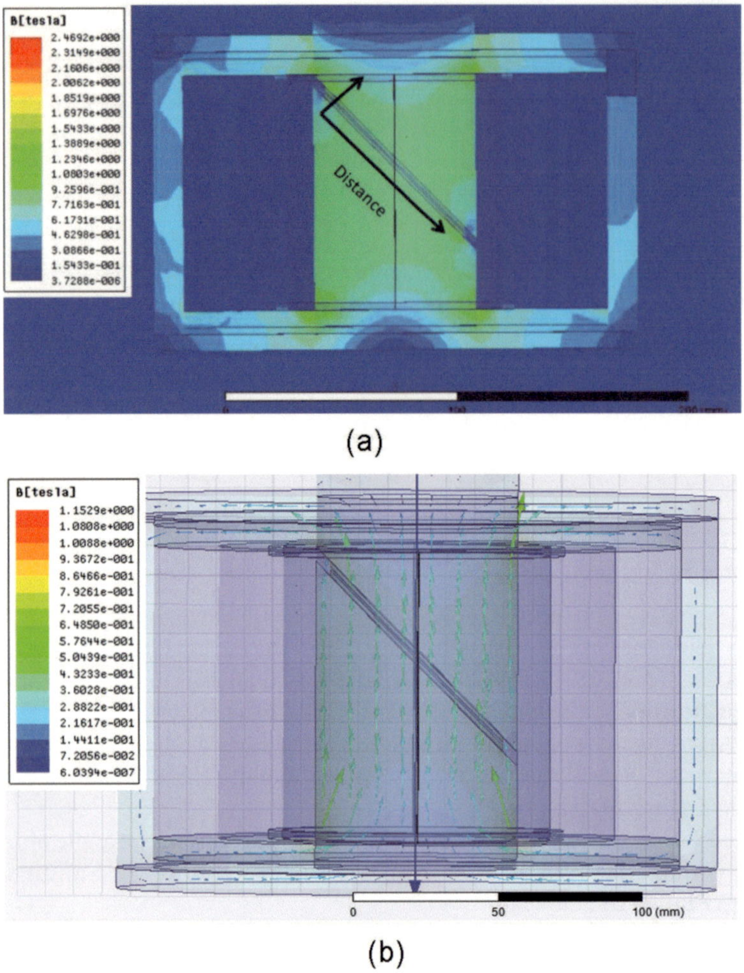

Figure 10.
Simulation results of a 45°: Vector magnetic (a) distribution of magnetic flux and (b) vector.

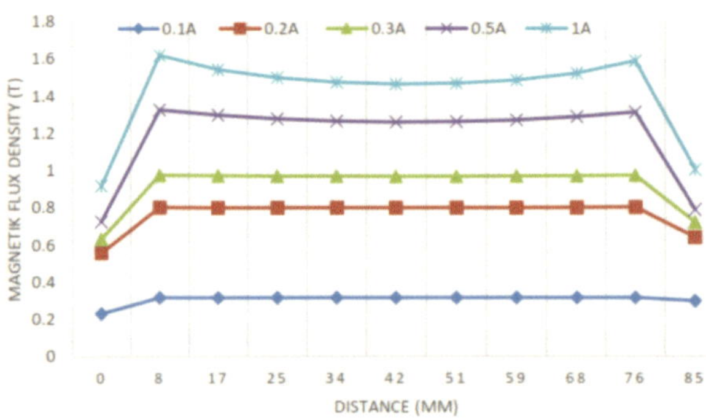

Figure 11.
Magnetic flux density distribution values in a 45° curing chamber at a current of 0.1; 0.2; 0.3; 0.5; and 1 A.

calculate shear stress based on Bingham Model that would generate braking torque. The braking torque generated on modeling torque and experiment was 1,51 Nm and 1,91 Nm at 1 A current with 20% difference, respectively. **Figure 12** shows an exploded design of M.R. brake.

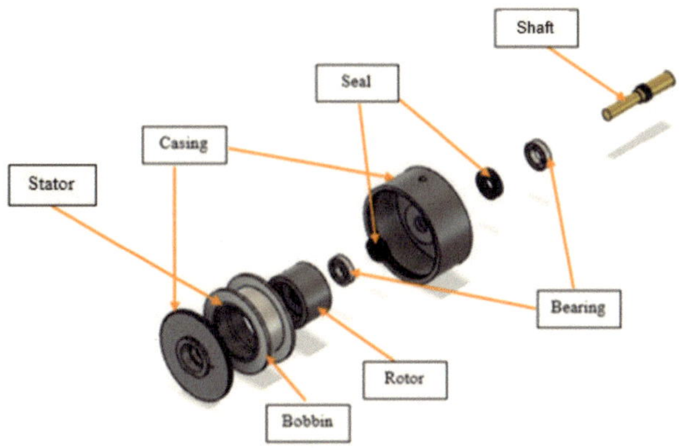

Figure 12.
Exploded design of M.R. brake.

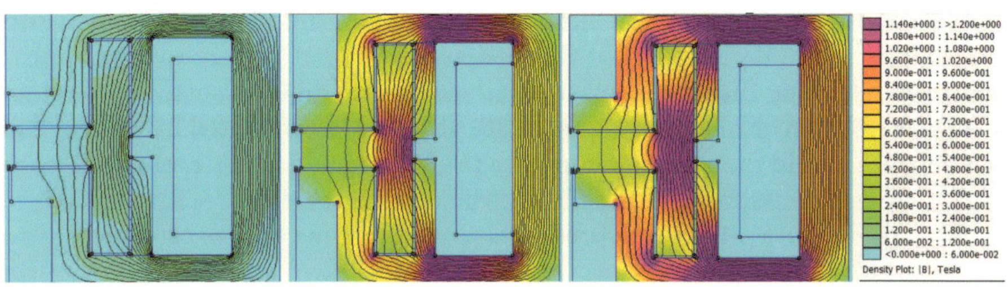

Figure 13.
Simulation results for magnetostatics: (a) 0.1 A; (b) 0.5 A and (c) 1 A.

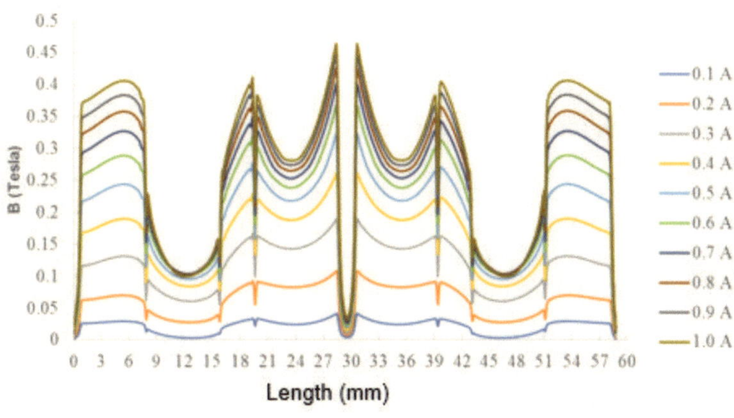

Figure 14.
Distribution of magnetic flux density along the MRF gap at variations of electric current 0.1–1 A.

The use of an electric current greatly affects the magnetic flux density. The greater the electric current used, the greater the magnetic flux density produced. It can be illustrated in **Figure 13**. The results given show the change in the resulting magnetic flux density, which is marked in a darker color, accompanied by more flux lines produced. The change has a limit point due to the ability of the material [32] as well as the flux that attaches to a particular component.

Figure 14 shows the distribution of magnetic flux density along the MRF gap with different variations of electric current. At current 1 A, the greatest magnetic flux density value exceeds 0.45 T, which is in the outer annular part. The higher the current applied, the lower the increase in magnetic flux density. This is because the direction of the magnetic flux is getting closer to the wall so that the resulting flux is limited. The ability of the copper wire to distribute magnetic flux also affects the result.

5.4 Performance prediction of magnetorheological damper for seismic

The new concept of a magnetorheological (M.R.) damping device used in the seismic building is discussed in this paper. The damper is aimed to deliver a comparable damping performance with the existing semi-active seismic damper design but with lower M.R. fluids volume requirement. This capability is achieved through the improvement in the M.R. valve performance using a meandering flow structure which was placed in the bypass line. **Figure 15** shows the sectional design of M.R. damper for seismic building and its valve.

This research is focused on the performance analysis of the M.R. valve pressure drop using an analytical approach. There are two main steps needed for the analytical approach, the magnetic field simulation, and the analytical pressure drop calculation. The simulation work of the M.R. valve magnetic circuit performance was carried out using finite element method magnetic (FEMM) software to calculate the distribution of magnetic flux density values. The simulated magnetic field density values would then be matched with the M.R. fluids characteristics data to predict the yield stress value of the fluids to be used in the pressure drop calculation. As a result, the M.R. valve is predicted to generate maximum off-state pressure drop of 5.35 MPa and a piston speed of 0.184 m/s. Meanwhile, at on-state condition (1.4 A), the valve is generating pressure drop up to 9.13 MPa at a piston speed of 0.184 m/s. The generated total pressure drop of the M.R. valve reaches 16.39 MPa. The MR fluids that are used in this design are only 1.5×10^{-4} m^3. From the generated total pressure drop, the peak of the damping force is obtained with 1.4 A, which is 32.19 kN. Meanwhile, the calculation result of the seismic force is 125.3 kN. Thus, it can be concluded that with the peak generated damping force, this seismic damper design will be capable of providing a damping performance which is appropriate to the seismic force with four parallel devices.

In this study, the FEMM simulation was used to obtain the magnetic flux density value in the valve section. The magnetic flux density value is used to calculate the predicted yield stress value, which is then used to predict the value of the pressure drop and the damping force. Yield stress is obtained through magnetic simulation using FEMM software which aims to obtain a magnetic flux density graph. Then the

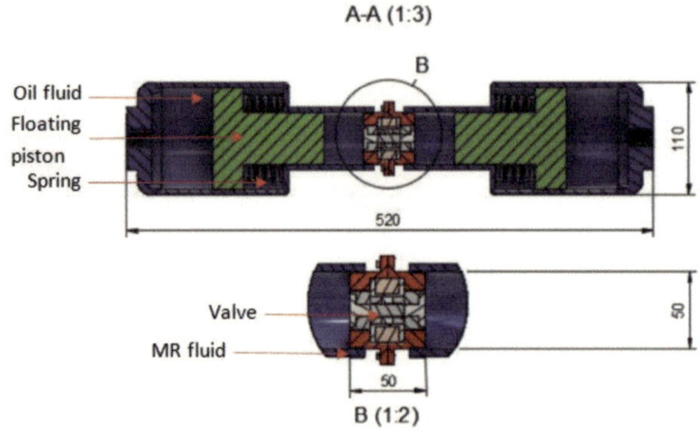

Figure 15.
M.R. damper for seismic building design.

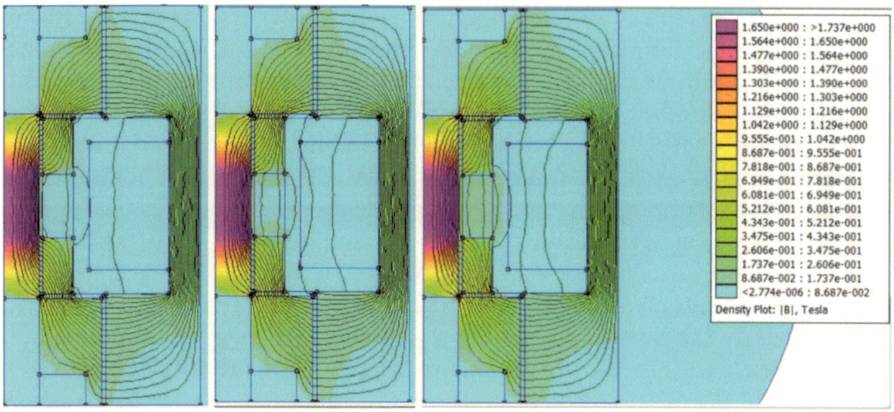

Figure 16.
Result of FEMM simulation.

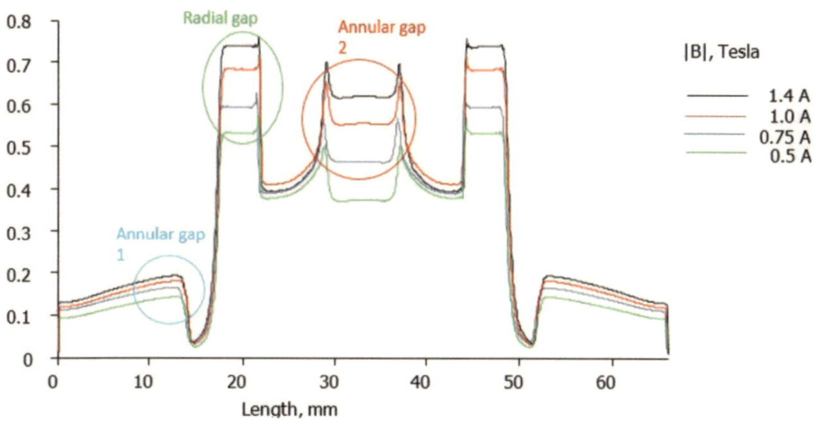

Figure 17.
Magnetic flux density result.

resulting magnetic flux density value is included in the calculation to get the yield stress value. The simulation process used is a magnetic simulation of the working fluid with a viscosity of 0.112 Pa.s which is obtained from the MRF132-DG property data by Lord Corp [33]. **Figure 16** shows the results of 2D magnetic simulations with FEMM and magnetic flux density graphs obtained through the simulation process.

Figure 17 above is obtained from a FEMM simulation based on a 2D design with an MRF132-DG working fluid and 900 coils. The wire used uses copper wire 28 A.W.G. with a diameter of 0.3211 mm with a resistance of 213 Ω/km. The graph shows the results of the magnetic flux density at the annular and radial channel against the variation of current input 0.5 A; 0.75 A; 1.0 A; 1.4 A.

6. Conclusion

Finite element magnetic is a method that can be used to facilitate an intricate work or may not even be completed by other methods. In this case, the field of magnetorheology is an example of a problem by using the finite element method solution. The magnetorheological device is a device that uses iron particle materials whose working principle is to change its rheological properties due to the influence of a magnetic field. This magnetic field gives rise to a magnetic flux density in the device whose magnitude can be determined by solving the finite element method. To find out the magnitude of the magnetic flux density value, some finite element method magnetics software can be used, such as FEMM and Ansoft Maxwell. The solution to these problems can be resolved with the existence of boundary conditions and initial setups that are following the procedure, such as the type of problem, the use of materials, and a clear design configuration. Thus, problems requiring the finite element method can be resolved with good accuracy. Problem-solving with the finite element method simulation is considered more accurate than other methods. In the case of magnetorheological devices, magnetic simulation with the finite element method is very helpful to achieve the research objectives.

Acknowledgements

The authors gratefully thank the ministry of research and technology (RISTEK/BRIN) for the funding of Hibah World-Class Research 2020-2021.

Conflict of interest

The authors have no conflict of interest.

Notes/thanks/other declarations

The authors thank Ariyo Nurachman Satiya Permata, Ilham Rizkia Nyaubit, Ilham Bagus Wiranto, and Rivananda Rama S that have been doing the simulations for finite element magnetic in magnetorheological devices so the authors could do this book chapter as well.

Author details

Ubaidillah* and Bhre Wangsa Lenggana
Universitas Sebelas Maret, Surakarta, Indonesia

*Address all correspondence to: ubaidillah_ft@staff.uns.ac.id

References

[1] K.B. Baltzis, The finite element method magnetics (FEMM) freeware package: May it serve as an educational tool in teaching electromagnetics?, Educ. Inf. Technol. 15 (2010) 19-36. doi:10.1007/s10639-008-9082-8.

[2] D. Meeker, Finite Element Method Magnetics Version 4.2 User Manual (2015)

[3] W.H. Li, H. Du, N.Q. Guo, Finite element analysis and simulation evaluation of a magnetorheological valve, Int. J. Adv. Manuf. Technol. 21 (2003) 438-445. doi:10.1007/s001700300051.

[4] C. Jędryczka, FE analysis of electromagnetic field coupled with fluid dynamics in an M.R. clutch, COMPEL - Int. J. Comput. Math. Electr. Electron. Eng. 26 (2007) 1028-1036. doi:10.1108/03321640710756357.

[5] M. Barham, D. White, Finite element simulation of permanent magnetoelastic thin films, IEEE Trans. Magn. 47 (2011) 1402-1405. doi:10.1109/TMAG.2010.2088382.

[6] A. Boczkowska, L. Czechowski, M. Jaroniek, T. Niezgoda, Analysis of magnetic field effect on ferromagnetic spheres embedded in elastomer pattern, J. Theor. Appl. Mech. 48 (2010) 659-676.

[7] Z. Parlak, T. Engin, I. Çalli, Optimal design of M.R. damper via finite element analyses of fluid dynamic and magnetic field, Mechatronics. 22 (2012) 890-903. doi:10.1016/j.mechatronics.2012.05.007.

[8] T.A. Voronina, G.M. Molodavkin, S.A. Sergeeva, O.I. Epstein, Anxiolytic Effect of Proproten under Conditions of Punished and Unpunished Behavior, Bull. Exp. Biol. Med. 135-136 (2003) 120-122. doi:10.1023/A:1024771906306.

[9] J. Morozionkov, J.A. Virbalis, Investigation of electric reactor magnetic field using finite element method, Elektron. Ir Elektrotechnika. (2008) 9-12. doi:10.5755/j01.eee.85.5.11150.

[10] H. Zhou, M. Bai, Y. He, S. Ren, F.E.M. analysis of magnetorheological fluid seal of circular cooler and its experimental research, Int. J. Appl. Electromagn. Mech. 41 (2013) 419-431. doi:10.3233/JAE-121622.

[11] V. Pendefunda, A.C. Pendefunda, N. Ioanid, A. Apostu, Finite element analysis of periodontal stresses in fixed prosthodontics, 5 (2013) 82-87.

[12] Y. Li, J. Li, Finite element design and analysis of adaptive base isolator utilizing laminated multiple magnetorheological elastomer layers, J. Intell. Mater. Syst. Struct. 26 (2015) 1861-1870. doi:10.1177/1045389X15580654.

[13] Z. Chao, F. Tyan, S. Tu, W.S. Jeng, Finite Element Analysis of a Magnetorheological Fluid Damper, 24th CSME Conf. CYCU, Chungli, Taiwan, R.O.C. (2015) 3145-3149.

[14] M.Y. Salloom, Z. Samad, Finite element modeling and simulation of proposed design magnetorheological valve, Int. J. Adv. Manuf. Technol. 54 (2011) 421-429. doi:10.1007/s00170-010-2963-1.

[15] B. Yang, J. Luo, L. Dong, Magnetic circuit F.E.M. analysis and optimum design for M.R. damper, Int. J. Appl. Electromagn. Mech. 33 (2010) 207-216. doi:10.3233/JAE-2010-1115.

[16] Ubaidillah, J. Sutrisno, A. Purwanto, S.A. Mazlan, Recent progress on magnetorheological solids: Materials, fabrication, testing, and applications,

Adv. Eng. Mater. 17 (2015) 563-597. doi:10.1002/adem.201400258.

[17] P. Skalski, K. Kalita, Role of Magnetorheological Fluids and Elastomers in Today's World, Acta Mech. Autom. 11 (2017) 267-274. doi:10.1515/ama-2017-0041.

[18] M. Schwartz, Smart Materials, Taylor & Francis Group, Boca Raton (2009)

[19] D. Baranwal, T.S. Deshmukh, MR-Fluid Technology and Its Application - A Review, Int. J. Emerg. Technol. Adv. Eng. 2 (2012) 563-569.

[20] K.D. Weiss, J.D. Carlson, D.A. Nixon, Viscoelastic properties of magneto- and electro-rheological fluids, J. Intell. Mater. Syst. Struct. 5 (1994) 772-775. doi:10.1177/1045389X9400500607.

[21] M.R. Jolly, J.W. Bender, J.D. Carlson, Properties and applications of commercial magnetorheological fluids, J. Intell. Mater. Syst. Struct. 10 (1999) 5-13. doi:10.1106/R9AJ-XYT5-FG0J-23G1.

[22] F. Imaduddin, S.A. Mazlan, H. Zamzuri, A design and modelling review of rotary magnetorheological damper, Mater. Des. 51 (2013) 575-591. doi:10.1016/j.matdes.2013.04.042.

[23] M. Avraam, M. Horodinca, I. Romanescu, A. Preumont, Computer controlled rotational MR-brake for wrist rehabilitation device, J. Intell. Mater. Syst. Struct. 21 (2010) 1543-1557. doi:10.1177/1045389X10362274.

[24] G. M. Webb, Exercise apparatus and associated method including rheological fluid brake (1998)

[25] A.G. Olabi, A. Grunwald, Design and application of magnetorheological fluid, Mater. Des. 28 (2007) 2658-2664. doi:10.1016/j.matdes.2006.10.009.

[26] Z. Rigbi, L. Jilkén, The response of an elastomer filled with soft ferrite to mechanical and magnetic influences, J. Magn. Magn. Mater. 37 (1983) 267-276. doi:10.1016/0304-8853(83)90055-0.

[27] M. Lokander, B. Stenberg, Improving the magnetorheological effect in isotropic magnetorheological rubber materials, Polym. Test. 22 (2003) 677-680. doi:10.1016/S0142-9418(02)00175-7.

[28] L.C. Balbas, G. Borstel, J.A. Alonso, Nonlocal density functional calculation of the electron affinity of atoms, Phys. Lett. A. 114 (1986) 236-240. doi:10.1016/0375-9601(86)90214-8.

[29] S. Mantripragada, X. Wang, F. Gordaninejad, B. Hu, A. Fuchs, Rheological properties of novel magnetorheological fluids, Int. J. Mod. Phys. B. 21 (2007) 4849-4857. doi:10.1142/s021797920704575x.

[30] E.C. Sekaran, Magnetic circuits and power transformers, Elsevier Inc., 2016. doi:10.1016/B978-0-12-804448-3.00011-6.

[31] S.A. Afsari, Optimal Design and Analysis of a Novel Reluctance Axial Flux Magnetic Gear, (n.d.).

[32] K. Karakoc, E.J. Park, A. Suleman, Design considerations for an automotive magnetorheological brake, Mechatronics. 18 (2008) 434-447. doi:10.1016/j.mechatronics.2008.02.003.

[33] C. Load, MRF-132DG Magneto-Rheological Fluid, Lord Prod. Sel. Guid. Lord Magnetorheol. Fluids. 54 (2011) 11. doi:)D DS7015 (Rev.1 11/11).